ORCCA

Open Resources for Community College Algebra

ORCCA

Open Resources for Community College Algebra

A Textbook Created by
Portland Community College Faculty

Contributing Authors

Ann Cary

Alex Jordan

Ross Kouzes

Scot Leavitt

Cara Lee

Carl Yao

Ralf Youtz

MTH 60 Winter 2018

Project Leads: Ann Cary and Alex Jordan
Technology Engineer: Alex Jordan
Cover Image: Ralf Youtz

Edition: Beta (Winter 2018 MTH 60/65 Pilot)

Website: spot.pcc.edu/math/orcca/orcca.html

© 2016–2018 Portand Community College

Permission is granted to copy, distribute and/or modify this document under the terms of the Creative Commons Attribution 4.0 International License.

Acknowledgements

This book has been made possible through Portland Community College's Strategic Investment Funding, approved by PCC's Budget Planning Advisory Council and the Board of Directors. Without significant funding to provide the authors with the adequate time, an ambitious project such as this one would not be possible.

The technology that makes it possible to create synced print, eBook, and WeBWorK content is PreTeXt, created by Rob Beezer. Additionally, David Farmer and the American Institute of Mathematics have worked to make the PreTeXt eBook layout functional, yet simple. A grant from OpenOregon funded the original bridge between WeBWorK and PreTeXt.

This book uses WeBWorK to provide most of its exercises, which may be used for online homework. WeBWorK was created by Mike Gage and Arnie Pizer, and has benefited from over 25 years of contributions from open source developers. In 2013, Chris Hughes, Alex Jordan, and Carl Yao programmed most of the WeBWorK questions in this book with a PCC curriculum development grant.

The javascript library MathJax, created and maintained by David Cervone, Volker Sorge, Christian Lawson-Perfect, and Peter Krautzberger allows math content to render nicely on screen in the eBook. Additionally, MathJax makes web accessible mathematics possible.

The print edition (PDF) is built using the typesetting software LaTeX, created by Donald Knuth and enhanced by Leslie Lamport.

Each of these open technologies, along with many that we use but have not listed here, has been enhanced by many additional contributors spanning the past 40 years. We are grateful for all of these contributions.

To All

HTML and PDF This book is avaialble as an eBook, a free PDF, or printed and bound. All versions offer the same content and are synchronized such that cross-refernces match across versions.

- A web version is available at `http://spot.pcc.edu/math/orcca/orcca.html`, and this version is recommended. It offers interactive elements and easier navigation than print versions.
- A PDF is available at `http://spot.pcc.edu/~ajordan/orcca.pdf`. Some content is in color, but most of the colorized content from the eBook has been converted to black and white to ensure adequate contrast when printing. The exceptions are the graphs generated by WeBWorK.
- Printed and bound copies are available online, through various merchants. Contact the authors if you have trouble finding the latest version online. For each online sale, all royalties go to a PCC Foundation account, where roughly half will fund student scholarships, and half will fund continued maintenance of this book and other OER.

Copying Content The graphs and other images that appear in this manual may be copied in various file formats using the eBook version. Below each image are links to `.png`, `.eps`, `.svg`, `.pdf`, and `.tex` files that contain the image.

Similarly, tables can be copied from the eBook version and pasted into applications like *MS Word*. However, mathematical content within tables will not always paste correctly without a little extra effort as described below.

Mathematical content can be copied from the eBook. To copy math content into *MS Word*, right-click or control-click over the math content, and click to `Show Math As MathML Code`. Copy the resulting code, and `Paste Special` into *Word*. In the `Paste Special` menu, paste it as `Unformatted Text`. To copy math content into LaTeX source, right-click or control-click over the math content, and click to `Show Math As TeX Commands`.

Accessibility The HTML version is intended to meet or exceed all web accessibility standards. If you encounter an accessibility issue, please report it to the editor.

- All graphs and images will eventually have meaningful alt text that communicates what a sighted person would see, without necessarily giving away anything that is intended to be deduced from the image. Construction of alt text is underway, and will be complete by Summer 2018.
- All math content is rendered using MathJax. MathJax has a contextual menu that can be accessed in several ways, depending on what operating system and browser you are using. The most common way is to right-click or control-click on some piece of math content.

- In the MathJax contextual menu, you may set options for triggering a zoom effect on math content, and also by what factor the zoom will be.
- If you change the MathJax renderer to MathML, then a screen reader will generally have success verbalizing the math content.

Tablets and Smartphones PreTeXt documents like this lab manual are "mobile-friendly". When you view the HTML version, the display adapts to whatever screen size or window size you are using. A math teacher will always recommend that you do not study from the small screen on a phone, but if it's necessary, this manual gives you that option.

WeBWorK for Online Homework Most exercises are available in a ready-to-use collection of WeBWorK problem sets. Visit `https://webwork.pcc.edu/webwork2/orcca-demonstration` to see a demonstration WeBWorK course where guest login is enabled. Anyone faculty interested in using these problem sets should contact the authors.

Pedagogical Decisions

The authors have taken various stances on certain pedagogical and notational questions that arise in basic algebra instruction. We attempt to catalog these decisions here, although this list will certainly be incomplete. If you find something in the book that runs contrary to these decisions, please let us know.

- Interleaving is our preferred approach, compared to a proficiency-based approach. To us, this means that once the book covers a topic, that topic will be appear in subsequent sections and chapters in indirect ways.
- Chapter 1 is written as a *review*, and is not intended to teach these topics from first principles.
- We round decimal results to four significant digits, or possibly fewer leaving out trailing zeros. We do this to maintain consistency with the most common level of precision that WeBWorK uses to assess decimal answers. We *round*, not *truncate*. And we use the $\approx$ symbol. For example $\pi \approx 3.142$ and Portland's population is ≈ 609500.
- We intend to offer *alternative* video lessons associated with each section. These are intended to provide readers with an alternative to whatever we have written on a topic. We have produced videos for Chapters 1–4. In later chapters we sometimes use videos from YouTube, but intend to produce videos at some point in the future. The YouTube videos more than likely do not cover 100% of what our written content covers. And such videos may use notation and approaches that differ from ours.
- Traditionally, a math textbook has "examples" throughout each section. This textbook generally uses two different types of such "examples".

Static These are labeled "Example." Static examples may or may not be subdivided into a "statement" followed by a walk-through solution. This is basically what traditional examples from math textbooks do.

Active These are labeled "Exercise," not to be confused with the exercises that come at the end of a section that might be assigned for homework, etc. In the HTML output, Active examples have WeBWorK answer blanks where a reader could try submitting an answer. In the PDF output, Active examples are almost indistinguishable from Static examples. Generally, a walk-through solution is provided immediately following the answer blank.

Some HTML readers will skip the opportunity to try an Active example and go straight to its solution. Some readers will try an active example once and then move on to the solution. Some readers will tough it out for a period of time and resist reading the

solution.

For readers of the PDF, it is expected that they would read the example and its solution just as they would read a Static example.

It is important to understand that a reader is *not* required to try submitting an answer to an Active example before moving on. It is also important to understand that a reader *is* expected to read the solution to an Active exercise, even if they succeed on their own at finding an answer.

Interspersed through a section there are usually several exercises that are intended as active reading exercises. A reader can work these examples and submit answers to WeBWorK to see if they are correct. The important thing is to keep the reader actively engaged instead of providing another static written example. In most cases, it is expected that a reader will read the solutions to these exercises just as they would be expected to read a more traditional static example.

- We believe in nearly always opening a topic with some level of application rather than abstract examples. From applications and practical questions, we move to motivate more abstract definitions and notation. This approach is perhaps absent in the first chapter, which is intended to be a review only. At first this may feel backwards to some instructors, with some "easier" exmaples (with no context) coming later than some of the contextual examples.
- Linear inequalities are not strictly separated from linear equations. The same section that teaches how to solve $2x+3=8$ will also teach how to solve $2x+3<8$. There will be sufficient subdivisions within sections so that an instructor may focus on equations only or inequalities only if they so choose.

  Our aim is to not treat inequalities as an add-on optional topic, but rather to show how intimately related they are to corresponding equations.
- When issues of "proper formatting" of student work arise, we value that the reader understand *why* such things help the reader to communicate outwardly. We believe that mathematics is about more than understanding a topic, but also about understanding it well enough to communicate results to others.

  For example we promote progression of equations like

  $$\begin{aligned} 1+1+1 &= 2+1 \\ &= 3 \end{aligned}$$

  instead of

  $$1+1+1=2+1=3.$$

  And we want students to *understand* that the former method makes their work easier for a reader to read. It is not simply a matter of "this is the standard and this is how it's done."
- When soliving equations (or systems), every example will come with a check, intended to communicate to students that checking is part of the process. In Chapters 1–4, these checks will be complete simplifications using order of operations one step at a time. The later sections will often have more summary checks where either order of operations steps are skipped in

groups, or we promote entering expressions into a calculator. Occasionally in later sections the checks will still have finer details, especially when there are issues like with negative numbers squared.

- Within a section, any first example of solving some equation (or system) should summarize with some variant of both "the solution is..." and "the solution set is...". Later examples can mix it up, but always offer at least one of these.
- There is a section on very basic arithmetic (five operations on natural numbers) in an appendix, not in the first chapter.
- With applications of linear equations (as opposed to linear systems), we limit applications to situations where the setup will be in the form $x + f(x) = C$ and also certain rate problems where the setup will be in the form $5t + 4t = C$. There are other classes of application problem (mixing problems, interest problems, ...) which can be handled with a system of two equations, and we reserve these until linear systems are covered.
- With simplifications of rational expressions, we always include domain restrictions that are lost in the simplification. For example, we would write $\frac{x(x+1)}{x+1} = x$, for $x \neq -1$.

Contents

Acknowledgements **v**

To All **vii**

Pedagogical Decisions **ix**

1 Basic Math Review **1**

1.1 Arithmetic with Negative Numbers 1
1.2 Fractions and Fraction Arithmetic 14
1.3 Absolute Value and Square Root 28
1.4 Order of Operations 36
1.5 Set Notation and Types of Numbers 49
1.6 Comparison Symbols 58
1.7 Notation for Intervals 64
1.8 Chapter Review 70

2 Variables, Expressions, and Equations **79**

2.1 Variables and Evaluating Expressions 79
2.2 Equations and Inequalities as True/False Statements 99
2.3 Solving One-Step Equations 109
2.4 Solving One-Step Inequalities 125
2.5 One-Step Equations With Percentages 133
2.6 Modeling with Equations and Inequalities 146
2.7 Introduction to Exponent Rules 156
2.8 Simplifying Expressions 162
2.9 Chapter Review 177

3 Linear Equations and Inequalities **187**

3.1 Solving Multistep Linear Equations and Inequalities 187
3.2 Linear Equations and Inequalities with Fractions 211
3.3 Isolating a Linear Variable 223
3.4 Ratios and Proportions 230
3.5 Special Solution Sets 246
3.6 Chapter Review 251

4 Graphing Lines **259**

4.1 Cartesian Coordinates 259
4.2 Graphing Equations 267
4.3 Exploring Two-Variable Data and Rate of Change 277

4.4 Slope 288
4.5 Slope-Intercept Form 303
4.6 Point-Slope Form 327
4.7 Standard Form 340
4.8 Horizontal, Vertical, Parallel, and Perpendicular Lines 355
4.9 Summary of Graphing Lines 372
4.10 Linear Inequalities in Two Variables 381
4.11 Graphing Lines Chapter Review 386

Index **403**

CHAPTER 1

Basic Math Review

This chapter is *mostly* intended to *review* topics from a basic math course, especially Sections 1.1–1.4. These topics are covered differently than they would be covered for a student seeing them for the very first time ever.

1.1 Arithmetic with Negative Numbers

Adding, subtracting, multiplying, dividing, and raising to powers all have their own peculiarities when negative numbers are involved. This section reviews arithmetic with signed (both positive and negative) numbers.

1.1.1 Signed Numbers

Is it valid to subtract a large number from a smaller one? It may be hard to imagine what it would mean physically to subtract 8 cars from your garage if you only have 1 car in there in the first place. Nevertheless, mathematics has found a way to give meaning to expressions like $1-8$ using **signed numbers**.

In daily life, the signed numbers we might see most often are temperatures. Most people on Earth use the Celsius scale; if you're not familiar with the Celsius temperature scale, think about these examples:

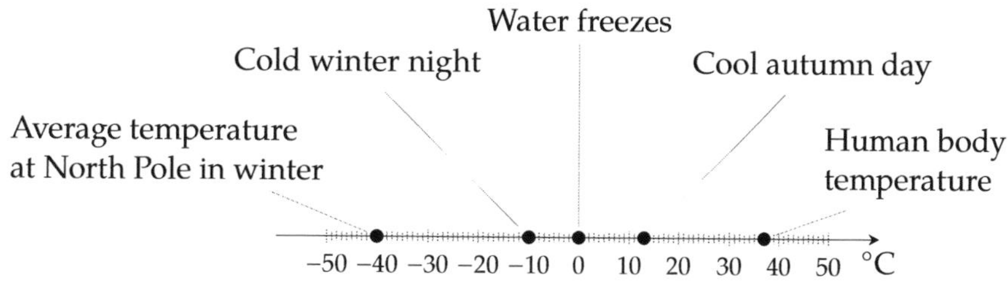

Figure 1.1.1: Number line with interesting Celsius temperatures

CC BY

Figure 1.1.1 uses a **number line** to illustrate these positive and negative numbers. A number line is a useful device for visualizing how numbers relate to each other and combine with each other. Values to the right of 0 are called **positive** numbers and values to the left of 0 are called **negative numbers**.

Warning 1.1.2 (Subtraction Sign versus Negative Sign). Unfortunately, the symbol we use for subtraction looks just like the symbol we use for marking a negative number. It will help to identify when a "minus" sign means "subtract" or means "negative." The key is to see if there is a number to its left, not counting anything farther left than an open parenthesis. Here are some examples.

- -13 has one negative sign and no subtraction sign.
- $20 - 13$ has no negative signs and one subtraction sign.
- $-20 - 13$ has a negative sign and then a subtraction sign.
- $(-20)(-13)$ has two negative signs and no subtraction sign.

Exercise 1.1.3. Identify "minus" signs.

In each expression, how many negative signs and subtraction signs are there?

a. $1 - 9$

has ☐ negative signs and ☐ subtraction signs.

b. $-12 + (-50)$

has ☐ negative signs and ☐ subtraction signs.

c. $\dfrac{-13 - (-15) - 17}{23 - 4}$

has ☐ negative signs and ☐ subtraction signs.

1.1.2 Adding

An easy way to think about adding two numbers with the *same sign* is to simply (at first) ignore the signs, and add the numbers as if they were both positive. Then make sure your result is either positive or negative, depending on what the sign was of the two numbers you started with.

Example 1.1.4 (Add Two Negative Numbers). If you needed to add -18 and -7, note that both are negative. Maybe you have this expression in front of you:

$$-18 + -7$$

but that "plus minus" is awkward, and in this book you are more likely to have this expression:

$$-18 + (-7)$$

with extra parentheses. (How many subtraction signs do you see? How many negative signs?)

Since *both* our terms are *negative*, we can add 18 and 7 to get 25 and immediately realize that our final result should be negative. So our result is -25:

$$-18 + (-7) = -25$$

This approach works because adding numbers is like having two people tugging on a rope in one direction or the other, with strength indicated by each number. In Example 1.1.4 we have two people pulling to the left, one with strength 18, the other with strength 7. Their forces combine to pull *left* with strength 25, giving us our total of -25, as illustrated in Figure 1.1.5.

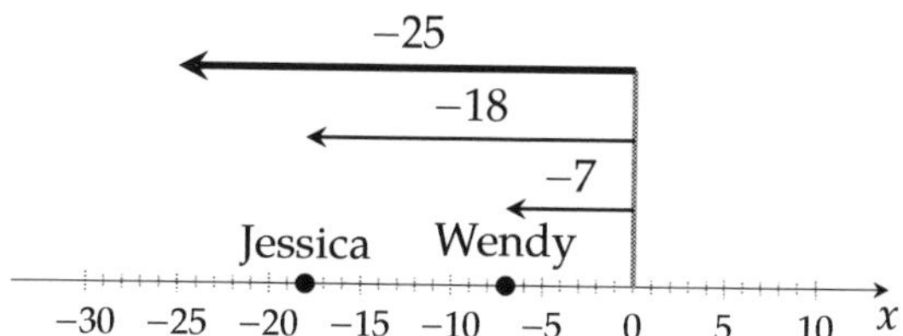

Figure 1.1.5: Working together

If we are adding two numbers that have *opposite* signs, then the two people tugging the rope are opposing each other. If either of them is using more strength, then the overall effect will be a net pull in that person's direction. And the overall pull on the rope will be the *difference* of the two strengths. This is illustrated in Figure 1.1.6.

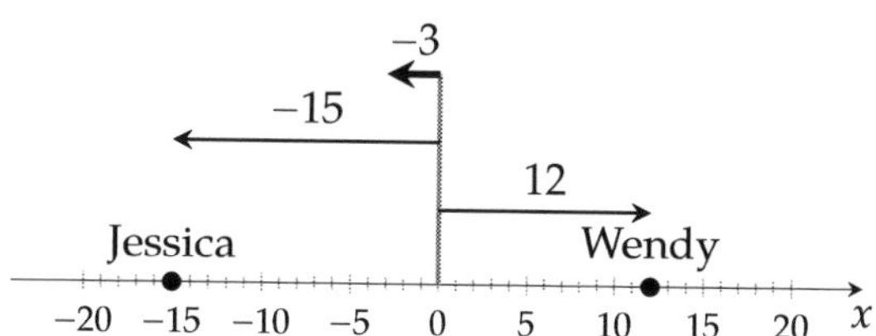

Figure 1.1.6: Working in opposition

Example 1.1.7 (Adding One Number of Each Sign). Here are four examples of addition where one number is positive and the other is negative.

a. $-15 + 12$

We have one number of each sign, with sizes 15 and 12. Their difference is 3. But of the two numbers, the negative number dominated. So the result from adding these is -3:

$$-15 + 12 = -3$$

b. $200 + (-100)$

We have one number of each sign, with sizes 200 and 100. Their difference is 100. But of the two numbers, the positive number dominated. So the result from adding these is 100:

$$200 + (-100) = 100$$

c. $12.8 + (-20)$

We have one number of each sign, with sizes 12.8 and 20. Their difference is 7.2. But of the two numbers, the negative number dominated. So the result from adding these is -7.2:

$$12.8 + (-20) = -7.2$$

d. $-87.3 + 87.3$

We have one number of each sign, both with size 87.3. The opposing forces cancel each other, leaving a result of 0:

$$-87.3 + 87.3 = 0$$

Exercise 1.1.8. Take a moment to practice adding when at least one negative number is involved. The expectation is that readers can make these calculations here without a calculator.

a. Add -1 to 9.

b. Add $-12 + (-98)$.

c. Add $123 + (-100)$.

d. Find the sum $-2.1 + (-2.1)$.

e. Find the sum $-34.67 + 81.53$.

1.1.3 Subtracting

Perhaps you can handle a subtraction such as $18 - 5$, where a small positive number is subtracted from a larger number. There are other instances of subtraction that might leave you scratching your head. In such situations, we recommend that you view each subtraction as *adding* the opposite number.

	Original	Adding the Opposite
Subtracting an even larger positive number:	$12 - 30$	$12 + (-30)$
Subtracting from a negative number:	$-8.1 - 17$	$-8.1 + (-17)$
Subtracting a negative number:	$42 - (-23)$	$42 + 23$

The benefit is that perhaps you already mastered addition with positive and negative numbers, and this strategy that you convert subtraction to addition means you don't have all that much more to learn. These examples might be computed as follows:

$12 - 30$	$-8.1 - 17$	$42 - (-23)$
$= 12 + (-30)$	$= -8.1 + (-17)$	$= 42 + 23$
$= -18$	$= -25.1$	$= 65$

Exercise 1.1.9. Take a moment to practice subtracting when at least one negative number is involved. The expectation is that readers can make these calculations here without a calculator.

a. Subtract -1 from 9.

b. Subtract $32 - 50$.

c. Subtract $108 - (-108)$.

d. Find the difference $-5.9 - (-3.1)$.

e. Find the difference $-12.04 - 17.2$.

1.1.4 Multiplying

Making sense of multiplication of negative numbers isn't quite so straightforward, but it's possible. Should the product of 3 and -7 be a positive number or a negative number? Remembering that we can view multiplication as repeated addition, we can see this result on a number line:

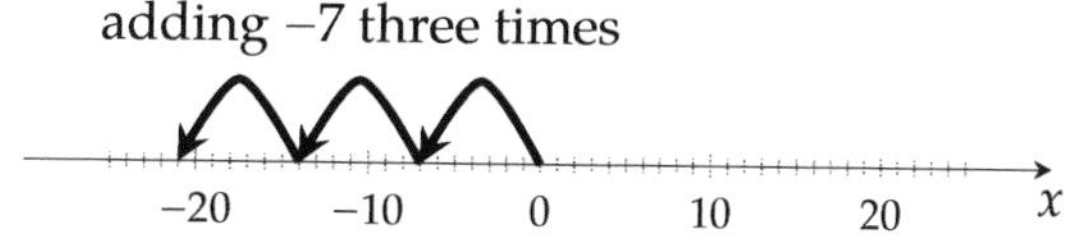

Figure 1.1.10: Viewing $3 \cdot (-7)$ as repeated addition

Figure 1.1.10 illustrates that $3 \cdot (-7) = -21$, and so it would seem that a positive number times a negative number will always give a negative result. (Note that it would not change things if the negative number came first in the product, since the order of multiplication doesn't affect the result.)

What about the product $-3 \cdot (-7)$, where both factors are negative? Should the product be positive or negative? If $3 \cdot (-7)$ can be seen as adding -7 three times as in Figure 1.1.11, then it isn't too crazy to interpret $-3 \cdot (-7)$ as *subtracting* -7 three times, as in Figure 1.1.11.

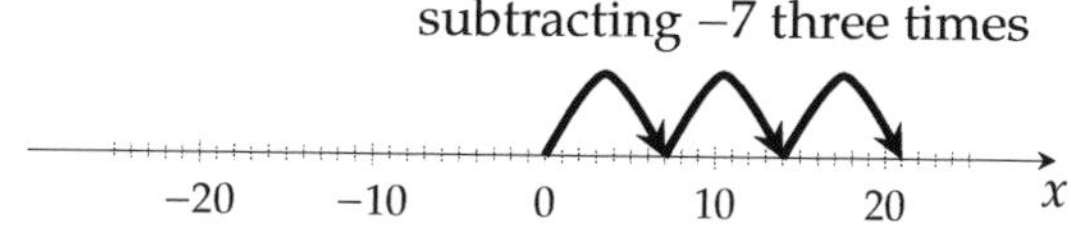

Figure 1.1.11: Viewing $-3 \cdot (-7)$ as repeated subtraction

This illustrates that $-3 \cdot (-7) = 21$, and it would seem that a negative number times a negative number always gives a positive result.

Exercise 1.1.12. Here are some practice exercises with multiplication and signed numbers. The expectation is that readers can make these calculations here without a calculator.

a. Multiply $-13 \cdot 2$.

b. Find the product of 30 and -50.

c. Compute $-12(-7)$.

d. Find the product $-285(0)$.

1.1.5 Powers

For early sections of this book the only exponents you will see will be the **natural numbers**: $\{1, 2, 3, \ldots\}$. But negative numbers can and will arise as the *base* of a power.

An exponent is a shorthand for how many times to multiply by the base. For example,

$$(-2)^5 \text{ means } \overbrace{(-2) \cdot (-2) \cdot (-2) \cdot (-2) \cdot (-2)}^{\text{5 instances}}$$

Will the result here be positive or negative? Since we can view $(-2)^5$ as repeated multiplication, and we now understand that multiplying two negatives gives a positive result, this expression can be thought of this way:

$$\underbrace{\underbrace{(-2) \cdot (-2)}_{\text{positive}} \cdot \underbrace{(-2) \cdot (-2)}_{\text{positive}}}_{\text{positive}} \cdot (-2)$$

and that lone last negative number will be responsible for making the final product negative.

More generally, if the base of a power is negative, then whether or not the result is positive or negative depends on if the exponent is even or odd. It depends on whether or not the factors can all be paired up to "cancel" negative signs, or if there will be a lone factor left by itself.

Once you understand whether the result is positive or negative, for a moment you may forget about signs. Continuing the example, you may calculate that $2^5 = 32$, and then since we know $(-2)^5$ is negative, you can report

$$(-2)^5 = -32$$

Warning 1.1.13 (Negative signs and Exponents). Expressions like -3^4 may not mean what you think they mean. What base do you see here? The correct answer is 3. The exponent 4 *only* applies to the 3, not to -3. So this expression, -3^4, is actually the same as $-(3^4)$, which is -81. Be careful not to treat -3^4 as having base -3. That would make it equivalent to $(-3)^4$, which is *positive* 81.

Exercise 1.1.14. Here is some practice with natural exponents on negative bases. The expectation is that readers can make these calculations here without a calculator.

a. Compute $(-8)^2$.

b. Calculate the power $(-1)^{203}$.

c. Find $(-3)^3$.

d. Calculate -5^2.

1.1.6 Summary

The various arithmetic combinations involving negative numbers and how to think about them are summarized here.

Addition Add two negative numbers: add their positive counterparts and make the result negative.

Add a positive with a negative: find their difference using subtraction, and keep the sign of the dominant number.

Subtraction Any subtraction can be converted to addition of the opposite number. For all but the most basic subtractions, this is a useful strategy.

Multiplication Multiply two negative numbers: multiply their positive counterparts and make the result positive.

Multiply a positive with a negative: multiply their positive counterparts and make the result negative.

Multiply any number by 0: the result will be 0.

Division (not discussed in this section) Division by some number is the same as multiplication by its reciprocal. So the multiplication rules can be adopted.

Division of 0 by any nonzero number always results in 0.

Division of any number by 0 is always undefined.

Powers Raise a negative number to an even power: raise the positive counterpart to that power.

Raise a negative number to an odd power: raise the positive counterpart to that power, then make the result negative.

Expressions like -2^4 mean $-(2^4)$, not $(-2)^4$.

1.1.7 Exercises

These skills practice exercises simply ask you to calculate something.

1. Perform the given addition and subtraction:

a. $-13 - 9 + (-10) =$ ☐

b. $1 - (-13) + (-17) =$ ☐

2. Perform the given addition and subtraction:

a. $-12 - 6 + (-6) =$ ☐

b. $8 - (-13) + (-11) =$ ☐

3. Add the following:

a. $-8 + (-3) =$ ☐

b. $-4 + (-5) =$ ☐

c. $-1 + (-9) =$ ☐

4. Add the following:

a. $-8 + (-1) =$ ☐

b. $-6 + (-7) =$ ☐

c. $-1 + (-7) =$ ☐

5. Add the following:

a. $1 + (-8) =$ ☐

b. $9 + (-1) =$ ☐

c. $7 + (-7) =$ ☐

6. Add the following:

a. $2 + (-10) =$ ☐

b. $5 + (-2) =$ ☐

c. $7 + (-7) =$ ☐

7. Add the following:

a. $-6 + 2 =$ ☐

b. $-3 + 8 =$ ☐

c. $-4 + 4 =$ ☐

8. Add the following:

a. $-8 + 3 =$ ☐

b. $-4 + 10 =$ ☐

c. $-4 + 4 =$ ☐

9. Subtract the following integers:

a. $3 - 6 =$ ☐

b. $7 - 4 =$ ☐

c. $5 - 14 =$ ☐

10. Subtract the following integers:

a. $4 - 9 =$ ☐

b. $9 - 3 =$ ☐

c. $4 - 19 =$ ☐

11. Subtract the following integers:

a. $-2 - 3 =$ ☐

b. $-10 - 2 =$ ☐

c. $-6 - 6 =$ ☐

12. Subtract the following integers:

a. $-1 - 4 =$ ☐

b. $-6 - 3 =$ ☐

c. $-4 - 4 =$ ☐

13. Subtract the following integers:

a. $-5 - (-8) =$ ☐

b. $-5 - (-2) =$ ☐

c. $-4 - (-4) =$ ☐

14. Subtract the following integers:

a. $-1 - (-7) =$ ☐

b. $-7 - (-1) =$ ☐

c. $-4 - (-4) =$ ☐

15. Multiply the following integers.

a. $(-10) \cdot (-3) =$ ☐

b. $(-4) \cdot 2 =$ ☐

c. $7 \cdot (-5) =$ ☐

d. $(-2) \cdot 0 =$ ☐

16. Multiply the following integers.

a. $(-10) \cdot (-1) =$ ☐

b. $(-6) \cdot 7 =$ ☐

c. $7 \cdot (-2) =$ ☐

d. $(-10) \cdot 0 =$ ☐

17. Multiply the following integers.

a. $(-2)\cdot(-6)\cdot(-3) =$ ☐

b. $5\cdot(-8)\cdot(-4) =$ ☐

c. $(-97)\cdot(-77)\cdot 0 =$ ☐

18. Multiply the following integers.

a. $(-2)\cdot(-4)\cdot(-5) =$ ☐

b. $3\cdot(-8)\cdot(-2) =$ ☐

c. $(-96)\cdot(-65)\cdot 0 =$ ☐

19. Multiply the following integers.

a. $(-2)(-2)(-2)(-1) =$ ☐

b. $(-1)(-1)(3)(-1) =$ ☐

20. Multiply the following integers.

a. $(-3)(-1)(-3)(-3) =$ ☐

b. $(-3)(-2)(-1)(-2) =$ ☐

21. Evaluate the following.

a. $\frac{-30}{-3} =$ ☐

b. $\frac{36}{-6} =$ ☐

c. $\frac{-28}{7} =$ ☐

22. Evaluate the following.

a. $\frac{-14}{-2} =$ ☐

b. $\frac{12}{-4} =$ ☐

c. $\frac{-63}{7} =$ ☐

23. Evaluate each of the following.

a. $\frac{-10}{-1} =$ ☐

b. $\frac{5}{-1} =$ ☐

c. $\frac{120}{-120} =$ ☐

d. $\frac{-10}{-10} =$ ☐

e. $\frac{9}{0} =$ ☐

f. $\frac{0}{-8} =$ ☐

24. Evaluate each of the following.

a. $\frac{-9}{-1} =$ ☐

b. $\frac{10}{-1} =$ ☐

c. $\frac{160}{-160} =$ ☐

d. $\frac{-13}{-13} =$ ☐

e. $\frac{9}{0} =$ ☐

f. $\frac{0}{-3} =$ ☐

25. Evaluate the following expressions that have integer exponents:

a. $3^2 =$ ☐

b. $4^3 =$ ☐

c. $(-5)^2 =$ ☐

d. $(-4)^3 =$ ☐

26. Evaluate the following expressions that have integer exponents:

a. $3^2 =$ ☐

b. $2^3 =$ ☐

c. $(-4)^2 =$ ☐

d. $(-5)^3 =$ ☐

27. Evaluate the following expressions that have integer exponents:

a. $1^8 =$ ☐

b. $(-1)^{15} =$ ☐

c. $(-1)^{18} =$ ☐

d. $0^{20} =$ ☐

28. Evaluate the following expressions that have integer exponents:

a. $1^8 =$ ☐

b. $(-1)^{13} =$ ☐

c. $(-1)^{14} =$ ☐

d. $0^{18} =$ ☐

29. Evaluate the following expressions that have integer exponents:

a. $(-3)^2 =$ ☐

b. $-10^2 =$ ☐

30. Evaluate the following expressions that have integer exponents:

a. $(-1)^2 =$ ☐

b. $-2^2 =$ ☐

31. Evaluate the following expressions that have integer exponents:

a. $(-1)^3 =$ ☐

b. $-2^3 =$ ☐

32. Evaluate the following expressions that have integer exponents:

a. $(-4)^3 =$ ☐

b. $-1^3 =$ ☐

33. Add these two decimals without using a calculator.

$-2.74 + (-22.3) =$ ☐

34. Add these two decimals without using a calculator.

$-3.47 + (-82.7) =$ ☐

35. Add these two decimals without using a calculator.

$-2.63 + 7.1 =$ ☐

36. Add these two decimals without using a calculator.

$-4.33 + 7.7 =$ ☐

37. Subtract these two decimals without using a calculator.

$-1.03 - (-8.4) =$ ☐

38. Subtract these two decimals without using a calculator.

$-4.73 - (-8.1) =$ ☐

39. It's given that $87 \cdot 85 = 7395$. Use this fact to calculate the following without using a calculator:

$8.7(-8.5) =$ ☐

40. It's given that $94 \cdot 32 = 3008$. Use this fact to calculate the following without using a calculator:

$9.4(-0.032) =$ ☐

41. It's given that $19 \cdot 79 = 1501$. Use this fact to calculate the following without using a calculator:

$(-1.9)(-7.9) =$ ☐

42. It's given that $26 \cdot 16 = 416$. Use this fact to calculate the following without using a calculator:

$(-2.6)(-0.016) =$ ☐

Apply your skills with arithmetic to solve some applied questions.

43. Consider the following situation in which you borrow money from your cousin:

- On June 1st, you borrowed 1100 dollars from your cousin.
- On July 1st, you borrowed 360 more dollars from your cousin.
- On August 1st, you paid back 610 dollars to your cousin.
- On September 1st, you borrowed another 880 dollars from your cousin.

How much money do you owe your cousin now?

44. Consider the following scenario in which you study your bank account.

- On Jan. 1, you had a balance of −270 dollars in your bank account.
- On Jan. 2, your bank charged 50 dollar overdraft fee.
- On Jan. 3, you deposited 990 dollars.
- On Jan. 10, you withdrew 530 dollars.

What is your balance on Jan. 11?

45. A mountain is 1300 feet *above* sea level.

A trench is 420 feet *below* sea level.

What is the difference in elevation between the mountain top and the bottom of the trench?

46. A mountain is 1300 feet *above* sea level.

A trench is 350 feet *below* sea level.

What is the difference in elevation between the mountain top and the bottom of the trench?

1.2 Fractions and Fraction Arithmetic

1.2.1 Breaking Apart Fractions

The word "fraction" comes from the Latin word *fractio*, which means "break into pieces." Ancient cultures all over the world use fractions to understand parts of wholes, but it took humanity thousands of years to develop the symbols we use today.

1.2.1.1 Parts of a Whole

One approach to understanding fractions is to think of them as counting parts of a whole.

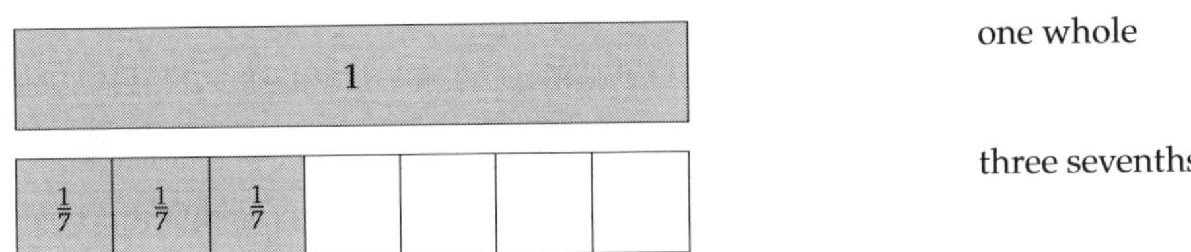

Figure 1.2.1: Representing $\frac{3}{7}$ as parts of a whole.

In Figure 1.2.1, we see 1 whole amount divided into 7 parts. Since 3 of the 7 parts are highlighted, we have an illustration of the fraction $\frac{3}{7}$. The **denominator** 7 lets us know how many equal parts of the whole amount we're considering; since we've got 7 parts here, they're called "sevenths." The **numerator** 3 tells us how many of those sevenths we're considering.

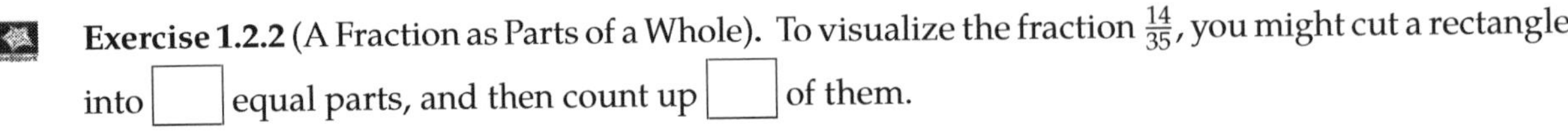

Exercise 1.2.2 (A Fraction as Parts of a Whole). To visualize the fraction $\frac{14}{35}$, you might cut a rectangle into ☐ equal parts, and then count up ☐ of them.

Instead of using rectangles, we can also locate fractions on number lines. When a number line is marked off with whole numbers, equal divisions of the unit 1 can represent the equal parts, as in Figure 1.2.3.

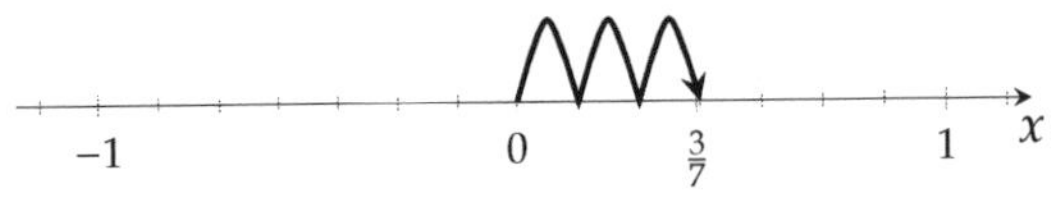

Figure 1.2.3: Representing $\frac{3}{7}$ on a number line.

Exercise 1.2.4 (A Fraction on a Number Line). In the given number line, what fraction is marked?

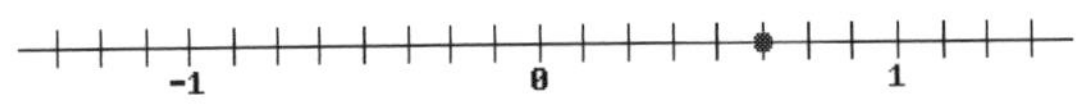

The mark represents the fraction ______.

Hint. Count how many subdivisions there are between 0 and 1.

1.2.1.2 Division

Another helpful way to understand fractions like $\frac{3}{7}$ is to see them as division of the numerator by the denominator. In this case, 3 is divided into 7 parts, as in Figure 1.2.5.

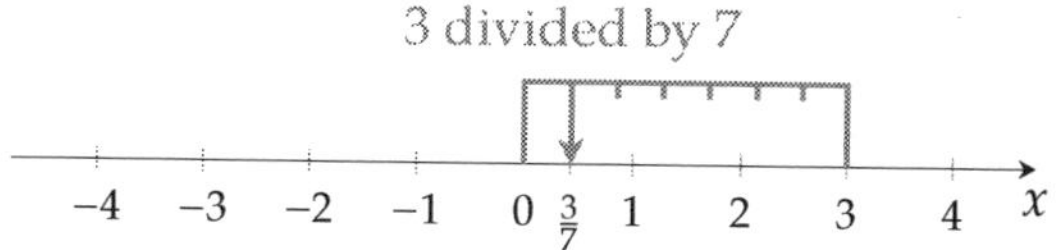

Figure 1.2.5: Representing $\frac{3}{7}$ on a number line.

Exercise 1.2.6 (Seeing a Fraction as Division Arithmetic). The fraction $\frac{21}{40}$ can be thought of as dividing ______ by ______. In other words, dividing the whole number ______ into ______ equal-sized parts.

1.2.2 Equivalent Fractions

It's common to have two fractions that represent the same amount. Consider $\frac{2}{5}$ and $\frac{6}{15}$ represented in various ways in Figures 1.2.7–1.2.9.

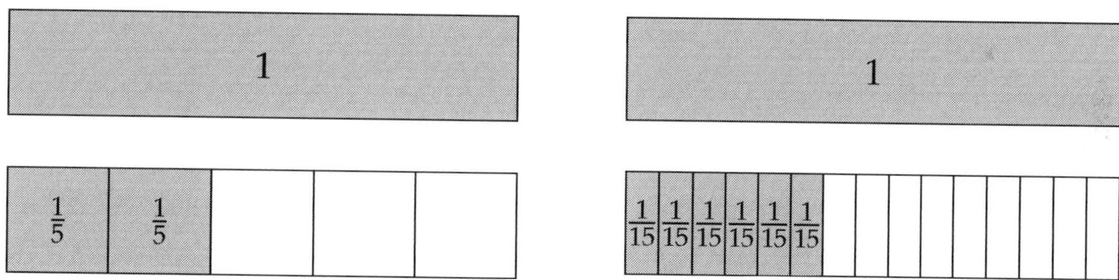

Figure 1.2.7: $\frac{2}{5}$ and $\frac{6}{15}$ as equal parts of a whole

Figure 1.2.8: $\frac{2}{5}$ and $\frac{6}{15}$ as equal on a number line

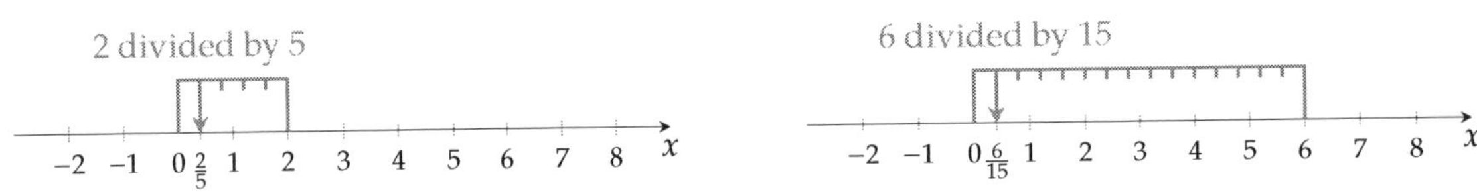

Figure 1.2.9: $\frac{2}{5}$ and $\frac{6}{15}$ as equal results from division

Those two fractions, $\frac{2}{5}$ and $\frac{6}{15}$ are equal, as those figures demonstrate. Also, because they each equal 0.4 as a decimal. If we must work with this number, the fraction that uses smaller numbers, $\frac{2}{5}$, is preferable. Working with smaller numbers decreases the likelihood of making a human arithmetic error. And it also increases the chances that you might make useful observations about the nature of that number.

So if you are handed a fraction like $\frac{6}{15}$, it is important to try to **reduce** it to "lowest terms." The most important skill you can have to help you do this is to know the multiplciation table very well. If you know it well, you know that $6 = 2 \cdot 3$ and $15 = 3 \cdot 5$, so you know

$$\begin{aligned}\frac{6}{15} &= \frac{2\cdot 3}{3\cdot 5}\\ &= \frac{2\cdot \cancel{3}\cdot 1}{1\cdot \cancel{3}\cdot 5}\\ &= \frac{2}{5}\end{aligned}$$

Both the numerator and denominator were divisible by 3, so they could be "factored out" and then as factors, canceled out.

Exercise 1.2.10. Reduce these fractions into lowest terms.

a. $\frac{14}{42} =$ ☐ b. $\frac{8}{30} =$ ☐ c. $\frac{70}{90} =$ ☐

Solution. With $\frac{14}{42}$, we have $\frac{2\cdot 7}{2\cdot 3\cdot 7}$, which reduces to $\frac{1}{3}$.

With $\frac{8}{30}$, we have $\frac{2\cdot 2\cdot 2}{2\cdot 3\cdot 5}$, which reduces to $\frac{4}{15}$.

With $\frac{70}{90}$, we have $\frac{7\cdot 10}{9\cdot 10}$, which reduces to $\frac{7}{9}$.

Sometimes it is useful to do the opposite of reducing a fraction, and **build up** the fraction to use larger numbers.

Exercise 1.2.11. Sayid scored $\frac{21}{25}$ on a recent exam. Build up this fraction so that the denominator is 100, so that Sayid can understand what percent score he earned.

Solution. To change the denominator from 25 to 100, it needs to be multiplied by 4. So we calculate

$$\begin{aligned}\frac{21}{25} &= \frac{21\cdot 4}{25\cdot 4}\\ &= \frac{84}{100}\end{aligned}$$

So the fraction $\frac{21}{25}$ is equivalent to $\frac{84}{100}$. (This means Sayid scored an 84%.)

1.2.3 Multiplying with Fractions

To double a recipe or cut it in half, we need to consider fractions of fractions.

Example 1.2.12. Say a recipe calls for $\frac{2}{3}$ cup of milk, but we'd like to double the recipe. One way to measure this out is to fill a measuring cup to $\frac{2}{3}$, two times:

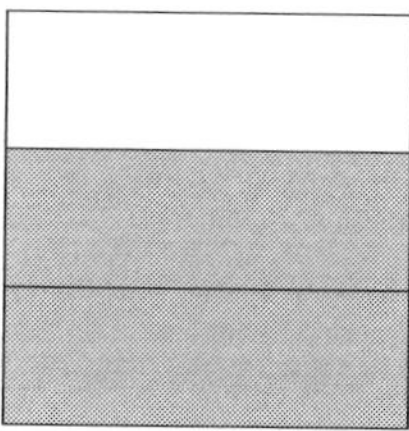

Altogether there are four thirds of a whole here. So $\frac{2}{3} \cdot 2 = \frac{4}{3}$. The figure shows $\frac{2}{3}$ of two wholes. Two wholes can be written as 2, or as the fraction $\frac{2}{1}$. So mathematically, our figure says

$$\frac{2}{3} \cdot 2 = \frac{2}{3} \cdot \frac{2}{1} = \frac{4}{3}.$$

Example 1.2.13. We could also use multiplication to decrease amounts. How much is $\frac{1}{2}$ of $\frac{2}{3}$ cup?

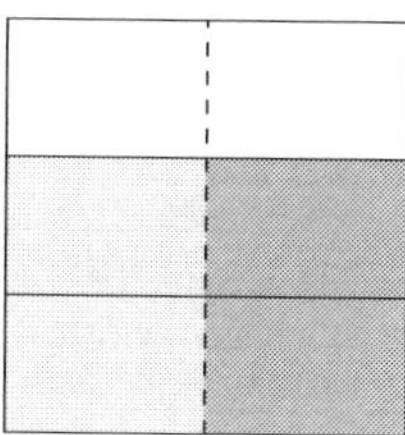

So $\frac{1}{2}$ of $\frac{2}{3}$ cup is $\frac{2}{6}$ cup. Mathematically, we can write

$$\frac{2}{3} \cdot \frac{1}{2} = \frac{2}{6}.$$

In our two examples, we have observed that

$$\frac{2}{3} \cdot \frac{2}{1} = \frac{2 \cdot 2}{3 \cdot 1} \qquad \frac{2}{3} \cdot \frac{1}{2} = \frac{2 \cdot 1}{3 \cdot 2}$$

This idea works generally, no matter what numbers are involved with the fractions.

Fact 1.2.14 (Multiplication with Fractions). *As long as b and d are not 0, then fractions multiply this way:*

$$\frac{a}{b} \cdot \frac{c}{d} = \frac{a \cdot c}{b \cdot d}$$

Try some fraction multiplications for practice:

Exercise 1.2.15. Simplify these fraction products.

a. $\frac{1}{3} \cdot \frac{10}{7} =$ ☐

b. $\frac{12}{3} \cdot \frac{15}{3} =$ ☐

c. $-\frac{14}{5} \cdot \frac{2}{3} =$ ☐

d. $\frac{70}{27} \cdot \frac{12}{-20} =$ ☐

1.2.4 Division with Fractions

How does division with fractions work? Are we able to compute/simplify each of these examples?

$$3 \div \frac{2}{7} \qquad \frac{18}{19} \div 5 \qquad \frac{14}{3} \div \frac{8}{9} \qquad \frac{\frac{2}{5}}{\frac{5}{2}}$$

We know that when we divide something by 2, this is the same as multiplying it by $\frac{1}{2}$. Conversely, dividing a number or expression by $\frac{1}{2}$ is the same as multiplying by $\frac{2}{1}$, or just 2. The more general property is that when we divide a number or expression by $\frac{a}{b}$, this is equivalent to multiplying by the reciprocal $\frac{b}{a}$.

Fact 1.2.16 (Division with Fractions). *As long as b, c and d are not 0, then division with fractions works this way:*

$$\frac{a}{b} \div \frac{c}{d} = \frac{a}{b} \cdot \frac{d}{c}$$

Example 1.2.17. With our examples from the beginning of this subsection:

$$\begin{aligned} 3 \div \frac{2}{7} &= 3 \cdot \frac{7}{2} \\ &= \frac{3}{1} \cdot \frac{7}{2} \\ &= \frac{21}{2} \end{aligned}$$

$$\begin{aligned} \frac{18}{19} \div 5 &= \frac{18}{19} \div \frac{5}{1} \\ &= \frac{18}{19} \cdot \frac{1}{5} \\ &= \frac{18}{95} \end{aligned}$$

$$\begin{aligned} \frac{14}{3} \div \frac{8}{9} &= \frac{14}{3} \cdot \frac{9}{8} \\ &= \frac{14}{1} \cdot \frac{3}{8} \end{aligned}$$

$$\begin{aligned} \frac{\frac{2}{5}}{\frac{5}{2}} &= \frac{2}{5} \div \frac{5}{2} \\ &= \frac{2}{5} \cdot \frac{2}{5} \end{aligned}$$

$$= \frac{7}{1} \cdot \frac{3}{4}$$
$$= \frac{21}{4}$$

$$= \frac{4}{25}$$

Try some divisions with fractions for practice:

Exercise 1.2.18. Simplify these fraction division expressions.

a. $\frac{1}{3} \div \frac{10}{7} =$ ☐

b. $\frac{12}{5} \div 5 =$ ☐

c. $-14 \div \frac{3}{2} =$ ☐

d. $\frac{70}{9} \div \frac{11}{-20} =$ ☐

1.2.5 Adding and Subtracting Fractions

With whole numbers and integers, operations of addition and subtraction are relatively straightforward. The situation is almost as straightforward with fractions *if the two fractions have the same denominator*. Consider

$$\frac{7}{2} + \frac{3}{2} = 7 \text{ halves} + 3 \text{ halves}$$

In the same way that 7 tacos and 3 tacos make 10 tacos, we have:

$$\begin{array}{ccccc} 7 \text{ halves} & + & 3 \text{ halves} & = & 10 \text{ halves} \\ \frac{7}{2} & + & \frac{3}{2} & = & \frac{10}{2} \\ & & & = & 5 \end{array}$$

Fact 1.2.19 (Adding/Subtracting with Fractions Having the Same Denominator). *To add or subtract two fractions having the same denominator, keep that denominator, and add or subtract the numerators.*

$$\frac{a}{b} + \frac{c}{b} = \frac{a+c}{b} \qquad \frac{a}{b} - \frac{c}{b} = \frac{a-c}{b}$$

If it's possible, useful, or required of you, simplify the result by reducing to lowest terms.

Exercise 1.2.20. Add or subtract these fractions.

a. $\frac{1}{3} + \frac{10}{3} =$ ☐

b. $\frac{13}{6} - \frac{5}{6} =$ ☐

Whenever we'd like to combine fractional amounts that don't represent the same number of parts of a whole (that is, when the denominators are different), finding sums and differences is more complicated.

Example 1.2.21 (Quarters and Dimes). Find the sum $\frac{3}{4} + \frac{2}{10}$. Does this seem intimidating? Consider this:

- $\frac{1}{4}$ of a dollar is a quarter, and so $\frac{3}{4}$ of a dollar is 75 cents.

- $\frac{1}{10}$ of a dollar is a dime, and so $\frac{2}{10}$ of a dollar is 20 cents.

So if you know what to look for, the expression $\frac{3}{4}+\frac{2}{10}$ is like adding 75 cents and 20 cents, which gives you 95 cents. As a fraction of one dollar, that is $\frac{95}{100}$. So we can report

$$\frac{3}{4}+\frac{2}{10}=\frac{95}{100}.$$

(Although we should probably reduce that last fraction to $\frac{19}{20}$.)

This example was not something you can apply to other fraction addition situations, because the denominators here worked especially well with money amounts. But there is something we can learn here. The fraction $\frac{3}{4}$ was equivalent to $\frac{75}{100}$, and the other fraction $\frac{2}{10}$ was equivalent to $\frac{20}{100}$. These *equivalent* fractions have the same denominator and are therefore "easy" to add. What we saw happen was:

$$\begin{aligned}\frac{3}{4}+\frac{2}{10}&=\frac{75}{100}+\frac{20}{100}\\&=\frac{95}{100}\end{aligned}$$

This realization gives us a strategy for adding (or subtracting) fractions.

Fact 1.2.22 (Adding/Subtracting Fractions with Different Denominators). *To add (or subtract) generic fractions together, use their denominators to find a* ***common denominator****. This means some whole number that is a whole multiple of both of the original denominators. Then rewrite the two fractions as equivalent fractions that use this common denominator. Write the result keeping that denominator and adding (or subtracting) the numerators. Reduce the fraction if that is useful or required.*

Example 1.2.23. Let's add $\frac{2}{3}+\frac{2}{5}$. The denominators are 3 and 5, so the number 15 would be a good common denominator.

$$\begin{aligned}\frac{2}{3}+\frac{2}{5}&=\frac{2\cdot5}{3\cdot5}+\frac{2\cdot3}{5\cdot3}\\&=\frac{10}{15}+\frac{6}{15}\\&=\frac{16}{15}\end{aligned}$$

Exercise 1.2.24. A chef had $\frac{2}{3}$ cups of flour and needed to use $\frac{1}{8}$ cup to thicken a sauce. How much flour is left? [ ]

Solution. We need to compute

$$\frac{2}{3}-\frac{1}{8}$$

The denominators are 3 and 8. One common denominator is 24, so we move to rewrite each fraction using 24 as the denominator:

$$\begin{aligned}\frac{2}{3}-\frac{1}{8}&=\frac{2\cdot 8}{3\cdot 8}-\frac{1\cdot 3}{8\cdot 3}\\&=\frac{16}{24}-\frac{3}{24}\\&=\frac{13}{24}\end{aligned}$$

The numerical result is $\frac{13}{24}$, but a pure number does not answer this question. The amount of flour remaining is $\frac{13}{24}$ *cups*.

1.2.6 Mixed Numbers and Improper Fractions

A simple recipe for bread contains only a few ingredients:

1 1/2	tablespoons yeast
1 1/2	tablespoons kosher salt
6 1/2	cups unbleached, all-purpose flour (more for dusting)

Table 1.2.25: Ingredients for simple crusty bread.

Each ingredient is listed as a **mixed number** that quickly communicates how many whole amounts and how many parts are needed. It's useful for quickly communicating a practical amount of something you are cooking with, measuring on a ruler, purchasing at the grocery store, etc. But it causes trouble in an algebra class. The number 1 1/2 means "one ***and*** one half." So really,

$$1\frac{1}{2}=1+\frac{1}{2}$$

The trouble is that with 1 1/2, you have two numbers written right next to each other. Normally with two math expressions written right next to each other, they should be *multiplied*, not *added*. But with a mixed number, they *should* be added.

Fortunately we just reviewed how to add fractions. If we need to do any arithmetic with a mixed number like 1 1/2, we can treat it as $1+\frac{1}{2}$ and simplify to get a "nice" fraction instead:

$$\begin{aligned}1\frac{1}{2}&=1+\frac{1}{2}\\&=\frac{1}{1}+\frac{1}{2}\\&=\frac{2}{2}+\frac{1}{2}\\&=\frac{3}{2}\end{aligned}$$

A fraction like $\frac{3}{2}$ is called an **improper fraction** because it's actually larger than 1. And a "proper" fraction would be something small that is only *part* of a whole instead of *more* than a whole.

1.2.7 Exercises

Multiplying/Dividing Fractions

1. Multiply these two fractions: $\frac{4}{9} \cdot \frac{5}{9}$
2. Multiply these two fractions: $\frac{2}{7} \cdot \frac{2}{9}$
3. Multiply these two fractions: $\frac{15}{7} \cdot \frac{5}{3}$
4. Multiply these two fractions: $\frac{3}{2} \cdot \frac{5}{12}$
5. Multiply these two fractions: $-\frac{8}{3} \cdot \frac{7}{22}$
6. Multiply these two fractions: $-\frac{14}{13} \cdot \frac{7}{6}$
7. Multiply the integer with the fraction: $15 \cdot \left(-\frac{4}{3}\right)$
8. Multiply the integer with the fraction: $45 \cdot \left(-\frac{5}{9}\right)$
9. Multiply these fractions: $\frac{7}{25} \cdot \frac{6}{49} \cdot \frac{5}{4}$
10. Multiply these fractions: $\frac{6}{25} \cdot \frac{7}{9} \cdot \frac{5}{49}$
11. Carry out the division: $\frac{5}{6} \div \frac{8}{5}$
12. Carry out the division: $\frac{5}{8} \div \frac{8}{5}$
13. Carry out the division: $\frac{2}{15} \div \left(-\frac{5}{9}\right)$
14. Carry out the division: $\frac{1}{6} \div \left(-\frac{8}{9}\right)$
15. Carry out the division: $-\frac{10}{9} \div (-25)$
16. Carry out the division: $-\frac{12}{7} \div (-8)$
17. Carry out the division: $12 \div \frac{3}{4}$
18. Carry out the division: $9 \div \frac{9}{4}$
19. Multiply the following: $1\frac{11}{25} \cdot 1\frac{7}{18}$
20. Multiply the following: $1\frac{3}{25} \cdot 2\frac{11}{12}$

Adding/Subtracting Fractions

21. Add these two fractions: $\frac{1}{20} + \frac{1}{20}$
22. Add these two fractions: $\frac{1}{24} + \frac{19}{24}$
23. Add these two fractions: $\frac{3}{10} + \frac{31}{60}$
24. Add these two fractions: $\frac{1}{9} + \frac{4}{27}$
25. Add these two fractions: $\frac{4}{9} + \frac{19}{45}$
26. Add these two fractions: $\frac{5}{6} + \frac{7}{30}$
27. Add these two fractions: $\frac{3}{8} + \frac{4}{9}$
28. Add these two fractions: $\frac{2}{9} + \frac{1}{6}$
29. Add these two fractions: $\frac{1}{5} + \frac{3}{10}$
30. Add these two fractions: $\frac{1}{6} + \frac{3}{10}$
31. Add these two fractions: $-\frac{1}{5} + \frac{3}{5}$
32. Add these two fractions: $-\frac{2}{5} + \frac{4}{5}$

33. Add these two fractions: $-\frac{2}{9}+\frac{16}{27}$

34. Add these two fractions: $-\frac{1}{10}+\frac{7}{60}$

35. Add these two fractions: $-\frac{7}{8}+\frac{2}{5}$

36. Add these two fractions: $-\frac{5}{8}+\frac{3}{5}$

37. Add these together: $3+\frac{3}{4}$

38. Add these together: $4+\frac{1}{10}$

39. Add these fractions: $\frac{2}{5}+\frac{1}{10}+\frac{1}{3}$

40. Add these fractions: $\frac{1}{3}+\frac{1}{5}+\frac{1}{6}$

41. Subtract one fraction from the other: $\frac{11}{12}-\frac{1}{12}$

42. Subtract one fraction from the other: $\frac{15}{14}-\frac{9}{14}$

43. Subtract one fraction from the other: $\frac{29}{50}-\frac{3}{10}$

44. Subtract one fraction from the other: $\frac{23}{60}-\frac{3}{10}$

45. Subtract one fraction from the other: $-\frac{3}{10}-\frac{1}{6}$

46. Subtract one fraction from the other: $-\frac{3}{10}-\frac{1}{6}$

47. Subtract one fraction from the other: $-\frac{3}{10}-\left(-\frac{4}{5}\right)$

48. Subtract one fraction from the other: $-\frac{5}{6}-\left(-\frac{3}{10}\right)$

49. Carry out the subtraction: $-5-\frac{11}{2}$

50. Carry out the subtraction: $-4-\frac{13}{8}$

Use your fraction arithmetic skills to solve some applications.

51. Jon walked $\frac{1}{6}$ of a mile in the morning, and then walked $\frac{4}{11}$ of a mile in the afternoon. How far did Jon walk altogether?

Jon walked a total of [ ] of a mile.

52. Kimball and Kenji are sharing a pizza. Kimball ate $\frac{1}{7}$ of the pizza, and Kenji ate $\frac{1}{6}$ of the pizza. How much of the pizza was eaten in total?

They ate [ ] of the pizza.

53. Find the perimeter of the rectangle.

Its perimeter is [ ] meters. (Use a fraction in your answer.)

54. The pie chart represents a school's student population.

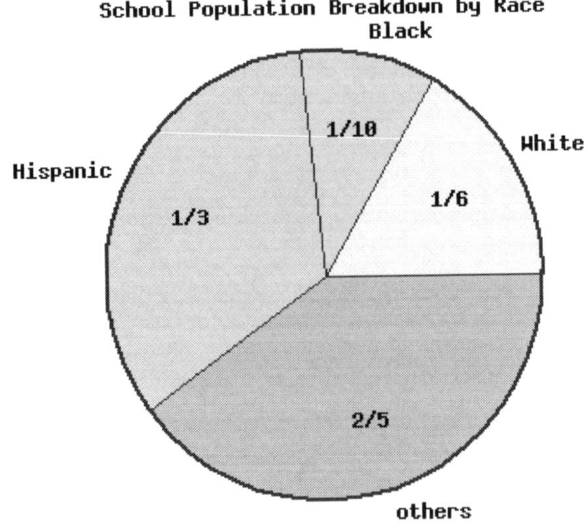

Fill in the blank with a fraction.

Together, white and black students make up ______ of the school's population.

55. A trail's total length is $\frac{25}{63}$ of a mile. It has two legs. The first leg is $\frac{2}{7}$ of a mile long. How long is the second leg?

The second leg is ______ of a mile in length.

56. Each page of a book is $5\frac{5}{6}$ inches in height, and consists of a header (a top margin), a footer (a bottom margin), and the middle part (the body). The header is $\frac{7}{9}$ of an inch thick and the middle part is $4\frac{1}{9}$ inches from top to bottom.

What is the thickness of the footer?

The footer is ______ of an inch thick.

57. Hayden and Eric are sharing a pizza. Hayden ate $\frac{3}{10}$ of the pizza, and Eric ate $\frac{1}{8}$ of the pizza. How much more pizza did Hayden eat than Eric?

Hayden ate ______ more of the pizza than Eric ate.

58. A school had a fund raising event. The revenue came from three resources: ticket sales, auction sales, and donations. Ticket sales account for $\frac{3}{5}$ of the total revenue; auction sales account for $\frac{3}{8}$ of the total revenue. What fraction of the revenue came from donations?

______ of the revenue came from donations.

59. The pie chart represents a school's student population.

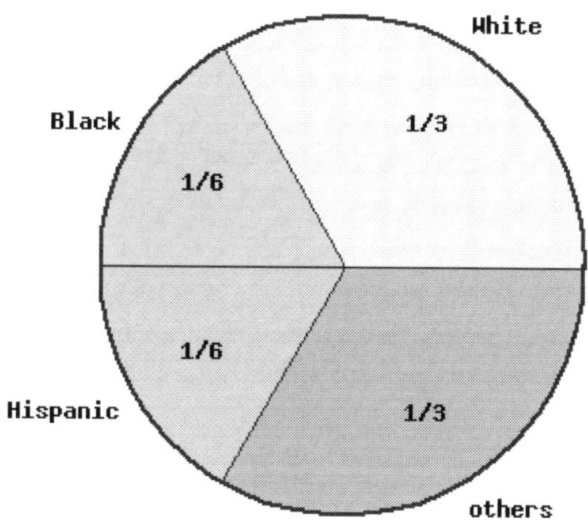

Answer the following question with fraction.

[ ] more of the school is white students than black students.

60. A few years back, a car was purchased for \$14,000. Today it is worth $\frac{1}{4}$ of its original value. What is the car's current value?

The car's current value is [ ].

61. A town has 200 residents in total, of which $\frac{3}{4}$ are Native Americans. How many Native Americans reside in this town?

There are [ ] Native Americans residing in this town.

62. A company received a grant, and decided to spend $\frac{5}{6}$ of this grant in research and development next year. Out of the money set aside for research and development, $\frac{2}{3}$ will be used to buy new equipment. What fraction of the grant will be used to buy new equipment?

[ ] of the grant will be used to buy new equipment.

63. Find the area of the rectangle.

Its area is [ ] square meters. (Use a fraction in your answer.)

64. A food bank just received 16 kilograms of emergency food. Each family in need is to receive $\frac{2}{5}$ kilograms of food. How many families can be served with the 16 kilograms of food?

[ ] families can be served with the 16 kilograms of food.

65. A construction team maintains a 75-mile-long sewage pipe. Each day, the team can cover $\frac{5}{9}$ of a mile. How many days will it take the team to complete the maintenance of the entire sewage pipe?

It will take the team [ ] days to complete maintaining the entire sewage pipe.

66. A child is stacking up tiles. Each tile's height is $\frac{2}{3}$ of a centimeter. How many layers of tiles are needed to reach 12 centimeters in total height?

To reach the total height of 12 centimeters, [ ] layers of tiles are needed.

67. A restaurant made 100 cups of pudding for a festival.

Customers at the festival will be served $\frac{1}{9}$ of a cup of pudding per serving. How many customers can the restaurant serve at the festival with the 100 cups of pudding?

The restaurant can serve [ ] customers at the festival with the 100 cups of pudding.

68. A 2×4 piece of lumber in your garage is $66\frac{1}{4}$ inches long. A second 2×4 is $38\frac{1}{8}$ inches long. If you lay them end to end, what will the total length be?

The total length will be [ ] inches.

69. A 2×4 piece of lumber in your garage is $42\frac{1}{8}$ inches long. A second 2×4 is $68\frac{1}{4}$ inches long. If you lay them end to end, what will the total length be?

The total length will be [ ] inches.

70. Each page of a book consists of a header, a footer and the middle part. The header is $\frac{7}{10}$ inches in height; the footer is $\frac{9}{20}$ inches in height; and the middle part is $6\frac{1}{10}$ inches in height.

What is the total height of each page in this book? Use mixed number in your answer if needed.

Each page in this book is [ ] inches in height.

71. When driving on a high way, noticed a sign saying exit to Johnstown is $1\frac{1}{2}$ miles away, while exit to Jerrystown is $3\frac{1}{4}$ miles away. How far is Johnstown from Jerrystown?

Johnstown and Jerrystown are [ ] miles apart.

72. A cake recipe needs $1\frac{1}{5}$ cups of flour. Using this recipe, to bake 6 cakes, how many cups of flour are needed?

To bake 6 cakes, [ ] cups of flour are needed.

73. Find the area of the triangle.

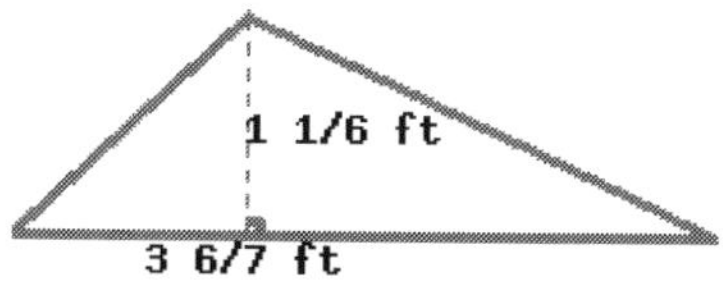

1 1/6 ft

3 6/7 ft

Its area is [ ] square feet. (Use a fraction in your answer.)

1.3 Absolute Value and Square Root

In this section, we will learn the basics of **absolute value** and **square root**. These are actions you can *do* to a given number, often changing the number into something else.

1.3.1 Introduction to Absolute Value

Definition 1.3.2. The **absolute value** of a number is the distance between that number and 0 on a number line. For the absolute value of x, we write $|x|$.

Let's look at $|2|$ and $|-2|$, the absolute value of 2 and -2.

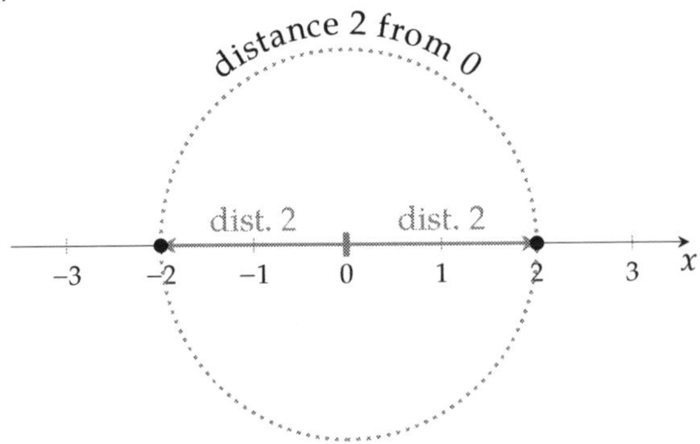

Figure 1.3.3: $|2|$ and $|-2|$

Since the distance between 2 and 0 on the number line is 2 units, the absolute value of 2 is 2. We write $|2| = 2$.

Since the distance between -2 and 0 on the number line is also 2 units, the absolute value of -2 is also 2. We write $|-2| = 2$.

Fact 1.3.4 (Absolute Value). *Taking the absolute value of a number results in whatever the "positive version" of that number is. This is because the real meaning of absolute value is its distance from zero.*

Exercise 1.3.5 (Calculating Absolute Value). Try calculating some absolute values.

a. $|57| =$ ____ b. $|-43| =$ ____ c. $\left|\frac{2}{-5}\right| =$ ____

Solution.

a. 57 is 57 units away from 0 on a number line, so $|57| = 57$. Another way to think about this is that the "positive version" of 57 is 57.

b. -43 is 43 units away from 0 on a number line, so $|-43| = 43$. Another way to think about this is that the "positive version" of -43 is 43.

c. $\frac{2}{-5}$ is $\frac{2}{5}$ units away from 0 on a number line, so $\left|\frac{2}{-5}\right| = \frac{2}{5}$. Another way to think about this is that the "positive version" of $\frac{2}{-5}$ is $\frac{2}{5}$.

Warning 1.3.6 (Absolute Value Does Not Exactly "Make Everything Positive"). Students may see an expression like $|2-5|$ and incorrectly think it is OK to "make everything positive" and write $2+5$. This is incorrect since $|2-5|$ works out to be 3, not 7, as we are actually taking the absolute value of -3 (the equivalent number inside the absolute value).

1.3.2 Square Root Facts

If you have learned your basic multiplication table, you know:

×	1	2	3	4	5	6	7	8	9
1	1	2	3	4	5	6	7	8	9
2	2	4	6	8	10	12	14	16	18
3	3	6	9	12	15	18	21	24	27
4	4	8	12	16	20	24	28	32	36
5	5	10	15	20	25	30	35	40	45
6	6	12	18	24	30	36	42	48	54
7	7	14	21	28	35	42	49	56	63
8	8	16	24	32	40	48	56	64	72
9	9	18	27	36	45	54	63	72	81

Table 1.3.7: Multiplication table with squares

The numbers along the diagonal are special; they are known as **perfect squares**. And for working with square roots, it will be helpful if you can memorize these first few perfect square numbers.

"Taking a square root" is the opposite action of squaring a number. For example, when you square 3, the result is 9. So when you take the square root of 9, the result is 3. Just knowing that 9 comes about as 3^2 lets us realize that 3 is the square root of 9. This is why memorizing the perfect squares from the multiplication table can be so helpful.

The notation we use for taking a square root is the **radical**, $\sqrt{}$. For example, "the square root of 9" is denoted $\sqrt{9}$. And now we know enough to be able to write $\sqrt{9} = 3$.

Tossing in a few extra special square roots, it's advisable to memorize the following:

$$\begin{array}{llll} \sqrt{0} = 0 & \sqrt{1} = 1 & \sqrt{4} = 2 & \sqrt{9} = 3 \\ \sqrt{16} = 4 & \sqrt{25} = 5 & \sqrt{36} = 6 & \sqrt{49} = 7 \\ \sqrt{64} = 8 & \sqrt{81} = 9 & \sqrt{100} = 10 & \sqrt{121} = 11 \\ \sqrt{144} = 12 & \sqrt{169} = 13 & \sqrt{196} = 14 & \sqrt{225} = 15 \end{array}$$

1.3.3 Calculating Square Roots with a Calculator

Most square roots are actually numbers with decimal places that go on forever. Take $\sqrt{5}$ as an example:

$$\sqrt{4} = 2 \qquad \sqrt{5} = ? \qquad \sqrt{9} = 3$$

Since 5 is between 4 and 9, then $\sqrt{5}$ must be somewhere between 2 and 3. There are no whole numbers between 2 and 3, so $\sqrt{5}$ must be some number with decimal places. If the decimal places eventually stopped, then squaring it would give you another number with decimal places that stop further out. But squaring it gives you 5 with no decimal places. So the only possibility is that $\sqrt{5}$ is a decimal between 2 and 3 that goes on forever. With a calculator, we can see

$$\sqrt{5} \approx 2.236.$$

Actually the decimal will not terminate, and that is why we used the $\approx$ symbol instead of an equals sign. To get 2.236 we rounded down slightly from the true value of $\sqrt{5}$. With a calculator, we can check that $2.236^2 = 4.999696$, a little shy of 5.

1.3.4 Square Roots of Fractions

We can calculate the square root of some fractions by hand, such as $\sqrt{\frac{1}{4}}$. The idea is the same: can you think of a number that you would square to get $\frac{1}{4}$? Being familiar with fraction multiplication, we know that

$$\frac{1}{2} \cdot \frac{1}{2} = \frac{1}{4}$$

and so $\sqrt{\frac{1}{4}} = \frac{1}{2}$.

Exercise 1.3.8 (Square Roots of Fractions). Try calculating some absolute values.

a. $\sqrt{\frac{1}{25}} = \square$ b. $\sqrt{\frac{4}{9}} = \square$ c. $\sqrt{\frac{81}{121}} = \square$

1.3.5 Square Root of Negative Numbers

Can we find the square root of a negative number, such as $\sqrt{-25}$? That would mean that there is some number out there that multiplies by itself to make -25. Would $\sqrt{-25}$ be positive or negative?

Either way, once you square it (multiply it by itself) the result would be positive. So it couldn't possibly square to −25. So there is no square root of −25 or of any negative number for that matter.

Imaginary Numbers Mathematicians imagined a new type of number, neither positive nor negative, that would square to a negative result. But that is beyond the scope of this chapter.

If you are confronted with an expression like $\sqrt{-25}$, or any other square root of a negative number, you can state that "there is no real square root," that the result "does not exist," or just "DNE" for short.

1.3.6 Exercises

These skills practice familiarity with absolute value.

1. Find the absolute value of this number.

$|-9| =$ ☐

2. Find the absolute value of this number.

$|-7| =$ ☐

3. Find the absolute value of the following numbers.

a. $|4| =$ ☐

b. $|-4| =$ ☐

c. $-|4| =$ ☐

d. $-|-4| =$ ☐

4. Find the absolute value of the following numbers.

a. $|5| =$ ☐

b. $|-5| =$ ☐

c. $-|5| =$ ☐

d. $-|-5| =$ ☐

5. Evaluate the following expressions which involve the absolute value:

a. $|5| =$ ☐

b. $|-7| =$ ☐

c. $|0| =$ ☐

d. $|19 + (-2)| =$ ☐

e. $|-5 - (-1)| =$ ☐

6. Evaluate the following expressions which involve the absolute value:

a. $|7| =$ ☐

b. $|-1| =$ ☐

c. $|0| =$ ☐

d. $|13 + (-4)| =$ ☐

e. $|-5 - (-3)| =$ ☐

7. Evaluate the following expressions which involve the absolute value:

a. $-|4 - 9| =$ ☐

b. $|-4 - 9| =$ ☐

c. $-2|9 - 4| =$ ☐

8. Evaluate the following expressions which involve the absolute value:

a. $-|2 - 6| =$ ☐

b. $|-2 - 6| =$ ☐

c. $-2|6 - 2| =$ ☐

These skills practice familiarity with square roots.

9. Which of the following are square numbers? There may be more than one correct answer.

☐ 130 ☐ 144 ☐ 121 ☐ 36
☐ 113 ☐ 137

10. Which of the following are square numbers? There may be more than one correct answer.

☐ 73 ☐ 1 ☐ 129 ☐ 144
☐ 66 ☐ 64

11. Find the square root of the following numbers:

a. $\sqrt{4} =$ ☐

b. $\sqrt{16} =$ ☐

c. $\sqrt{36} =$ ☐

12. Find the square root of the following numbers:

a. $\sqrt{16} =$ ☐

b. $\sqrt{121} =$ ☐

c. $\sqrt{81} =$ ☐

13. Find the square root of the following numbers.

a. $\sqrt{\frac{25}{49}} =$ ____

b. $\sqrt{-\frac{9}{100}} =$ ____

14. Find the square root of the following numbers.

a. $\sqrt{\frac{36}{25}} =$ ____

b. $\sqrt{-\frac{121}{16}} =$ ____

15. Find the square root of the following numbers without using a calculator.

a. $\sqrt{64} =$ ____

b. $\sqrt{0.64} =$ ____

c. $\sqrt{6400} =$ ____

16. Find the square root of the following numbers without using a calculator.

a. $\sqrt{81} =$ ____

b. $\sqrt{0.81} =$ ____

c. $\sqrt{8100} =$ ____

17. Find the square root of the following numbers without using a calculator.

a. $\sqrt{100} =$ ____

b. $\sqrt{10000} =$ ____

c. $\sqrt{1000000} =$ ____

18. Find the square root of the following numbers without using a calculator.

a. $\sqrt{144} =$ ____

b. $\sqrt{14400} =$ ____

c. $\sqrt{1440000} =$ ____

19. Find the square root of the following numbers without using a calculator.

a. $\sqrt{1} =$ ____

b. $\sqrt{0.01} =$ ____

c. $\sqrt{0.0001} =$ ____

20. Find the square root of the following numbers without using a calculator.

a. $\sqrt{4} =$ ____

b. $\sqrt{0.04} =$ ____

c. $\sqrt{0.0004} =$ ____

21. Without using a calculator, estimate the value of $\sqrt{28}$:

- ○ 5.29
- ○ 5.71
- ○ 4.29
- ○ 4.71

22. Without using a calculator, estimate the value of $\sqrt{38}$:

- ○ 6.16
- ○ 6.84
- ○ 5.16
- ○ 5.84

23. Simplify the radical expression or state that it is not a real number.

$\sqrt{\frac{36}{121}}$ is ______.

24. Simplify the radical expression or state that it is not a real number.

$\sqrt{\frac{49}{81}}$ is ______.

25. Simplify the radical expression or state that it is not a real number.

$-\sqrt{81}$ is ______.

26. Simplify the radical expression or state that it is not a real number.

$-\sqrt{121}$ is ______.

27. Simplify the radical expression or state that it is not a real number.

$\sqrt{-144}$ is ______.

28. Simplify the radical expression or state that it is not a real number.

$\sqrt{-4}$ is ______.

29. Simplify the radical expression or state that it is not a real number.

$\sqrt{-\frac{4}{121}}$ is ______.

30. Simplify the radical expression or state that it is not a real number.

$\sqrt{-\frac{9}{64}}$ is ______.

31. Simplify the radical expression or state that it is not a real number.

$-\sqrt{\frac{25}{81}}$ is ______.

32. Simplify the radical expression or state that it is not a real number.

$-\sqrt{\frac{36}{121}}$ is ______.

33. Simplify the radical expression or state that it is not a real number.

a. $\sqrt{100} - \sqrt{64} =$ ______

b. $\sqrt{100 - 64} =$ ______

34. Simplify the radical expression or state that it is not a real number.

a. $\sqrt{25} - \sqrt{9} =$ ______

b. $\sqrt{25 - 9} =$ ______

35. Rationalize the denominator and simplify the expression.

$\frac{9}{\sqrt{25}} = \boxed{}.$

36. Rationalize the denominator and simplify the expression.

$\frac{9}{\sqrt{16}} = \boxed{}.$

1.4 Order of Operations

Mathematical symbols are a means of communication, and it's important that when you write something, everyone else knows exactly what you intended. For example, if we say in English, "two times three squared," do we mean that:

- 2 is multiplied by 3, and then the result is squared?
- or that 2 is multiplied by the result of squaring 3?

English is allowed to have ambiguities like this. But mathematical language needs to be precise and mean the same thing to everyone reading it. For this reason, a standard **order of operations** has been adopted, which we review here.

1.4.1 Grouping Symbols

Consider the math expression $2 \cdot 3^2$. There are two mathematical operations here: a multiplication and an exponentiation. The result of this expression will change depending on which operation you decide to execute first: the multiplication or the exponentiation. If you multiply $2 \cdot 3$, and then square the result, you have 36. If you square 3, and then multiply 2 by the result, you have 18. If we want all people everywhere to interpret $2 \cdot 3^2$ in the same way, then only *one* of these can be correct.

The first tools that we have to tell readers what operations to execute first are grouping symbols, like parentheses and brackets. If you *intend* to execute the multiplication first, then writing

$$(2 \cdot 3)^2$$

clearly tells your reader to do that. And if you *intend* to execute the power first, then writing

$$2 \cdot \left(3^2\right)$$

clearly tells your reader to do that.

To visualize the difference between $2 \cdot \left(3^2\right)$ or $(2 \cdot 3)^2$, consider these garden plots:

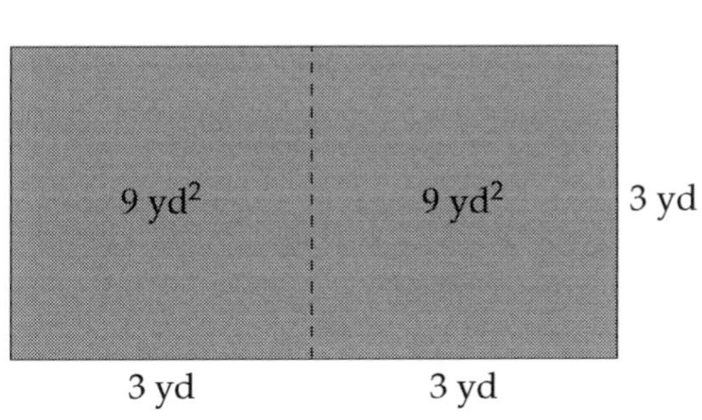

Figure 1.4.2: 3 yd is squared, then doubled: $2 \cdot (3^2)$

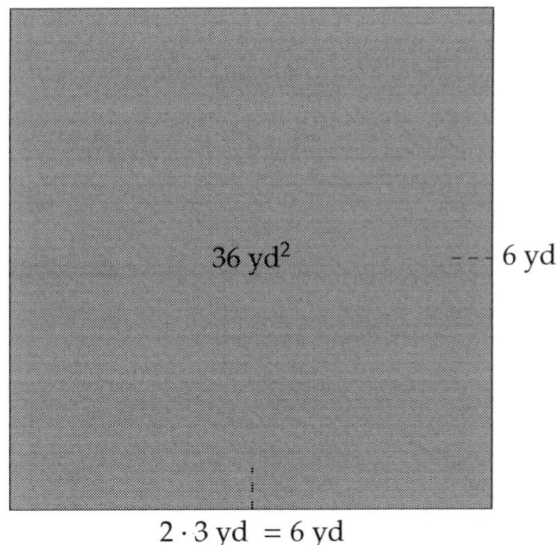

Figure 1.4.3: 3 yd is doubled, then squared: $(2 \cdot 3)^2$

If we calculate 3^2, we have the area of one of the small square garden plots on the left. If we then double that, we have $2 \cdot (3^2)$, the area of the left garden plot.

But if we calculate $(2 \cdot 3)^2$, then first we are doubling 3. So we are calculating the area of a square garden plot whose sides are twice as long. We end up with the area of the garden plot on the right.

The point is that these amounts are different.

Exercise 1.4.4. Try simplifying this expression, focusing on executing the content inside the grouping symbols first.

Calculate the value of $30 - ((2 + 3) \cdot 2)$, respecting the order that the grouping symbols are telling you to execute the arithmetic operations.

Solution. The grouping symbols tell us what to work on first. In this exercise, we have grouping symbols within grouping symbols, so any opearation in there (the addition) should be executed first:

$$\begin{aligned} 30 - ((2 + 3) \cdot 2) &= 30 - (5 \cdot 2) \\ &= 30 - 10 \\ &= 20 \end{aligned}$$

1.4.2 Beyond Grouping Symbols

If all math expressions used grouping symbols for each and every arithmetic operation, we wouldn't need to say anything more here. In fact, some computer systems work that way, *requiring* the use of grouping symbols all the time. But it is much more common to permit math expressions with no grouping symbols at all, like $5 + 3 \cdot 2$. Should the addition $5 + 3$ be executed first, or should the multiplication $3 \cdot 2$? We need what's known formally as the **order of operations** to tell us what to do.

The **order of operations** is nothing more than an agreement that we all have made to prioritize the arithmetic operations in a certain order.

(P)arentheses and other grouping symbols Grouping symbols should always direct you to the highest priority arithmetic first.

(E)xponentiation After grouping symbols, exponentiation has the highest priority. Excecute any exponentiation before other arithmetic operations.

(M)ultiplication, (D)ivision, and Negation After all exponentiation has been executed, start executing multiplications, divisions, and negations. These things all have equal priority. If there are more than one of them in your expression, the highest priority is the one that is leftmost (which comes first as you read it).

(A)ddition and (S)ubtraction After all other arithmetic has been executed, these are all that is left. Addition and subtraciton have equal priority. If there are more than one of them in your expression, the highest priority is the one that is leftmost (which comes first as you read it).

List 1.4.5: Order of Operations

A common acronym to help you remember this order of operations is PEMDAS. There are a handful of mnemonic devices for remembering this ordering (such as Please Excuse My Dear Aunt Sally, People Eat More Donuts After School, etc.).

We'll start with a few examples that only invoke a few operations each.

Example 1.4.6. Use the order of operations to simplify the following expressions.

a. $10 + 2 \cdot 3$. With this expression, we have the operations of addition and multiplication. The order of operations says the multiplication has higher priority, so execute that first:

$$\begin{aligned} 10 + 2 \cdot 3 &= 10 + \overbrace{2 \cdot 3} \\ &= 10 + 6 \\ &= 16 \end{aligned}$$

b. $4+10\div2-1$. With this expression, we have addition, division, and subtraction. According to the order of operations, the first thing we need to do is divide. After that, we'll apply the addition and subtraction, working left to right:

$$\begin{aligned} 4+10\div 2-1 &= 4+\overbrace{10\div 2}-1 \\ &= \overbrace{4+5}-1 \\ &= 9-1 \\ &= 8 \end{aligned}$$

c. $7-10+4$. This example *only* has subtraction and addition. While the acronym PEMDAS may mislead you to do addition before subtraction, remember that these operations have the same priority, and so we work left to right when executing them:

$$\begin{aligned} 7-10+4 &= \overbrace{7-10}+4 \\ &= -3+4 \\ &= 1 \end{aligned}$$

d. $20\div 4\cdot 7$. This expression has only division and multiplication. Again, remember that although PEMDAS shows "MD," the operations of multiplication and division have the same priority, so we'll apply them left to right:

$$\begin{aligned} 20\div 4\cdot 5 &= \overbrace{20\div 4}\cdot 5 \\ &= 5\cdot 5 \\ &= 25 \end{aligned}$$

e. $(6+7)^2$. With this expression, we have addition inside a set of parentheses, and an exponent of 2 outside of that. We must compute the operation inside the parentheses first, and after that we'll apply the exponent:

$$\begin{aligned} (6+7)^2 &= (\overbrace{6+7})^2 \\ &= 13^2 \\ &= 169 \end{aligned}$$

f. $4(2)^3$. This expression has multiplication and an exponent. There are parentheses too, but no operation inside them. Parentheses used in this manner make it clear that the 4 and 2 are separate numbers, not to be confused with 42. In other words, $4(2)^3$ and 42^3 mean very different things. Exponentiation has the higher priority, so we'll apply the exponent first, and then we'll multiply:

$$4(2)^3 = 4\ \overbrace{(2)^3}$$

$$= 4(8)$$
$$= 32$$

Remark 1.4.7. There are many different ways that we write multiplication. We can use the symbols $\cdot$, $\times$, and $*$ to denote multiplication. We can also use parentheses to denote multiplication, as we've seen in Example 1.4.6, Item f. Once we start working with variables, there is even another way.

No matter how multiplication is written, it does not change the priority that multiplication has in the order of operations.

Exercise 1.4.8 (Practice with order of operations). Simplify this expression one step at a time, using the order of operations.

$5 - 3(7 - 4)^2$

= ______

= ______

= ______

= ______

1.4.3 Absolute Value Bars, Radicals, and Fraction Bars are Grouping Symbols

When we first discussed grouping symbols, we only mentioned parentheses and brackets. Each of the following examples has an *implied* grouping symbol aside from parentheses and brackets: absolute value bars, radicals, and fraction bars.

Absolute Value Bars The absolute value bars, as in $|2 - 5|$, group the expression inside it just like a set of parentheses would.

Radicals The same is true of the radical symbol — everything inside the radical is grouped, as with $\sqrt{12 - 3}$.

Fraction Bars With a horizontal division bar, the numerator is treated as one group and the denominator as another, as with $\frac{2+3}{5-2}$.

We don't *need* parentheses for these three things since the absolute value bars, radical, and horizontal division bar each denote this grouping on their own. As far as priority in the order of operations goes, it's important to remember that these work just like our most familiar grouping symbols, parentheses.

With absolute value bars and radicals, these grouping symbols also *do* something to what's inside (but only *after* the operations inside the grouping symbols have been executed). For example, $|-2| = 2$, and $\sqrt{9} = 3$.

Example 1.4.9. Use the order of operations to simplify the following expressions.

a. $4-3|5-7|$. For this expression, we'll treat the absolute value bars just like we treat parentheses. This implies we'll simplify what's inside the bars first, and then compute the absolute value. After that, we'll multiply and then finally subtract:

$$\begin{aligned} 4-3|5-7| &= 4-3\left|\overbrace{5-7}\right| \\ &= 4-3\overbrace{|-2|} \\ &= 4-\overbrace{3(2)} \\ &= 4-6 \\ &= -2 \end{aligned}$$

We may not do $4-3=1$ first, because 3 is connected to the absolute value bars by multiplication (although implicitly), which has a higher order than subtraction.

b. $8-\sqrt{5^2-8\cdot 2}$. This expression has an expression inside the radical of $5^2-8\cdot 2$. We'll treat this radical like we would a set of parentheses, and simplify that internal expression first. We'll then apply the square root, and then our last step will be to subtract that expression from 8:

$$\begin{aligned} 8-\sqrt{5^2-8\cdot 2} &= 8-\sqrt{\overbrace{5^2}-8\cdot 2} \\ &= 8-\sqrt{25-\overbrace{8\cdot 2}} \\ &= 8-\sqrt{\overbrace{25-16}} \\ &= 8-\overbrace{\sqrt{9}} \\ &= 8-3 \\ &= 5 \end{aligned}$$

c. $\dfrac{2^4+3\cdot 6}{5-18\div 2}$. For this expression, the first thing we want to do is to recognize that the main fraction bar serves as a separator that groups the numerator and groups the denominator. Another way this expression could be written is $(2^4+3\cdot 6)\div(15-18\div 2)$. This implies we'll simplify the numerator and denominator separately according to the order of operations (since there are implicit parentheses around each of these). As a final step we'll simplify the resulting fraction (which is division).

$$\frac{2^4+3\cdot 6}{5-18\div 2} = \frac{\overbrace{2^4}+3\cdot 6}{5-\underbrace{18\div 2}}$$

$$= \frac{16 + \overbrace{3 \cdot 6}}{5 - 9}$$
$$= \frac{16 + 18}{-4}$$
$$= \frac{34}{-4}$$
$$= -\frac{17}{2}$$

Exercise 1.4.10 (More Practice with Order of Operations). Use the order of operations to evaluate $\dfrac{6 + 3\,|9 - 10|}{\sqrt{3 + 18 \div 3}}$.

Solution. We start by identifying the innermost, highest priority operations:

$$\frac{6 + 3\,|9 - 10|}{\sqrt{3 + 18 \div 3}} = \frac{6 + 3\left|\overbrace{9 - 10}\right|}{\sqrt{3 + \underbrace{18 \div 3}}}$$
$$= \frac{6 + 3\,\overbrace{|-1|}}{\sqrt{\underbrace{3 + 6}}}$$
$$= \frac{6 + \overbrace{3(1)}}{\sqrt{9}}$$
$$= \frac{6 + 3}{3}$$
$$= \frac{9}{3}$$
$$= 3$$

1.4.4 Negation and Distinguishing $(-a)^m$ from $-a^m$

We noted in the that using the negative sign to negate a number has the same priority as multiplication and division. To understand why this is, observe that $-1 \cdot 23 = -23$, just for one example. So negating 23 gives the same result as multiplying 23 by -1. For this reason, negation has the same priority in the order of operations as multiplication.

This can be a source of misunderstandings. How would you write a math expression that takes the number -4 and squares it?

$$-4^2? \qquad (-4)^2? \qquad \text{it doesn't matter?}$$

Well, it *does* matter. Certainly the second option, $(-4)^2$ is squaring the number -4. The parentheses empasize this.

But the expression -4^2 is something different. There are two actions in this expression: a negation and and exponentiation. According to the order of operations, the exponentiation has higher priority than the negation, so the exponent of 2 in -4^2 is going to apply to the 4 *before* the negative sign (multiplication by -1) ever gets taken into account. We would have:

$$\begin{aligned} -4^2 &= -\overbrace{4^2} \\ &= -16 \end{aligned}$$

and this is not the same as $(-4)^2$, which is *positive* 16.

Warning 1.4.11 (Negative Numbers Raised to Powers). You may find yourself needing to raise a negative number to a power, and using a calculator to do the work for you. If you do not understand the issue described here, then you may get incorrect results.

- For example, entering `-4^2` into a calculator will result in -16, the negative of 4^2.
- But entering `(-4)^2` into a calculator will result in 16, the square of -4.

Go ahead and try entering these into your own calculator.

Exercise 1.4.12 (Negating and Raising to Powers). Compute the following:

a. $-3^4 =$ ______ and $(-3)^4 =$ ______

b. $-4^3 =$ ______ and $(-4)^3 =$ ______

c. $-1.1^2 =$ ______ and $(-1.1)^2 =$ ______

Remark 1.4.13. You might observe in the previous example that there is no difference between -4^3 and $(-4)^3$. It's true that the results are the same, -64, but the two expressions still do say different things. With -4^3, you raise to a power first, then negate. With $(-4)^3$, you negate first, then raise to a power.

As was discussed in Subsection 1.1.5, if the base of a power is negative, then whether or not the result is positive or negative depends on if the exponent is even or odd. It depends on whether or not the factors can all be paired up to "cancel" negative signs, or if there will be a lone factor left by itself.

1.4.5 More Examples

Here are some example exercises that involve applying the order of operations to more complicated expressions. Try these exercises and read the steps given in each solution.

Exercise 1.4.14 (Complications). Simplify $10 - 4(5-7)^3$.

Solution. For the expression $10 - 4(5-7)^3$, we have simplify what's inside parentheses first, then we'll apply the exponent, then multiply, and finally subtract:

$$10 - 4(5-7)^3 = 10 - 4(\overbrace{5-7})^3$$

$$= 10 - 4\overbrace{(-2)^3}$$
$$= 10 - \overbrace{4(-8)}$$
$$= 10 - (-32)$$
$$= 10 + 32$$
$$= 42$$

Exercise 1.4.15 (Complications). Simplify $24 \div (15 \div 3 + 1) + 2$.

Solution. With the expression $24 \div (15 \div 3 + 1) + 2$, we'll simplify what's inside the parentheses according to the order of operations, and then take 24 divided by that expression as our last step:

$$\begin{aligned} 24 \div (15 \div 3 + 1) + 2 &= 24 \div (\overbrace{15 \div 3} + 1) + 2 \\ &= 24 \div \overbrace{(5 + 1)} + 2 \\ &= \overbrace{24 \div 6} + 2 \\ &= 4 + 2 \\ &= 6 \end{aligned}$$

Exercise 1.4.16 (Complications). Simplify $6 - (-8)^2 \div 4 + 1$.

Solution. To simplify $6 - (-8)^2 \div 4 + 1$, we'll first apply the exponent of 2 to -8, making sure to recall that $(-8)^2 = 64$. After this, we'll apply division. As a final step, we'll be have subtraction and addition, which we'll apply working left-to-right:

$$\begin{aligned} 6 - (-8)^2 \div 4 + 1 &= 6 - \overbrace{(-8)^2} \div 4 + 1 \\ &= 6 - \overbrace{(64) \div 4} + 1 \\ &= \overbrace{6 - 16} + 1 \\ &= -10 + 1 \\ &= -9 \end{aligned}$$

Exercise 1.4.17 (Complications). Simplify $(20 - 4^2) \div (4 - 6)^3$.

Solution. The expression $(20 - 4^2) \div (4 - 6)^3$ has two sets of parentheses, so our first step will be to simplify what's inside each of those first according to the order of operations. Once we've done that, we'll apply the exponent and then finally divide:

$$\begin{aligned}(20-4^2)\div(4-6)^3 &= (20-\overbrace{4^2})\div(4-6)^3\\ &= \overbrace{(20-16)}\div(4-6)^3\\ &= 4\div\overbrace{(4-6)}^3\\ &= 4\div\overbrace{(-2)^3}\\ &= 4\div(-8)\\ &= \frac{4}{-8}\\ &= \frac{1}{-2}\\ &= -\frac{1}{2}\end{aligned}$$

Exercise 1.4.18 (Complications). Simplify $\dfrac{2\,|9-15|+1}{\sqrt{(-5)^2+12^2}}$.

Solution. To simplify this expression, the first thing we want to recognize is the role of the main fraction bar, which groups the numerator and denominator. This implies we'll simplify the numerator and denominator separately according to the order of operations, and then reduce the fraction that results:

$$\begin{aligned}\frac{2\,|9-15|+1}{\sqrt{(-5)^2+12^2}} &= \frac{2\left|\overbrace{9-15}\right|+1}{\sqrt{\underbrace{(-5)^2}+12^2}}\\ &= \frac{2\,\overbrace{|-6|}+1}{\sqrt{25+\underbrace{12^2}}}\\ &= \frac{\overbrace{2(6)}+1}{\sqrt{\underbrace{25+144}}}\\ &= \frac{\overbrace{12+1}}{\underbrace{\sqrt{169}}}\\ &= \frac{13}{13}\\ &= 1\end{aligned}$$

1.4.6 Exercises

Practice order of operations by simplifying these expressions.

For the following exercises: Evaluate this expression:

1. $3 + 8(8) =$ ☐

2. $6 + 5(9) =$ ☐

3. $2(2 + 5) =$ ☐

4. $4(2 + 3) =$ ☐

5. $(2 \cdot 3)^2 =$ ☐

6. $(3 \cdot 2)^2 =$ ☐

7. $3 \cdot 3^3 =$ ☐

8. $4 \cdot 2^2 =$ ☐

9. $(14 - 4) \cdot 3 =$ ☐

10. $(12 - 5) \cdot 5 =$ ☐

11. $19 - 5 \cdot 3 =$ ☐

12. $18 - 2 \cdot 5 =$ ☐

13. $2 + 4 \cdot 6 =$ ☐

14. $3 + 2 \cdot 8 =$ ☐

15. $2 - 5 \cdot 10 =$ ☐

16. $3 - 4 \cdot 7 =$ ☐

17. $3 - 2(-10) =$ ☐

18. $4 - 5(-7) =$ ☐

19. $-[7 - (3 - 7)^2] =$ ☐

20. $-[8 - (4 - 7)^2] =$ ☐

21. $5 - 4[5 - (6 + 2 \cdot 3)] =$ ☐

22. $5 - 3[9 - (3 + 2 \cdot 5)] =$ ☐

23. $2 + 4(19 - 2 \cdot 2^3) =$ ☐

24. $2 + 4(45 - 5 \cdot 2^3) =$ ☐

25. $-3[5 - (2 - 4 \cdot 2)^2] =$ ☐

26. $-4[1 - (5 - 5 \cdot 2)^2] =$ ☐

27. $162 - 5[6^2 - (8 - 4)] =$ ☐

28. $29 - 4[3^2 - (7 - 5)] =$ ☐

29. $(17 - 4)^2 + 5(17 - 4^2)$ ☐

30. $(8 - 2)^2 + 2(8 - 2^2)$ ☐

31. $(2 \cdot 3)^2 - 2 \cdot 3^2 =$ ☐

32. $(2 \cdot 5)^2 - 2 \cdot 5^2 =$ ☐

33. $5 \cdot 4^2 - 27 \div 3^2 \cdot 2 + 5 =$ ☐

34. $6 \cdot 3^2 - 150 \div 5^2 \cdot 2 + 9 =$ ☐

35. $5(8 - 3)^2 - 5(8 - 3^2) =$ ☐

36. $5(8 - 4)^2 - 5(8 - 4^2) =$ ☐

37. $\dfrac{2 + 4}{7 - 5}$ ☐

38. $\dfrac{9 + 16}{11 - 6}$ ☐

39. $\dfrac{6^2 - 4^2}{7 + 3} =$ ☐

40. $\dfrac{6^2 - 2^2}{3 + 5} =$ ☐

41. $\dfrac{1-(-4)^3}{9-14} =$ ☐

42. $\dfrac{1-(-2)^3}{3-6} =$ ☐

43. $\dfrac{(-6)\cdot(-9)-(-4)\cdot 5}{(-2)^2+(-6)} =$ ☐

44. $\dfrac{(-6)\cdot(-1)-(-10)\cdot 3}{(-2)^2+(-6)} =$ ☐

45. $-|4-7| =$ ☐

46. $-|4-5| =$ ☐

47. $5-9|2-5|+8 =$ ☐

48. $1-5|4-5|+8 =$ ☐

49. $-8^2-|2\cdot(-9)| =$ ☐

50. $-7^2-|6\cdot(-5)| =$ ☐

51. $4-4\left|-1+(2-7)^3\right| =$ ☐

52. $5-2\left|-5+(4-7)^3\right| =$ ☐

53. $\dfrac{\left|27+(-4)^3\right|}{-1} =$ ☐

54. $\dfrac{\left|27+(-2)^3\right|}{-1} =$ ☐

55. $\left|\dfrac{27+(-4)^3}{-1}\right| =$ ☐

56. $\left|\dfrac{27+(-4)^3}{-1}\right| =$ ☐

57. $\dfrac{-2|1-2|}{23-(-5)^2} =$ ☐

58. $\dfrac{2|8-20|}{7-(-2)^2} =$ ☐

59. $\dfrac{4}{7}+4\cdot\dfrac{1}{14} =$ ☐

60. $\dfrac{4}{3}+8\cdot\dfrac{2}{3} =$ ☐

61. $\left(\dfrac{1}{2}-\dfrac{1}{10}\right)-4\left(\dfrac{1}{10}-\dfrac{1}{2}\right) =$ ☐

62. $\left(\dfrac{7}{6}-\dfrac{7}{12}\right)-4\left(\dfrac{7}{12}-\dfrac{7}{6}\right) =$ ☐

63. $\left|\dfrac{7}{2}-\dfrac{9}{10}\right|-4\left|\dfrac{9}{10}-\dfrac{7}{2}\right| =$ ☐

64. $\left|\dfrac{9}{10}-\dfrac{1}{8}\right|-4\left|\dfrac{1}{8}-\dfrac{9}{10}\right| =$ ☐

65. $\dfrac{4}{9}+10\left(\dfrac{5}{9}\right)^2 =$ ☐

66. $\dfrac{1}{3}+4\left(\dfrac{2}{3}\right)^2 =$ ☐

67. $\dfrac{1}{5}+\dfrac{4}{5}\div\dfrac{3}{2}-\dfrac{5}{4} =$ ☐

68. $\dfrac{2}{3}+\dfrac{1}{4}\div\dfrac{3}{4}-\dfrac{1}{2} =$ ☐

69. $3\sqrt{49-48} =$ ☐

70. $3\sqrt{91-27} =$ ☐

71. $4\sqrt{-12+4\cdot 7} =$ ☐

72. $4\sqrt{-34+7\cdot 5} =$ ☐

73. $2-5\sqrt{16+48} =$ ☐

74. $9-5\sqrt{58-42} =$ ☐

75. $\sqrt{4}-4\sqrt{-7+128} =$ ☐

76. $\sqrt{9}-3\sqrt{-4+68} =$ ☐

77. $\sqrt{77+2^2} =$ ☐

78. $\sqrt{-75+10^2} =$ ☐

79. $\sqrt{12^2 + 9^2} =$ ☐

80. $\sqrt{9^2 + 12^2} =$ ☐

81. $\dfrac{\sqrt{25} + 10}{\sqrt{25} - 10} =$ ☐

82. $\dfrac{\sqrt{64} + 6}{\sqrt{64} - 6} =$ ☐

83. $\dfrac{\sqrt{69 + 3 \cdot 4} + |-16 - 3|}{-129 - (-5)^3}$ ☐

84. $\dfrac{\sqrt{9 + 3 \cdot 9} + |-11 - 3|}{-130 - (-5)^3}$ ☐

85. $2[18 - 3(7 + 4)] =$ ☐

86. $2[16 - 5(5 + 4)] =$ ☐

87. $-4^2 - 9[10 - (2 - 2^3)] =$ ☐

88. $-6^2 - 7[5 - (7 - 2^3)] =$ ☐

1.5 Set Notation and Types of Numbers

When we talk about *how many* or *how much* of something we have, it often makes sense to use different types of numbers. For example, if we are counting dogs in a shelter, the possibilities are only $0, 1, 2, \ldots$. (It would be difficult to have $\frac{1}{2}$ of a dog.) On the other hand if you were weighing a dog in pounds, it doesn't make sense to only allow yourself to work with whole numbers. The dog might weigh something like 28.35 pounds. These examples highlight how certain kinds of numbers are appropriate for certain situations. We'll classify various types of numbers in this section.

1.5.1 Set Notation

What is the mathematical difference between these three "lists?"

$$28, 31, 30 \qquad \{28, 31, 30\} \qquad (28, 31, 30)$$

To a mathematician, the last one, $(28, 31, 30)$ is an *ordered* triple. What matters is not merely the three numbers, but *also* the order in which they come. The ordered triple $(28, 31, 30)$ is not the same as $(30, 31, 28)$; they have the same numbers in them, but the order has changed. For some context, February has 28 days; *then* March has 31 days; *then* April has 30 days. The order of the three numbers is meaningful in that context.

With curly braces and $\{28, 31, 30\}$, a mathematician sees a collection of three numbers and does not particularly care about the order they are in. Such a collection is called a **set**. All that matters is that these three numbers are part of a collection. They've been *written* in some particular order because that's necessary to write them down. But you might as well have put the three numbers in a bag and shaken up the bag. For some context, maybe your favorite three NBA players have jersey numbers 30, 31, and 28, and you like them all equally well. It doesn't really matter what order you use to list them.

So we can say:

$$\{28, 31, 30\} = \{30, 31, 28\} \qquad\qquad (28, 31, 30) \neq (30, 31, 28)$$

What about just writing $28, 31, 30$? This list of three numbers is ambiguous. Without the curly braces or parentheses, it's unclear to a reader if the order is important. **Set notation** is the use of curly braces to surround a list/collection of numbers, and we will use set notation frequently in this section.

Exercise 1.5.2 (Set Notation). Practice using (and not using) set notation.

According to Google, the three most common error codes from visiting a web site are `403`, `404`, and `500`. Without knowing which error code is most common, express this set mathematically.

Error code `500` is the most common. Error code `403` is the least common of these three. And that leaves `404` in the middle. Express the error codes in a mathematical way that appreciates how frequently they happen, from most often to least often.

1.5.2 Different Number Sets

In the introduction, we mentioned how different sets of numbers are appropriate for different situations. Here are the basic sets of numbers that are used in basic algebra.

Natural Numbers When we count, we begin: $1, 2, 3, \ldots$ and continue on in that pattern. These numbers are known as **natural numbers**.

$\mathbb{N} = \{1, 2, 3, \ldots\}$

Whole Numbers If we include zero, then we have the set of **whole numbers**.

$\{0, 1, 2, 3, \ldots\}$ has no standard symbol, but some options are $\mathbb{N}_0$, $\mathbb{N} \cup \{0\}$, and $\mathbb{Z}_{\geq 0}$.

Integers If we include the negatives of whole numbers, then we have the set of **integers**.

$\mathbb{Z} = \{\ldots, -3, -2, -1, 0, 1, 2, 3, \ldots\}$.

A $\mathbb{Z}$ is used because one word in German for "numbers" is "Zahlen".

Rational Numbers A **rational number** is any number that *can* be written as a fraction of integers, where the denominator is nonzero. Alternatively, a **rational number** is any number that *can* be written with a decimal that terminates or that repeats.

$\mathbb{Q} = \left\{0, 1, -1, 2, \frac{1}{2}, -\frac{1}{2}, -2, 3, \frac{1}{3}, -\frac{1}{3}, -3, \frac{3}{2}, \frac{2}{3} \ldots\right\}$

$\mathbb{Q} = \left\{0, 1, -1, 2, 0.5, -0.5, -2, 3, 0.\overline{3}, -0.\overline{3}, -3, 1.5, 0.\overline{6} \ldots\right\}$

A $\mathbb{Q}$ is used because fractions are *q*uotients of integers.

Irrational Numbers Any number that *cannot* be written as a fraction of integers belongs to the set of **irrational numbers**. Another way to say this is that any number whose decimal places goes on forever without repeating is an **irrational number**. Some examples include $\pi \approx 3.1415926\ldots$, $\sqrt{15} \approx 3.87298\ldots$, $e \approx 2.71828\ldots$

There is no standard symbol for the set of irrational numbers.

Real Numbers Any number that can be marked somewhere on a number line is a **real number**. Real numbers might be the only numbers you are familiar with. For a number to *not* be real, you have to start considering things called *complex numbers*, which are not our concern right now.

The set of real numbers can be denoted with $\mathbb{R}$ for short.

Warning 1.5.3 (Rational Numbers in Other Forms). Any number that *can* be written as a ratio of integers is rational, even if it's not written that way at first. For example, these numbers might not look rational to you at first glance: -4, $\sqrt{9}$, 0π, and $\sqrt[3]{\sqrt{5}+2} - \sqrt[3]{\sqrt{5}-2}$. But they are all rational, because they can respectively be written as $\frac{-4}{1}$, $\frac{3}{1}$, $\frac{0}{1}$, and $\frac{1}{1}$.

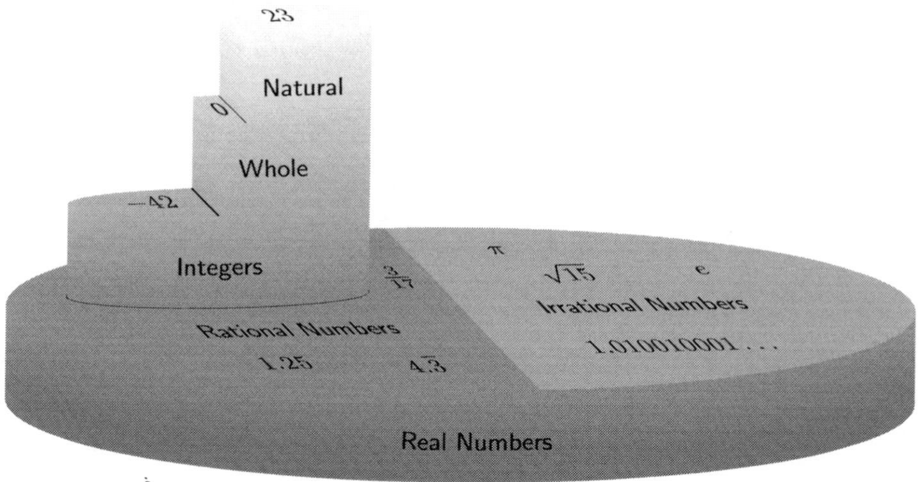

Figure 1.5.4: Types of Numbers

Example 1.5.5 (Determine if Numbers are This Type or That Type). Determine which numbers from the set $\left\{-102, -7.25, 0, \frac{\pi}{4}, 2, \frac{10}{3}, \sqrt{19}, \sqrt{25}, 10.\overline{7}\right\}$ are natural numbers, whole numbers, integers, rational numbers, irrational numbers, and real numbers.

Solution. All of these numbers are real numbers, becasue all of these numbers can be positioned on the real number line.

Each real number is either rational or irrational, and not both. -102, -7.25, 0, and 2 are rational because we can see directly that their decimal expressions terminate. $10.\overline{7}$ is also rational, because its decimal expression repeats. $\frac{10}{3}$ is rational because it is a ratio of integers. And last but not least, $\sqrt{25}$ is rational, because that's the same thing as 5.

This leaves only $\frac{\pi}{4}$ and $\sqrt{19}$ as irratinal numbers. Their decimal expressions go on forever without entering a repetetive cycle.

Only -102, 0, 2, and $\sqrt{25}$ (which is really 5) are integers.

Of these, only 0, 2, and $\sqrt{25}$ are whole numbers, because whole numbers excludes the negative integers.

Of these, only 2 and $\sqrt{25}$ are natural numbers, because the natural numbers exclude 0.

Exercise 1.5.6. Give examples of each type of number. If no such number exists, enter DNE or NONE.

a. Give an example of a whole number that is not an integer.

b. Give an example of an integer that is not a whole number.

c. Give an example of a rational number that is not an integer.

d. Give an example of a irrational number.

e. Give an example of a irrational number that is also an integer.

Solution.

a. Give an example of a whole number that is not an integer.

Since all whole numbers belong to integers, we cannot write any whole number which is not an integer. Type DNE (does not exist) for this question.

b. Give an example of an integer that is not a whole number. If no such number exists, enter DNE or NONE.

Any negative integer, like -1, is not a whole number, but is an integer.

c. Give an example of a rational number that is not an integer. If no such number exists, enter DNE or NONE.

Any terminating decimal, like 1.2, is a rational number, but is not an integer.

d. Give an example of a irrational number.

π is the easiest number to remember as an irrational number. Another constant worth knowing is $e \approx 2.718$. Finally, the square root of most integers are irrational, like $\sqrt{2}$ and $\sqrt{3}$.

e. Give an example of a irrational number that is also an integer. If no such number exists, enter DNE or NONE.

All irrational numbers are non-repeating and non-terminating decimals. No irrational numbers are integers. Type DNE (does not exist) for this question.

Exercise 1.5.7. In the introduction, we mentioned that the different types of numbers are appropriate in different situation. Which number set do you think is most appropriate in each of the following situations?

a. The number of people in a math class that play the ukulele.

This number is best considered as a (□ natural number □ whole number □ integer □ rational number □ irrational number □ real number) .

b. The hypotenuse's length in a given right triangle.

This number is best considered as a (□ natural number □ whole number □ integer □ rational number □ irrational number □ real number) .

c. The proportion of people in a math class that have a cat.

This number is best considered as a (□ natural number □ whole number □ integer □ rational number □ irrational number □ real number) .

d. The number of people in the room with you who have the same birthday as you.

This number is best considered as a (□ natural number □ whole number □ integer □ rational number ☐ irrational number ☐ real number) .

e. The total revenue (in dollars) generated for ticket sales at a Timbers soccer game.

   This number is best considered as a (☐ natural number ☐ whole number ☐ integer ☐ rational number ☐ irrational number ☐ real number) .

Solution.

a. The number of people who play the ukulele could be $0, 1, 2, \ldots$, so the whole numbers are the appropriate set.

b. A hypotenuse's length could be 1, 1.2, $\sqrt{2}$ (which is irrational), or any other positive number. So the real numbers are the appropriate set.

c. This proportion will be a ratio of integers, as both the total number of people in the class and the number of people who have a cat are integers. So the rational numbers are the appropriate set.

d. We know that the number of people must be a counting number, and since *you* are in the room with yourself, there is at least one person in that room with your brithday. So the natural numbers are the appropriate set.

e. The total revenue will be some number of dollars and cents, such as $631,897.15, which is a terminating decimal and thus a rational number. So the rational numbers are the appropriate set.

1.5.3 Converting Repeating Decimals to Fractions

We have learned that a terminating decimal number is a rational number. It's easy to convert a terminating decimal number into a fraction of integers: you just need to multiply and divide by one of the numbers in the set $\{10, 100, 1000, \ldots\}$. For example, when we say the number 0.123 out loud, we say "one hundred and twenty-three thousandths." While that's a lot to say, it makes it obvious that this number can be written as a ratio:

$$0.123 = \frac{123}{1000}.$$

Similarly,

$$21.28 = \frac{2128}{100} = \frac{532 \cdot 4}{25 \cdot 4} = \frac{532}{25},$$

demonstrating how *any* terminating decimal can be written as a fraction.

Repeating decimals can also be written as a fraction. To understand how, use a calculator to find the decimal for, say, $\frac{73}{99}$ and $\frac{189}{999}$ You will find that

$$\frac{73}{99} = 0.73737373\ldots = 0.\overline{73} \qquad \frac{189}{999} = 0.189189189\ldots = 0.\overline{189}.$$

The pattern is that diving a number by a number from $\{9, 99, 999, \ldots\}$ with the same number of digits will create a repeating decimal that starts as "0." and then repeats the numerator. We can use this observation to reverse engineer some fractions from repeating decimals.

Exercise 1.5.8.

a. Write the rational number 0.772772772... as a fraction.

b. Write the rational number 0.69696969... as a fraction.

Solution.

a. The *three*-digit number 772 repeats after the decimal. So we will make use of the *three*-digit denominator 999. And we have $\frac{772}{999}$.

b. The *two*-digit number 69 repeats after the decimal. So we will make use of the *two*-digit denominator 99. And we have $\frac{69}{99}$. But this fraction can be reduced to $\frac{23}{33}$.

Converting a repeating decimal to a fraction is not always quite this straightforward. There are complications if the number takes a few digits before it begins repeating. For your interest, here is one example on how to do that.

Example 1.5.9. Can we convert the repeating decimal $9.134343434\ldots = 9.1\overline{34}$ to a fraction? The trick is to separate its terminating part from its repeating part, like this:

$$9.1 + 0.034343434\ldots.$$

Now note that the terminating part is $\frac{91}{10}$, and the repeating part is almost like our earlier examples, except it has an extra 0 right after the decimal. So we have:

$$\frac{91}{10} + \frac{1}{10} \cdot 0.34343434\ldots.$$

With what we learned in the earlier examples and basic fraction arithmetic, we can continue:

$$\begin{aligned} 9.134343434\ldots &= \frac{91}{10} + \frac{1}{10} \cdot 0.34343434\ldots \\ &= \frac{91}{10} + \frac{1}{10} \cdot \frac{34}{99} \\ &= \frac{91}{10} + \frac{34}{990} \\ &= \frac{91 \cdot 99}{10 \cdot 99} + \frac{34}{990} \\ &= \frac{9009}{990} + \frac{34}{990} = \frac{9043}{990} \end{aligned}$$

Check that this is right by entering $\frac{9043}{990}$ into a calculator and seeing if it returns the decimal we started with, 9.134343434....

1.5.4 Exercises

These exercises examine set notation.

1. There are two numbers that you can square to get 1. Express this collection of two numbers using set notation.

2. There are four positive, even, one-digit numbers. Express this collection of four numbers using set notation.

3. There are six two-digit perfect square numbers. Express this collection of six numbers using set notation.

4. There is a set of three small positive integers where you can square all three numbers, then add the results, and get 50. Express this collection of three numbers using set notation.

These exercises examine different types of numbers.

5. Which of the following are whole numbers? There may be more than one correct answer.

□ 0 □ 38 □ $\frac{7}{68}$ □ $\sqrt{7}$ □ -1.191
□ -35646 □ $-1.101001000100001\ldots$
□ -9

6. Which of the following are whole numbers? There may be more than one correct answer.

□ $-1.6\overline{16}$ □ -74272 □ $-7.101001000100001\ldots$
□ 0 □ $\frac{5}{61}$ □ -3 □ 57941
□ $\sqrt{11}$

7. Which of the following are integers? There may be more than one correct answer.

□ -9 □ $\sqrt{36}$ □ -17860
□ $-6.7\overline{51}$ □ π □ $\sqrt{11}$ □ $-\frac{3}{73}$
□ 0

8. Which of the following are integers? There may be more than one correct answer.

□ -4.341 □ $6.101001000100001\ldots$
□ -4 □ $\sqrt{2}$ □ -8967 □ $\frac{6}{59}$
□ 41337 □ 0

9. Which of the following are rational numbers? There may be more than one correct answer.

□ 98 □ 0 □ $-\frac{6}{79}$ □ $-1.5\overline{62}$
□ -74 □ π □ -9 □ $\sqrt{11}$

10. Which of the following are rational numbers? There may be more than one correct answer.

□ 24733 □ $\frac{7}{32}$ □ 0 □ $-5.101001000100001\ldots$
□ -11772 □ π □ -7.161
□ -9

11. Which of the following are irrational numbers? There may be more than one correct answer.

□ $\frac{1}{47}$ □ -9 □ $0.101001000100001\ldots$
□ 0 □ $\sqrt{4}$ □ 4.429 □ $\sqrt{13}$
□ -82287

12. Which of the following are irrational numbers? There may be more than one correct answer.

□ -73394 □ -4 □ $-1.9\overline{31}$
□ $\sqrt{7}$ □ $-\frac{10}{83}$ □ 8129
□ $-2.101001000100001\ldots$ □ 0

13. Which of the following are real numbers? There may be more than one correct answer.

□ $-6.101001000100001\ldots$ □ 0 □ $3.4\overline{96}$
□ -9 □ π □ -64501 □ $-\frac{1}{73}$
□ 59

14. Which of the following are real numbers? There may be more than one correct answer.

□ $5.101001000100001\ldots$ □ 7.401
□ $\sqrt{6}$ □ -4 □ $\sqrt{25}$ □ -55608
□ $6.6\overline{41}$ □ 0

15. Determine the validity of each statement by selecting True or False.

(a) The number $\sqrt{29}$ is rational

(b) The number 0 is an integer

(c) The number $\frac{\pi}{2}$ is rational

(d) The number $\frac{-\sqrt{2}}{37\sqrt{2}}$ is irrational

(e) The number 0 is a natural number

16. Determine the validity of each statement by selecting True or False.

(a) The number 0.7405505005000500005... is rational

(b) The number $\sqrt{17}$ is a real number, but not an irrational number

(c) The number $\sqrt{29^2}$ is a real number, but not a rational number

(d) The number $\sqrt{20}$ is a real number, but not a rational number

(e) The number $\frac{17}{13}$ is rational, but not a natural number

17. Give examples of each type of number. If no such number exists, enter `DNE` or `NONE`.

a. Give an example of a whole number that is not an integer.
b. Give an example of an integer that is not a whole number.
c. Give an example of a rational number that is not an integer.
d. Give an example of a irrational number.
e. Give an example of a irrational number that is also an integer.

18. In each situation, which number set do you think is most appropriate?

a. The number of dogs a student has owned throughout their lifetime.
This number is best considered as a (□ natural number □ whole number □ integer □ rational number □ irrational number □ real number) .

b. The difference between the projected annual expenditures and the actual annual expenditures for a given company.
This number is best considered as a (□ natural number □ whole number □ integer □ rational number □ irrational number □ real number) .

c. The length around swimming pool in the shape of a half circle with radius 10 ft.
This number is best considered as a (□ natural number □ whole number □ integer □ rational number □ irrational number □ real number) .

d. The proportion of students at a college who own a car.
This number is best considered as a (□ natural number □ whole number □ integer □ rational number □ irrational number □ real number) .

e. The width of a sheet of paper, in inches.
This number is best considered as a (□ natural number □ whole number □ integer □ rational number □ irrational number □ real number) .

f. The number of people eating in a non-empty restaurant.
This number is best considered as a (□ natural number □ whole number □ integer □ rational number □ irrational number □ real number) .

Convert decimal numbers into fractions.

19. Write the rational number 9.95 as a fraction.

20. Write the rational number 19.682 as a fraction.

21. Write the rational number $0.\overline{28} = 0.282828\ldots$ as a fraction.

22. Write the rational number $0.\overline{335} = 0.335335335\ldots$ as a fraction.

23. Write the rational number $6.5\overline{42} = 6.5424242\ldots$ as a fraction.

24. Write the rational number $1.2\overline{497} = 1.2497497497\ldots$ as a fraction.

1.6 Comparison Symbols

As you know, 8 is larger than 3; that's a specific comparison between two numbers. We can also make a comparison between two less specific numbers, like if we say that average rent in Portland in 2016 is larger than it was in 2009. That makes a comparison using unspecified amounts. This section will go over the mathematical shorthand notation for making these kinds of comparisons.

In Oregon, only people who are 18 years old or older can vote in statewide elections.[1] Does that seem like a statement about the number 18? Maybe. But it's also a statement about numbers like 37 and 62: it says that people of these ages may vote as well. This section will also get into the mathematical notation for large collections of numbers like this.

In everyday language you can say something like "8 is larger than 3." In mathematical writing, it's not convenient to write that out in English. Instead the symbol ">" has been adopted, and it's used like this:

$$8 > 3$$

and read out loud as "8 is greater than 3." The symbol ">" is called the **greater-than symbol.**

Exercise 1.6.1.

a. Use mathematical notation to write "11.5 is greater than 4.2."

b. Use mathematical notation to write "age is greater than 20."

Solution.

a. $11.5 > 4.2$

b. We can just write the word age to represent age, and write $\text{age} > 20$. Or we could use an abbreviation like a for age, and write $a > 20$. Or, it is common to use x as a generic abbreviation, and we could write $x > 20$.

At some point in history, someone felt that $>$ was a good symbol for "is greater than." In "$8 > 3$," the tall side of the symbol is with the larger of the two numbers, and the small pointed side is with the smaller of the two numbers. That seems like a good system.

> **Alligator Jaws** Grade school teachers sometimes teach children that "the alligator wants to eat the larger number" as a way of remembering which direction to write the symbol.

We have to be careful when negative numbers are part of the comparison though. Is -8 larger or smaller than -3? In some sense -8 is larger, because if you owe someone 8 dollars, that's *more* than owing them 3 dollars. But that is not how the $>$ symbol works. This symbol is meant to tell you which number is farther to the right on a number line. And if that's how it goes, then $-3 > -8$.

---

[1]Some other states like Washington allow 17-year-olds to vote in primary elections provided they will be 18 by the general election.

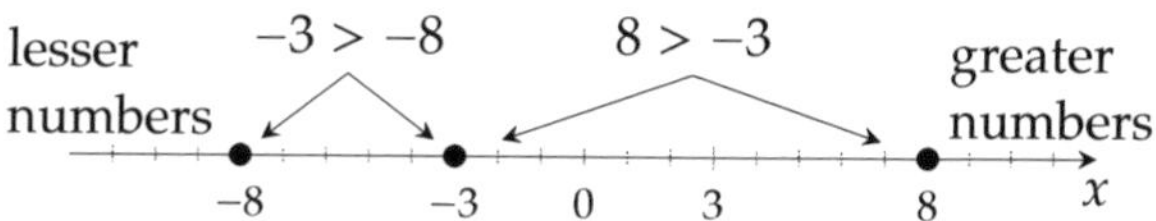

Figure 1.6.2: How the > symbol works.

Exercise 1.6.3. Use the > symbol to arrange the following numbers in order from greatest to least. For example, your answer might look like 4>3>2>1>0.

$$-7.6 \quad 6 \quad -6 \quad 9.5 \quad 8$$

Solution. We can order these numbers by placing these numbers on a number line.

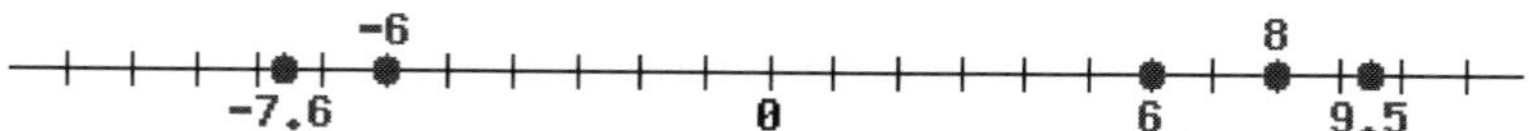

And so we see the answer is $9.5 > 8 > 6 > -6 > -7.6$.

Exercise 1.6.4. Use the > symbol to arrange the following numbers in order from greatest to least. For example, your answer might look like 4>3>2>1>0.

$$-5.2 \quad \pi \quad \frac{10}{3} \quad 4.6 \quad 8$$

Solution. We can order these numbers by placing these numbers on a number line. Knowing or computing their decimals helps with this.

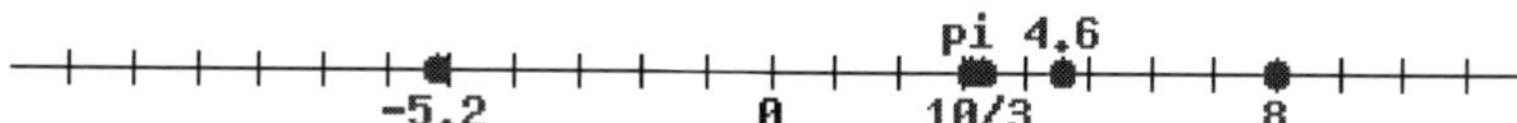

And so we see the answer is $8 > 4.6 > 3.33333 > 3.14159 > -5.2$.

The greater-than symbol has a close relative, the **greater-than-or-equal-to symbol**, "$\geq$." It means just like it sounds: the first number is either greater than, or equal to, the second number. These are all true statements:

$$8 \geq 3 \qquad 3 \geq -8 \qquad 3 \geq 3$$

but one of these three statements is false:

$$8 > 3 \qquad 3 > -8 \qquad 3 \overset{\text{no}}{>} 3$$

Remark 1.6.5. While it may not be that useful that we can write $3 \geq 3$, this symbol is quite useful when specific numbers aren't explicitly used on at least one side, like in these examples:

$$\text{(hourly pay rate)} \geq \text{(minimum wage)}$$

$$\text{(age of a voter)} \geq 18$$

Sometimes you want to emphasize that one number is *less than* another number instead of emphasizing which number is greater. To do this, we have symbols that are reversed from $>$ and $\geq$. The symbol "$<$" is the **less-than symbol** and it's used like this:

$$3 < 8$$

and read out loud as "3 is less than 8."

Table 1.6.6 gives the complete list of all six comparison symbols. Note that we've only discussed three in this section so far, but you already know the equals symbol, and we don't want to beat a dead horse with a full discussion of the last two.

Symbol	Means	Examples	
$=$	equals	$13 = 13$	$\frac{5}{4} = 1.25$
$>$	is greater than	$13 > 11$	$\pi > 3$
$\geq$	is greater than or equal to	$13 \geq 11$	$3 \geq 3$
$<$	is less than	$-3 < 8$	$\frac{1}{2} < \frac{2}{3}$
$\leq$	is less than or equal to	$-3 \leq 8$	$3 \leq 3$
$\neq$	is not equal to	$10 \neq 20$	$\frac{1}{2} \neq 1.2$

Table 1.6.6: Comparison Symbols

1.6.1 Exercises

1. Use the $>$ symbol to arrange the following numbers in order from greatest to least. For example, your answer might look like `4>3>2>1>0`.

$$10 \quad -3 \quad 5 \quad -9 \quad 8$$

2. Use the $>$ symbol to arrange the following numbers in order from greatest to least. For example, your answer might look like `4>3>2>1>0`.

$$-8 \quad -10 \quad -5 \quad 7 \quad 9$$

3. Use the $>$ symbol to arrange the following numbers in order from greatest to least. For example, your answer might look like `4>3>2>1>0`.

$$-5.73 \quad 3.67 \quad 1.68 \quad 1.11 \quad 7.48$$

4. Use the > symbol to arrange the following numbers in order from greatest to least. For example, your answer might look like `4>3>2>1>0`.

$$-3.2 \quad -2.16 \quad -9.14 \quad -4.98 \quad -2.06$$

5. Use the > symbol to arrange the following numbers in order from greatest to least. For example, your answer might look like `4>3>2>1>0`.

$$-\frac{1}{3} \quad 10 \quad -\frac{20}{7} \quad -10 \quad 7$$

6. Use the > symbol to arrange the following numbers in order from greatest to least. For example, your answer might look like `4>3>2>1>0`.

$$-\frac{4}{3} \quad -\frac{38}{7} \quad 3 \quad -7 \quad \frac{61}{8}$$

7. Use the > symbol to arrange the following numbers in order from greatest to least. For example, your answer might look like `4>3>2>1>0`.

$$\sqrt{2} \quad -5 \quad \frac{3}{5} \quad \sqrt{3} \quad 3 \quad \frac{2}{7}$$

8. Use the > symbol to arrange the following numbers in order from greatest to least. For example, your answer might look like `4>3>2>1>0`.

$$\frac{1}{2} \quad 6 \quad \pi \quad \sqrt{2} \quad 9 \quad \frac{2}{7}$$

For the following exercises: Decide if each comparison is true or false.

9.

$-2 = -1$	(□ True	□ False)
$-2 < 7$	(□ True	□ False)
$-9 \geq -9$	(□ True	□ False)
$-10 \neq 9$	(□ True	□ False)
$2 \leq -3$	(□ True	□ False)
$1 > 1$	(□ True	□ False)

10.

$5 \leq 5$	(□ True	□ False)
$-1 = 7$	(□ True	□ False)
$8 \neq 8$	(□ True	□ False)
$6 \geq -8$	(□ True	□ False)
$-6 \leq -4$	(□ True	□ False)
$6 > -4$	(□ True	□ False)

11.

$\frac{39}{4} \leq -\frac{18}{4}$	(□ True	□ False)
$\frac{25}{6} = -\frac{30}{8}$	(□ True	□ False)
$\frac{7}{3} \geq \frac{14}{6}$	(□ True	□ False)
$\frac{18}{7} < \frac{73}{9}$	(□ True	□ False)
$\frac{46}{6} \geq -\frac{34}{5}$	(□ True	□ False)
$-\frac{9}{4} = -\frac{18}{8}$	(□ True	□ False)

12.

$\frac{9}{9} \neq \frac{27}{27}$	(□ True	□ False)
$-\frac{6}{7} = -\frac{18}{21}$	(□ True	□ False)
$-\frac{39}{4} = \frac{22}{9}$	(□ True	□ False)
$-\frac{9}{2} < -\frac{9}{2}$	(□ True	□ False)
$-\frac{3}{2} \geq -\frac{3}{2}$	(□ True	□ False)
$\frac{6}{3} \leq \frac{12}{6}$	(□ True	□ False)

13. Choose <, >, or = to make a true statement.

$-\dfrac{7}{3}$ (□ < □ > □ =) $-\dfrac{3}{8}$

14. Choose <, >, or = to make a true statement.

$-\dfrac{6}{7}$ (□ < □ > □ =) $-\dfrac{1}{8}$

15. Choose <, >, or = to make a true statement.

$\frac{3}{4} + \frac{3}{2}$ (□ < □ > □ =) $\frac{4}{3} \div \frac{5}{3}$

16. Choose <, >, or = to make a true statement.

$\frac{4}{5} + \frac{4}{3}$ (□ < □ > □ =) $\frac{4}{5} \div \frac{5}{3}$

17. Choose <, >, or = to make a true statement.

$\frac{15}{16} \div \frac{15}{16}$ (□ < □ > □ =) $\frac{16}{28} - \frac{4}{7}$

18. Choose <, >, or = to make a true statement.

$\frac{17}{10} \div \frac{17}{10}$ (□ < □ > □ =) $\frac{32}{20} - \frac{8}{5}$

19. Compare these two numbers:

$-9\frac{1}{2}$ (□ < □ > □ =) -9

20. Compare these two numbers:

$-1\frac{2}{3}$ (□ < □ > □ =) -1

21. Compare these two numbers:

$-5\frac{1}{2}$ (□ < □ > □ =) 1

22. Compare these two numbers:

$-5\frac{1}{2}$ (□ < □ > □ =) 2

23. Compare the following numbers:

$\left|-\dfrac{7}{8}\right|$ (□ < □ > □ =) $|0.875|$

24. Compare the following numbers:

$\left|-\dfrac{5}{8}\right|$ (□ < □ > □ =) $|0.625|$

25. True or false?

$$-4 \geq -10$$

○ True

○ False

26. True or false?

$$-3 \geq -4$$

○ True

○ False

1.7 Notation for Intervals

If you say

$$(\text{age of a voter}) \geq 18$$

and have a particular voter in mind, what is that person's age? There's no way to know for sure. *Maybe* they are 18, but maybe they are older. It's helpful to visualize the possibilities with a number line, as in Figure 1.7.1.

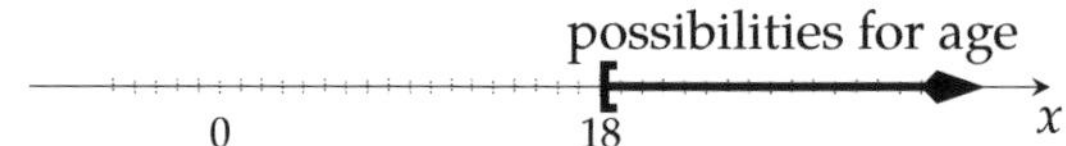

Figure 1.7.1: (age of a voter) ≥ 18

The shaded portion of the number line in Figure 1.7.1 is a mathematical **interval**. For now, that just means a collection of certain numbers. In this case, it's all the numbers 18 and above.

It's one thing to say (age of a voter) ≥ 18, and another thing to discuss *all the shaded numbers* in the interval in Figure 1.7.1. In mathematics,

$$(\text{age of a voter}) \geq 18$$

is saying that there is *one* age under consideration and all we know is that it's *18* or larger. It's subtle, but this is not the same thing as the *collection* of *all* numbers that are 18 or larger. Mathematics has two common ways to write down these kinds of collections.

Definition 1.7.2 (Set-Builder Notation). Set-builder notation attempts to directly say the condition that numbers in the interval satisfy. In general, write set-builder notation like:

$$\{x \mid \text{condition on } x\}$$

and read it out loud as "the set of all x such that …." For example,

$$\{x \mid x \geq 18\}$$

is read out loud as "the set of all x such that x is greater than or equal to 18." The breakdown is as follows.

$\{x \mid x \geq 18\}$	the set of
$\{x \mid x \geq 18\}$	all x
$\{x \mid x \geq 18\}$	such that
$\{x \mid x \geq 18\}$	x is greater than or equal to 18

Definition 1.7.3 (Interval Notation). Interval notation tries to just say the numbers where the interval starts and stops. For example, in Figure 1.7.1, the interval starts at 18. To the right, the interval extends forever and has no end, so we use the ∞ symbol (meaning "infinity"). This particular interval is denoted:

$$[18, \infty)$$

Why use "[" on one side and ")" on the other? The square bracket tells us that 18 *is* part of the interval and the round parenthesis tells us that ∞ is *not* part of the interval.[1]

In general there are four types of infinite intervals. Take note of the different uses of round parentheses and square brackets.

Figure 1.7.4: An **open, infinite** interval denoted by (a, ∞) means all numbers a or larger, *not* including a.

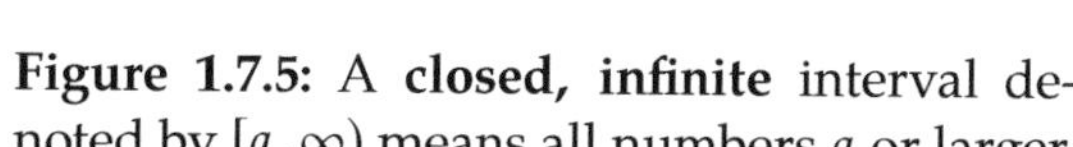

Figure 1.7.5: A **closed, infinite** interval denoted by $[a, \infty)$ means all numbers a or larger, *including* a.

Figure 1.7.6: An **open, infinite** interval denoted by $(-\infty, a)$ means all numbers a or smaller, *not* including a.

Figure 1.7.7: A **closed, infinite** interval denoted by $(-\infty, a]$ means all numbers a or smaller, *including* a.

Exercise 1.7.8 (Interval and Set-Builder Notation from Number Lines). For each interval expressed in the number lines, give the interval notation and set-builder notation.

a. −5 −4 −3 −2 −1 0 1 2 3 4 5

In set-builder notation:

In interval notation:

b. −5 −4 −3 −2 −1 0 1 2 3 4 5

In set-builder notation:

In interval notation:

c. −5 −4 −3 −2 −1 0 1 2 3 4 5

In set-builder notation:

In interval notation:

[1]And how could it be, since ∞ is not even a number?

1.7.1 Exercises

1. For each interval expressed in the number lines, give the interval notation and set-builder notation.

a.

−5 −4 −3 −2 −1 0 1 2 3 4 5

In set-builder notation:

In interval notation:

b.

−5 −4 −3 −2 −1 0 1 2 3 4 5

In set-builder notation:

In interval notation:

c.

−5 −4 −3 −2 −1 0 1 2 3 4 5

In set-builder notation:

In interval notation:

2. For each interval expressed in the number lines, give the interval notation and set-builder notation.

a.

−5 −4 −3 −2 −1 0 1 2 3 4 5

In set-builder notation:

In interval notation:

b.

−5 −4 −3 −2 −1 0 1 2 3 4 5

In set-builder notation:

In interval notation:

c.

−5 −4 −3 −2 −1 0 1 2 3 4 5

In set-builder notation:

In interval notation:

3. Here is a graph of an interval.

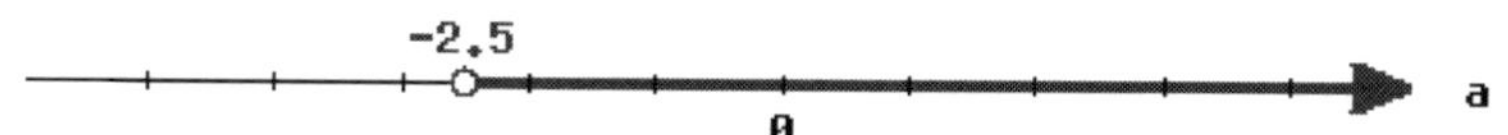

Write this inequality in set-builder notation:

Write this inequality in interval notation:

4. Here is a graph of an interval.

Write this inequality in set-builder notation:

Write this inequality in interval notation:

5. Here is a graph of an interval.

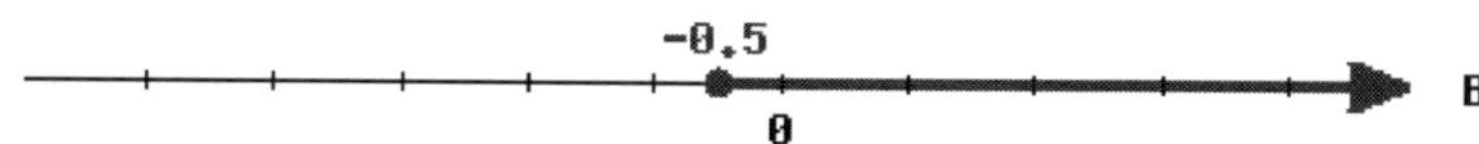

Write this inequality in set-builder notation:

Write this inequality in interval notation:

6. Here is a graph of an interval.

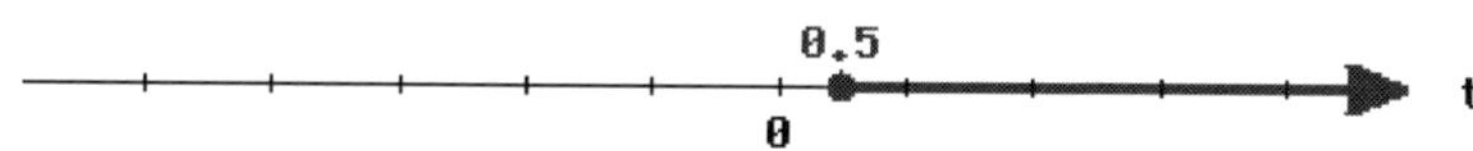

Write this inequality in set-builder notation:

Write this inequality in interval notation:

7. Here is a graph of an interval.

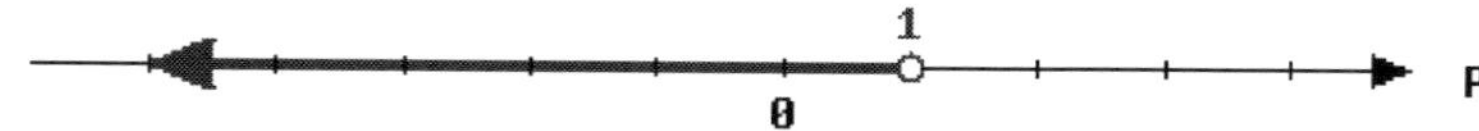

Write this inequality in set-builder notation:

Write this inequality in interval notation:

8. Here is a graph of an interval.

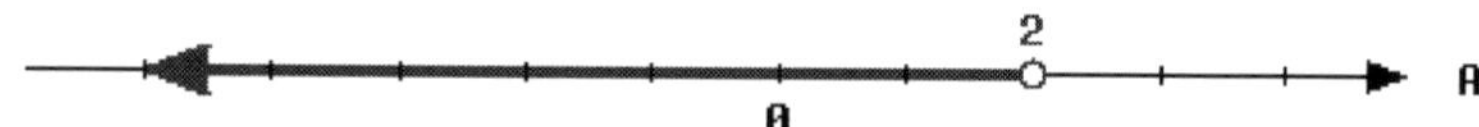

Write this inequality in set-builder notation:

Write this inequality in interval notation:

9. Here is a graph of an interval.

Write this inequality in set-builder notation: [ ]

Write this inequality in interval notation: [ ]

10. Here is a graph of an interval.

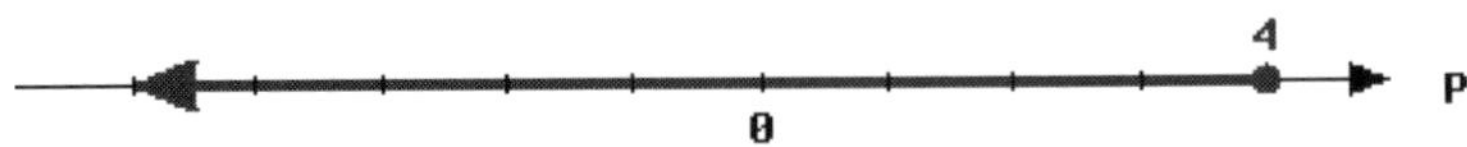

Write this inequality in set-builder notation: [ ]

Write this inequality in interval notation: [ ]

Convert set-builder notation to interval notation.

For the following exercises: Change the following inequality from set-builder notation to interval notation:

11. $\{x \mid x \leq -9\}$

The interval notation is [ ].

12. $\{x \mid x \leq -6\}$

The interval notation is [ ].

13. $\{x \mid x \geq -4\}$

The interval notation is [ ].

14. $\{x \mid x \geq -2\}$

The interval notation is [ ].

15. $\{x \mid x < 1\}$

The interval notation is [ ].

16. $\{x \mid x < 3\}$

The interval notation is [ ].

17. $\{x \mid x > 5\}$

The interval notation is [ ].

18. $\{x \mid x > 8\}$

The interval notation is [ ].

19. $\{x \mid 10 > x\}$

The interval notation is [ ].

20. $\{x \mid -9 > x\}$

The interval notation is [ ].

21. $\{x \mid -6 \geq x\}$

The interval notation is [ ].

22. $\{x \mid -4 \geq x\}$

The interval notation is [ ].

23. $\{x \mid -2 \leq x\}$

The interval notation is [ ].

24. $\{x \mid 1 \leq x\}$

The interval notation is [ ].

25. $\{x \mid 3 < x\}$

The interval notation is $\boxed{}$.

26. $\{x \mid 5 < x\}$

The interval notation is $\boxed{}$.

27. $\left\{x \mid \frac{9}{10} < x\right\}$

The interval notation is $\boxed{}$

28. $\left\{x \mid \frac{10}{7} < x\right\}$

The interval notation is $\boxed{}$

29. $\left\{x \mid x \leq -\frac{1}{4}\right\}$

The interval notation is $\boxed{}$

30. $\left\{x \mid x \leq -\frac{2}{7}\right\}$

The interval notation is $\boxed{}$

31. $\{x \mid x \leq 0\}$

The interval notation is $\boxed{}$.

32. $\{x \mid 0 < x\}$

The interval notation is $\boxed{}$.

1.8 Chapter Review

1.8.1 Review of Arithmetic with Negative Numbers

Adding Real Numbers with the Same Sign When adding two numbers with the same sign, we can ignore the signs, and simply add the numbers as if they were both positive.

Examples

a. $5 + 2 = 7$

b. $-5 + (-2) = -7$

Adding Real Numbers with Opposite Signs When adding two numbers with opposite signs, we find those two numbers' difference. The sum has the same sign as the number with the bigger value. If those two numbers have the same value, the sum is 0.

Examples

a. $5 + (-2) = 3$

b. $(-5) + 2 = -3$

Subtracting a Positive Number When subtracting a positive number, we can change the problem to adding the *opposite* number, and then apply the methods of adding numbers.

Examples

a.

$$\begin{aligned} 5 - 2 &= 5 + (-2) \\ &= 3 \end{aligned}$$

b.

$$\begin{aligned} 2 - 5 &= 2 + (-5) \\ &= -3 \end{aligned}$$

c.

$$\begin{aligned} -5 - 2 &= -5 + (-2) \\ &= -7 \end{aligned}$$

Subtracting a Negative Number When subtracting a negative number, we can change those two negative signs to a positive sign, and then apply the methods of adding numbers.

Examples

a.

$$\begin{aligned} 5 - (-2) &= 5 + 2 \\ &= 7 \end{aligned}$$

b.

$$\begin{aligned} -5 - (-2) &= -5 + 2 \\ &= -3 \end{aligned}$$

c.

$$\begin{aligned} -2 - (-5) &= -2 + 5 \\ &= 3 \end{aligned}$$

Multiplication and Division of Real Numbers When multiplying and dividing real numbers, each pair of negative signs cancel out each other (becoming a positive sign). If there is still one negative sign left, the result is negative; otherwise the result is positive.

Examples

a. $(6)(-2) = -12$

b. $(-6)(2) = -12$

c. $(-6)(-2) = 12$

d. $(-6)(-2)(-1) = -12$

e. $(-6)(-2)(-1)(-1) = 12$

f. $\frac{12}{-2} = -6$

g. $\frac{-12}{2} = -6$

h. $\frac{-12}{-2} = 6$

Powers When we raise a negative number to a certain power, apply the rules of multiplying real numbers: each pair of negative signs cancel out each other.

Examples

a.

$$\begin{aligned}(-2)^2 &= (-2)(-2)\\ &= 4\end{aligned}$$

b.

$$\begin{aligned}(-2)^3 &= (-2)(-2)(-2)\\ &= -8\end{aligned}$$

c.

$$\begin{aligned}(-2)^4 &= (-2)(-2)(-2)(-2)\\ &= 16\end{aligned}$$

Difference between $(-a)^n$ and $-a^n$ For the exponent expression 2^3, the number 2 is called the **base**, and the number 3 is called the **exponent**. The base of $(-a)^n$ is $-a$, while the base of $-a^n$ is a. This makes a difference in the result when the power is an even number.

Examples

a.

$$\begin{aligned}(-4)^2 &= (-4)(-4)\\ &= 16\end{aligned}$$

b.

$$\begin{aligned}-4^2 &= -(4)(4)\\ &= -16\end{aligned}$$

c.

$$\begin{aligned}(-4)^3 &= (-4)(-4)(-4)\\ &= -64\end{aligned}$$

d.

$$\begin{aligned}-4^3 &= -(4)(4)(4)\\ &= -64\end{aligned}$$

1.8.2 Fraction Arithmetic Review

Multiplying Fractions

When multiplying two fractions, we simply multiply their numerators and denominators. To avoid big numbers, we should reduce fractions before multiplying. If one number is an integer, we can change the integer to a fraction with a denominator of 1. For example, $2 = \frac{2}{1}$.

Example

$$\begin{aligned}\frac{1}{2} \cdot \frac{3}{4} &= \frac{1 \cdot 3}{2 \cdot 4} \\ &= \frac{3}{8}\end{aligned}$$

Dividing Fractions

When dividing two fractions, we "flip" the second number, and then do multiplication.

Example

$$\begin{aligned}\frac{1}{2} \div \frac{4}{3} &= \frac{1}{2} \cdot \frac{3}{4} \\ &= \frac{3}{8}\end{aligned}$$

Adding/Subtracting Fractions

Before adding/subtracting fractions, we need to change each fraction's denominator to the same number, called the **common denominator**. Then, we add/subtract the numerators, and the denominator remains the same.

Example

$$\begin{aligned}\frac{1}{2} - \frac{1}{3} &= \frac{1 \cdot 3}{2 \cdot 3} - \frac{1 \cdot 2}{3 \cdot 2} \\ &= \frac{3}{6} - \frac{2}{6} \\ &= \frac{1}{6}\end{aligned}$$

1.8.3 Absolute Value and Square Root Review

Absolute Value

The absolute value of a number is the distance from that number to 0 on the number line. An absolute value is always positive or 0.

Examples

a. $|2| = 2$

b. $\left|-\frac{1}{2}\right| = \frac{1}{2}$

c. $|0| = 0$

Square Root

The symbol $\sqrt{b}$ has meaning when $b \geq 0$. It means the positive number that can be squared to result in b.

Examples

a. $\sqrt{9} = 3$

b. $\sqrt{2} = 1.414\ldots$

c. $\sqrt{\frac{9}{16}} = \frac{3}{4}$

d. $\sqrt{-1}$ is undefined

1.8.4 Order of Operations Review

Order of Operations

When evaluating an expression with multiple operations, we must follow the order of operations:

1. (P)arentheses and other grouping symbols
2. (E)xponentiation
3. (M)ultiplication, (D)ivision, and Negation
4. (A)ddition and (S)ubtraction

Example

$$\begin{aligned} 4-2\left(3-(2-4)^2\right) &= 4-2\left(3-(\overbrace{2-4})^2\right) \\ &= 4-2\left(3-\overbrace{(-2)^2}\right) \\ &= 4-2\left(\overbrace{3-4}\right) \\ &= 4-\overbrace{2(-1)} \\ &= 4-(-2) \\ &= 6 \end{aligned}$$

1.8.5 Types of Numbers Review

Types of Numbers Real numbers are categorized into the following sets: natural numbers, whole numbers, integers, rational numbers and irrational numbers.

Examples Here are some examples of numbers from each set of numbers:

Natural Numbers $1, 251, 3462$

Whole Numbers $0, 1, 42, 953$

Integers $-263, -10, 0, 1, 834$

Rational Numbers $\frac{1}{3}, -3, 1.1, 0, 0.\overline{73}$

Irrational Numbers $\pi, e, \sqrt{2}$

1.8.6 Comparison Symbols Review

The following are symbols used to compare numbers.

Symbol	Meaning	Examples	
$=$	equals	$13 = 13$	$\frac{5}{4} = 1.25$
$>$	is greater than	$13 > 11$	$\pi > 3$
$\geq$	is greater than or equal to	$13 \geq 11$	$3 \geq 3$
$<$	is less than	$-3 < 8$	$\frac{1}{2} < \frac{2}{3}$
$\leq$	is less than or equal to	$-3 \leq 8$	$3 \leq 3$
$\neq$	is not equal to	$10 \neq 20$	$\frac{1}{2} \neq 1.2$

Table 1.8.1: Comparison Symbols

1.8.7 Notation for Intervals Review

The following are some examples of set-builder notation and interval notation.

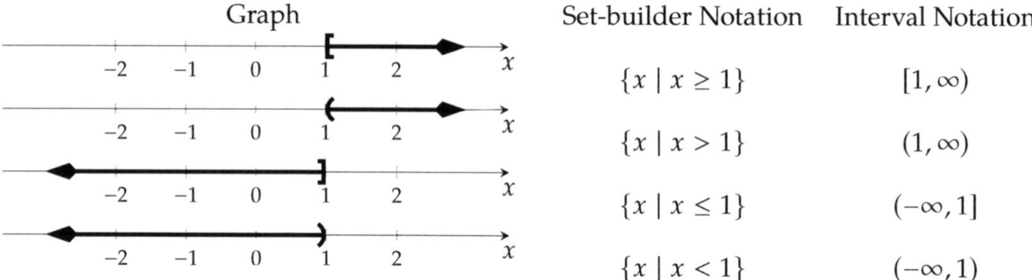

1.8.8 Exercises

1. Perform the given addition and subtraction:

a. $-15 - 9 + (-7) =$ ☐

b. $8 - (-19) + (-11) =$ ☐

2. Multiply the following integers.

a. $(-2) \cdot (-5) \cdot (-2) =$ ☐

b. $4 \cdot (-9) \cdot (-4) =$ ☐

c. $(-100) \cdot (-62) \cdot 0 =$ ☐

3. Evaluate the following.

a. $\dfrac{-12}{-4} =$ ☐

b. $\dfrac{21}{-7} =$ ☐

c. $\dfrac{-45}{5} =$ ☐

4. Evaluate the following expressions that have integer exponents:

a. $(-1)^2 =$ ☐

b. $-2^2 =$ ☐

5. Evaluate the following expressions that have integer exponents:

a. $(-1)^3 =$ ☐

b. $-2^3 =$ ☐

6. Add these two fractions: $-\dfrac{1}{6} + \dfrac{7}{10}$

7. Subtract one fraction from the other: $-\dfrac{1}{6} - \left(-\dfrac{9}{10}\right)$

8. Carry out the subtraction: $-2 - \dfrac{11}{5}$

9. Multiply these two fractions: $-\dfrac{2}{7} \cdot \dfrac{5}{26}$

10. Multiply the integer with the fraction: $-3 \cdot \dfrac{7}{18}$

11. Carry out the division: $\dfrac{3}{16} \div \left(-\dfrac{5}{12}\right)$

12. Carry out the division: $14 \div \dfrac{7}{4}$

13. Evaluate the following expressions which involve the absolute value:

a. $-|4-7| =$ ______

b. $|-4-7| =$ ______

c. $-2|7-4| =$ ______

14. Find the square root of the following numbers:

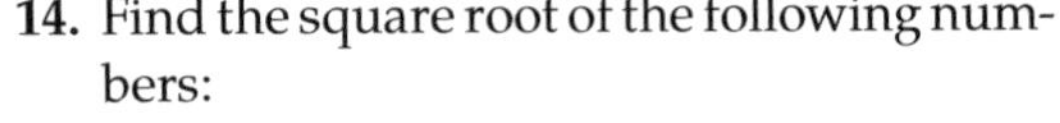

a. $\sqrt{144} =$ ______

b. $\sqrt{16} =$ ______

c. $\sqrt{81} =$ ______

15. Find the square root of the following numbers.

a. $\sqrt{\dfrac{1}{144}} =$ ______

b. $\sqrt{-\dfrac{9}{100}} =$ ______

16. Evaluate this expression:

$-2^2 - 7[7-(5-4^3)] =$ ______

17. Evaluate this expression:

$\dfrac{1-(-2)^3}{1-4} =$ ______

18. Evaluate this expression:

$5 - 8\left|-5+(2-7)^3\right| =$ ______

19. Compare the following integers:

7 (□ < □ > □ =) −4

−2 (□ < □ > □ =) −4

−4 (□ < □ > □ =) 0

20. Change the following inequality from set-builder notation to interval notation:

$\{x \mid x > 3\}$

The interval notation is ______.

21. Give examples of each type of number. If no such number exists, enter `DNE` or `NONE`.

a. Give an example of a whole number that is not an integer.

b. Give an example of an integer that is not a whole number.

c. Give an example of a rational number that is not an integer.

d. Give an example of a irrational number.

e. Give an example of a irrational number that is also an integer.

22. Here is a graph of an interval.

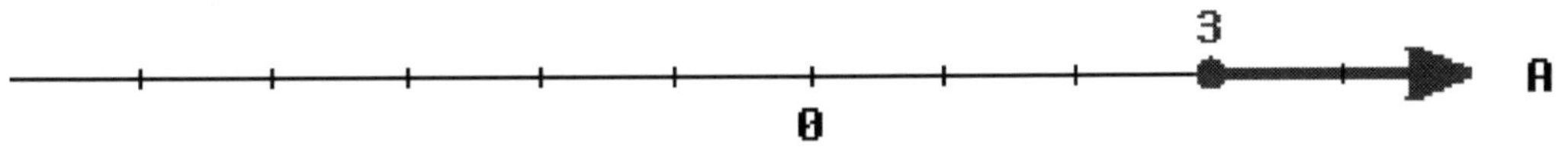

3
0
A

Write this inequality in set-builder notation:

Write this inequality in interval notation:

CHAPTER 2

Variables, Expressions, and Equations

2.1 Variables and Evaluating Expressions

A key sign that you are moving on from *arithmetic* to *algebra* is if you are using **variables**, discussed in this section. Any combination of numbers and variables using mathematical operations is called a mathematical **expression**. Some expressions are simple, and some are complicated. Some expressions are abstract, whereas some have context and meaning. One example of a simple expression with context is $220 - a$, which has one variable, a, and is the expression for the maximum heart rate of a person who is a years old.

Along with variables, and expressions, mathematical **equations** and **inequalities** are very important in algebra. In this section, we'll focus on expressions, and the remainder of this chapter is devoted to equations and inequalities.

2.1.1 Introduction to Variables

When we want to represent an unknown or changing numerical quantity, we use a **variable** to do so. For example, if you'd like to analyze the gas mileage of various cars, you could let the symbol "g" represent a car's gas mileage. The mileage might be 25 mpg, 30 mpg, or some other quantity. If we agree to use mpg for the units of measure, g might be a place holder for 25, 30, or some other number. Since we are using a variable and not a specific number, we can analyze gas mileage for Honda Civics at the same time we analyze gas mileage for Ford Explorers.

When using variables to stand for actual physical quantities, it's good practice to use letters that clearly correspond to the quantity they represent. For example, it's wise to use g to represent gas mileage. This helps the people who might read your work in the future to understand it better.

At the same time you identify what variable you would like to use, it is very important to identify what units of measurement will go along with that variable, and clearly tell your reader this. For example, suppose you are working with $g = 25$. A car whose gas mileage is 25 mpg is very different from a car whose gas mileage is 25 kpg (kilometer per gallon). So it would be important to state that g represents gas mileage *in miles per gallon*.

Exercise 2.1.3. Identify a variable you might use to represent each quantity. And identify what units would be most appropriate. 

a. Let ☐ be the age of a student, measured in ☐.

b. Let ☐ be the amount of time passed since a driver left Portland, Oregon, bounded for Boise, Idaho, measured in ☐.

c. Let ☐ be the area of a two-bedroom apartment, measured in ☐.

Solution.

a. The unknown quantity is age, which we generally measure in years. So we could define this variable as:

"Let a be the age of a student, measured in years."

b. The amount of time passed is the unknown quantity. Since this is a drive from Portland to Boise, it would make sense to measure this in hours. So we could define this variable as:

Let t be the amount of time passed since a driver left Portland, Oregon, bounded for Boise, Idaho, measured in hours.

c. The unknown quantity is area. Apartment area is usually measured in square feet. So we'll define this variable as:

Let A be the area of a two-bedroom apartment, measured in ft^2.

Remark 2.1.4. Note that unless an algebra problem specifies which letter(s) to use, we have a choice as to which letter we choose to represent our variable(s). However *without* any context to a problem, x, y, and z are the most common letters used as variables, and you may see these variables (especially x) a lot.

Also note that the units we use are often determined indirectly by other quantitative information given in an algebra problem. For example, if we're told that a car has used so many gallons of gas after traveling so many miles, then it suggests we should measure gas mileage in mpg.

2.1.2 Mathematical Expressions

A mathematical **expression** is any combination of variables and numbers using arithmetic operations. The following are all examples of mathematical expressions:

$$x + 1 \qquad 2\ell + 2w \qquad \frac{\sqrt{x}}{y+1} \qquad nRT$$

Note that this definition of "mathematical expression" does *not* include anything with signs like these in them: $=$, $<$, $\leq$, etc.

Example 2.1.5. The expression:

$$\frac{5}{9}(F - 32)$$

can be used to convert from degrees Fahrenheit to degrees Celsius. To do this, we need a Fahrenheit temperature, F. Then we can **evaluate** the expression. This means replacing its variable(s) with specific numbers and calculating the result. In this case, we can replace F with a specific number.

Let's convert the temperature 89 °F to the Celsius scale by evaluating the expression.

$$\begin{aligned} \frac{5}{9}(F-32) &= \frac{5}{9}(89-32) \\ &= \frac{5}{9}(57) \\ &= \frac{285}{9} \approx 31.67 \end{aligned}$$

This shows us that 89 °F is equivalent to approximately 31.67 °C.

Warning 2.1.6 (Vocabulary Usage). The steps in Example 2.1.5 are not considered to be "solving an equation," which is language you might be tempted to use. Instead, this process was "evaluating an expression." There is a special algebra meaning for words like "solve" and "solution" that will come soon.

In Example 2.1.5, we *did* use equals signs, but that was so we could assert that one expression equals another expression, which equals another expression, and so on. In the end, this example showed us that $\frac{5}{9}(F-32)$ *evaluates* to ≈ 31.67 when F is 89. It did not "solve" anything in a technical mathematical sense.

Exercise 2.1.7. Try evaluating the temperature expression for yourself.

Use the expression $\frac{5}{9}(F-32)$ to evaluate some Celsius temperatures.

a. If a temperature is 50°F what is that temperature measured in Celsius?

b. If a temperature is −20°F what is that temperature measured in Celsius?

Example 2.1.8 (Target heart rate). According to the American Heart Association, maximum heart rate in beats per minute (bpm) is given by $220 - a$, where a is age in years.

a. Determine the maximum heart rate for someone who is 31 years old.

b. *Target* heart rate for moderate exercise is 50% to 70% of maximum heart rate. If we want to reach 65% of an individual's maximum heart rate during moderate exercise, we'd use the expression $0.65(220 - a)$, where a is their age in years. Determine the target heart rate at this 65% level for someone who is 31 years old.

Solution. Both of these parts ask us to evaluate an expression.

a. Since a is defined to be age in years, we will evaluate this expression by substituting a with 31:

$$\begin{aligned} 220 - a &= 220 - 31 \\ &= 189 \end{aligned}$$

This tells us that the maximum heart rate for someone who is 31 years old is 189 bpm.

b. We'll again substitute a with 31, but this time using the target heart rate expression:

$$\begin{aligned} 0.65(220 - a) &= 0.65(220 - 31) \\ &= 0.65(189) \\ &= 122.85 \end{aligned}$$

This tells us that the target heart rate for someone who is 31 years old undertaking moderate exercise is 122.85 bpm.

Exercise 2.1.9. The target heart rate for moderate exercise is 50% to 70% of maximum heart rate. We can use the expression $\frac{p}{100}(220 - a)$ to represent a person's target heart rate when their target rate is p% of their maximum heart rate, and they are a years old.

Determine the target heart rate at the 53% level for moderate exercise for someone who is 56 years old.

At the 53% level, the target heart rate for moderate exercise for someone who is 56 years old is ☐ beats per minute.

Example 2.1.10 (Geometry).

An expression for the area of a trapezoid (shown in Figure 2.1.11) is $\frac{1}{2}(a + b)h$, where a and b are the parallel side lengths and h is the height. This expression is valid if a, b, and h are all measured in centimeters, inches, or any other unit for length. The result will be measured in square units, such as cm^2 or in^2.

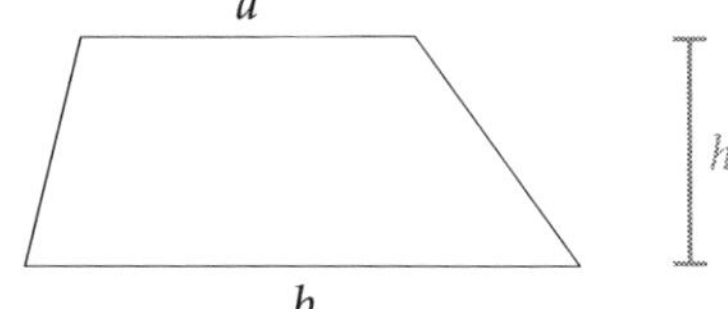

Figure 2.1.11: Trapezoid

Determine the area of a trapezoid for which $a = 3\,\text{cm}$, $b = 5\,\text{cm}$, and $h = 2\,\text{cm}$.

Solution. Replacing each variable with the given value:

$$\begin{aligned}\frac{1}{2}(a+b)h &= \frac{1}{2}(3\text{ cm}+5\text{ cm})(2\text{ cm})\\ &= \frac{1}{2}(8\text{ cm})(2\text{ cm})\\ &= 8\text{ cm}^2\end{aligned}$$

The area of this trapezoid is 8 cm^2.

Exercise 2.1.12 (Rising Rents). An expression estimating the average rent of a one-bedroom apartment in Portland, Oregon, from January, 2011 to October, 2016, is given by $10.173x + 974.78$, where x is the number of months since January, 2011. 

a. According to this model, what was the average rent of a one-bedroom apartment in Portland in January, 2011?

b. According to this model, what was the average rent of a one-bedroom apartment in Portland in January, 2016?

Solution.

a. This model uses x as the number of months after January, 2011. So in January, 2011, x is 0:

$$\begin{aligned}10.173x + 974.48 &= 10.173(0) + 974.48\\ &\approx 974.48\end{aligned}$$

According to this model, the average monthly rent for a one-bedroom apartment in Portland, Oregon, in January, 2011, was \$974.78.

b. The date we are given is January, 2016, which is 5 years after January, 2011. Recall that x is the number of *months* since January, 2011. So we need to use $x = 60$:

$$\begin{aligned}10.173x + 974.48 &= 10.173(60) + 974.48\\ &\approx 1584.86\end{aligned}$$

According to this model, the average monthly rent for a one-bedroom apartment in Portland, Oregon, was \$1584.86 in January 2016.

Terms and Factors Some expressions consist of **terms** — parts that are added together. For example, the expression $2x^2 + 3x - 4$ has three terms: $2x^2$, $3x$ and -4.

Some expressions consist of **factors** — parts that are multiplied together. For example, the term $2x^2$ has two factors: 2 and x^2 (with the multiplication symbol implied between them). The term $3x$ has two factors: 3 and x. The expression $x(x+1)(x-2)$ has three factors: x, $(x+1)$, and $(x - 2)$.

2.1.3 Evaluating Expressions with Exponents, Absolute Value, and Radicals

Mathematical expressions will often have exponents, absolute value bars, and radicals. This does not change the basic approach to evaluating them.

Example 2.1.13 (Cylinder Volume). The volume V, in cm^3, of a cylinder with height h and radius r, both in cm, is given by $V = \pi r^2 h$. Assume that a can of tuna has a radius of 4 cm and a height of 2 cm.

a. Calculate the volume of this can in terms of π.

b. Calculate the volume of this can and round the answer to four significant digits. Use the π button on your calculator, or use $\pi \approx 3.1415927$.

Solution.

a. Substitute $r = 4$ and $h = 2$ into the expression $\pi r^2 h$, we have:

$$\begin{aligned} V = \pi r^2 h &= \pi(4\text{ cm})^2(2\text{ cm}) \\ &= \pi(16\text{ cm}^2)(2\text{ cm}) \\ &= \pi(32\text{ cm}^3) \\ &= 32\pi\text{ cm}^3 \end{aligned}$$

The volume of this can is $32\pi\text{ cm}^3$.

b. For this part, we substitute $\pi \approx 3.1415927$ into out first result, 32π, and we have:

$$\begin{aligned} V = 32\pi &\approx 32(3.1415927) \\ &\approx 100.5309\ldots \\ &\approx \overbrace{100.5}^{\text{four}}309\ldots \end{aligned}$$

The volume of this can is about 100.5 cm^3.

Example 2.1.14 (Tsunami Speed). The speed of a tsunami (in meters per second) can be modeled by $\sqrt{9.8d}$, where d is the depth of the tsunami (in meters). Determine the speed of a tsunami that has a depth of 30 m to four significant digits.

Solution. Using $d = 30$, we find:

$$\sqrt{9.8d} = \sqrt{9.8(30)}$$

$$= \sqrt{294}$$
$$\approx \overbrace{17.14}^{\text{four}}6428\ldots$$

The speed of tsunami with a depth of 30 m is about 17.15 $\frac{\text{m}}{\text{s}}$.

Exercise 2.1.15 (Tent Height). While camping, the height (in feet) inside a tent when you are d ft from the north side of the tent is given by $-2|d-3|+6$.

a. When you are 5 ft from the north side, the height will be ______.

b. When you are 2.5 ft from the north side, the height will be ______.

Solution.

a. When $d = 5$, we have:

$$\begin{aligned} -2|d-3|+6 &= -2|5-3|+6 \\ &= -2|2|+6 \\ &= -2(2)+6 \\ &= -4+6 \\ &= 2 \end{aligned}$$

Thus when you are 5 ft from the north side, the height in the tent is 2 ft.

b. When $d = 2.5$, we have:

$$\begin{aligned} -2|d-3|+6 &= -2|2.5-3|+6 \\ &= -2|-0.5|+6 \\ &= -2(0.5)+6 \\ &= -1+6 \\ &= 5 \end{aligned}$$

Thus when you are 2.5 ft from the north side, the height in the tent is 5 ft.

Exercise 2.1.16 (Mortgage Payments). If we borrow L dollars for a home mortgage loan at an annual interest rate r, and intend to pay off the loan after n months, then the amount we should pay each month is given by the expression

$$\frac{rL\left(1+\frac{r}{12}\right)^n}{12\left(\left(1+\frac{r}{12}\right)^n-1\right)}$$

If we borrow \$200,000 at an interest rate of 6% with the intent to pay off the loan in 30 years, what should our monthly payment be? (Using a calculator is appropriate here.)

Solution. We must use $L = 200000$. Because the interest rate is a percentage, $r = 0.06$ (not 6). The variable n is supposed to be a number on months, but we will pay off the loan in 30 years.

Therefore we take $n = 360$.

$$\begin{aligned}
\frac{rL\left(1+\frac{r}{12}\right)^n}{12\left(\left(1+\frac{r}{12}\right)^n-1\right)} &= \frac{(0.06)(200000)\left(1+\frac{0.06}{12}\right)^{360}}{12\left(\left(1+\frac{0.06}{12}\right)^{360}-1\right)} \\
&= \frac{(0.06)(200000)(1+0.005)^{360}}{12\left((1+0.005)^{360}-1\right)} \\
&\approx \frac{(0.06)(200000)(6.022575\ldots)}{12\left(6.022575\ldots-1\right)} \\
&\approx \frac{(0.06)(200000)(6.022575\ldots)}{12(5.022575\ldots)} \\
&\approx \frac{72270.90\ldots}{60.2709\ldots} \\
&\approx 1199.10
\end{aligned}$$

Our monthly payment should be \$1,199.10.

2.1.4 Evaluating Expressions with Negative Numbers

When we substitute negative numbers into an expression, it's important to use parentheses around them. Let's look at some examples.

Example 2.1.17. Evaluate x^2 if $x = -2$.

We substitute:

$$\begin{aligned}
x^2 &= (-2)^2 \\
&= 4
\end{aligned}$$

If we don't use parentheses, we would have:

$$\begin{aligned}
x^2 &= -2^2 && \text{incorrect!} \\
&= -4
\end{aligned}$$

The original expression takes x and squares it. With $-2^2 = -4$, the number -2 is not being squared. Since the exponent has higher priority than the negation, it's just the number 2 that is being squared. With $(-2)^2 = 4$ the number -2 *is* being squared, which is what we would want given the expression x^2.

So it is wise to always use some parentheses when substituting in any negative number.

Exercise 2.1.18. Evaluate and simplify the following expressions for $x = -5$ and $y = -2$:

a. $x^3y^2 = \square$ b. $(-2x)^3 = \square$ c. $-3x^2y = \square$

Solution.

a.

$$\begin{aligned} x^3y^2 &= (-5)^3(-2)^2 \\ &= (-125)(4) \\ &= -600 \end{aligned}$$

b.

$$\begin{aligned} (-2x)^3 &= (-2(-5))^3 \\ &= (10)^3 \\ &= 1000 \end{aligned}$$

c.

$$\begin{aligned} -3x^2y &= -3(-5)^2(-2) \\ &= -3(25)(-2) \\ &= 150 \end{aligned}$$

2.1.5 Exercises

Evaluating Expressions

1. Evaluate the expression for $x = 5$.

$x + 6 = \square$

2. Evaluate the expression for $x = 7$.

$x - 1 = \square$

3. Evaluate the expression for $x = 10$.

$-9 - x = \square$

4. Evaluate the expression for $x = -9$.

$5 - x = \square$

5. Evaluate the expression for $x = -7$.

$-2x - 8 = \square$

6. Evaluate the expression for $x = -4$.

$-10x - 3 = \square$

7. Evaluate the expression for $b = 4$.

$-10b = \square$

8. Evaluate the expression for $A = -3$.

$5A = \square$

9. Evaluate the expression for $r = -10$.

$-3(r + 6) = \square$

10. Evaluate the expression for $r = 3$.

$-9(r + 3) = \square$

11. Evaluate the expression for $C = 2$ and $b = 10$.

$-6C - 4b = \square$

12. Evaluate the expression for $C = 4$ and $a = 2$.

$-7C - 4a = \square$

13. Evaluate the expression for $x = 2$ and $A = -4$.

$\frac{-3}{x} - \frac{9}{A} =$ ☐

14. Evaluate the expression for $r = 8$ and $b = -2$.

$\frac{-3}{r} - \frac{3}{b} =$ ☐

15. Evaluate the expression for $y = 9$.

$\frac{4y-2}{6y} =$ ☐

16. Evaluate the expression for $y = 1$.

$\frac{8y-5}{6y} =$ ☐

17. Evaluate the expression for $A = -8$ and $b = -4$.

$\frac{A-6b+9}{-A+3b} =$ ☐

18. Evaluate the expression for $B = 1$ and $C = -10$.

$\frac{B+5C-10}{7B+10C} =$ ☐

19. Evaluate the expression r^2:

a. When $r = 6$, $r^2 =$ ☐

b. When $r = -9$, $r^2 =$ ☐

20. Evaluate the expression t^2:

a. When $t = 3$, $t^2 =$ ☐

b. When $t = -4$, $t^2 =$ ☐

21. Evaluate the expression t^3:

a. When $t = 5$, $t^3 =$ ☐

b. When $t = -5$, $t^3 =$ ☐

22. Evaluate the expression x^3:

a. When $x = 3$, $x^3 =$ ☐

b. When $x = -2$, $x^3 =$ ☐

23. Evaluate the following expressions.

a. Evaluate $2x^2$ when $x = 4$. $2x^2 =$ ☐

b. Evaluate $(2x)^2$ when $x = 4$. $(2x)^2 =$ ☐

24. Evaluate the following expressions.

a. Evaluate $5y^2$ when $y = 2$. $5y^2 =$ ☐

b. Evaluate $(5y)^2$ when $y = 2$. $(5y)^2 =$ ☐

25. Evaluate the following expressions.

a. Evaluate y^2x^3 when $y = -2$ and $x = -1$. $y^2x^3 =$ ______

b. Evaluate y^3x^2 when $y = -2$ and $x = -1$. $y^3x^2 =$ ______

26. Evaluate the following expressions.

a. Evaluate r^2x^3 when $r = -3$ and $x = -1$. $r^2x^3 =$ ______

b. Evaluate r^3x^2 when $r = -3$ and $x = -1$. $r^3x^2 =$ ______

27. Evaluate the following expressions.

a. Evaluate $(-3r)^2$ when $r = -1$. $(-3r)^2 =$ ______

b. Evaluate $(-3r)^3$ when $r = -1$. $(-3r)^3 =$ ______

28. Evaluate the following expressions.

a. Evaluate $(-r)^2$ when $r = -2$. $(-r)^2 =$ ______

b. Evaluate $(-r)^3$ when $r = -2$. $(-r)^3 =$ ______

29. Evaluate the following expressions.

a. Evaluate $-4t^2x$ when $t = -1$ and $x = -3$. $-4t^2x =$ ______

b. Evaluate $(-4t)^2\,x$ when $t = -1$ and $x = -3$. $(-4t)^2\,x =$ ______

30. Evaluate the following expressions.

a. Evaluate $-4t^2r$ when $t = -2$ and $r = -1$. $-4t^2r =$ ______

b. Evaluate $(-4t)^2\,r$ when $t = -2$ and $r = -1$. $(-4t)^2\,r =$ ______

31. Evaluate each of the following expressions if $x = 3$, $y = 5$, $z = -6$.

$x + 12 =$ ______

$x + 3y + z =$ ______

$(x - 12) + 3(y + z) =$ ______

$100x + yz =$ ______

32. Evaluate each of the following expressions if $x = 4$, $y = 4$, $z = -4$.

$x + 12 =$ ______

$x + 3y + z =$ ______

$(x - 12) + 3(y + z) =$ ______

$100x + yz =$ ______

33. Evaluate the expression $\frac{1}{4}(x+2)^2 - 7$ when $x = -6$.

34. Evaluate the expression $\frac{1}{7}(x+2)^2 - 4$ when $x = -9$.

35. Evaluate the expression $\frac{1}{2}h(B + b)$ when $h = 10,\ B = 8,\ b = 6$.

36. Evaluate the expression $\frac{1}{2}h(B + b)$ when $h = 10,\ B = 6,\ b = 3$.

37. Evaluate the expression $-16t^2 + 64t + 128$ when $t = 3$.

38. Evaluate the expression $-16t^2 + 64t + 128$ when $t = 5$.

39. Evaluate each algebraic expression for the given value(s):

$\frac{y^2 + \sqrt{x+3}}{|2y - x|}$, for $x = 22$ and $y = -6$: ☐

40. Evaluate each algebraic expression for the given value(s):

$\frac{y^3 + \sqrt{x-2}}{|2x - y|}$, for $x = 51$ and $y = 7$: ☐

41. Evaluate each algebraic expression for the given value(s):

$\frac{\sqrt{x}}{y} - \frac{y}{x}$, for $x = 9$ and $y = 3$: ☐

42. Evaluate each algebraic expression for the given value(s):

$\frac{\sqrt{x}}{y} - \frac{y}{x}$, for $x = 16$ and $y = 6$: ☐

43. Evaluate

$$\frac{y_2 - y_1}{x_2 - x_1}$$

for $x_1 = -5$, $x_2 = -13$, $y_1 = 18$, and $y_2 = -17$:

44. Evaluate

$$\frac{y_2 - y_1}{x_2 - x_1}$$

for $x_1 = 14$, $x_2 = -6$, $y_1 = 13$, and $y_2 = -13$:

45. Evaluate

$$\sqrt{(x_2 - x_1)^2 + (y_2 - y_1)^2}$$

for $x_1 = 1$, $x_2 = -4$, $y_1 = -6$, and $y_2 = 6$:

46. Evaluate

$$\sqrt{(x_2 - x_1)^2 + (y_2 - y_1)^2}$$

for $x_1 = -5$, $x_2 = -17$, $y_1 = -6$, and $y_2 = -11$:

47. Evaluate the algebraic expression $7a + b$ for $a = \frac{8}{9}$ and $b = \frac{2}{3}$.

48. Evaluate the algebraic expression $-7a + b$ for $a = \frac{9}{5}$ and $b = \frac{1}{7}$.

49. Evaluate each algebraic expression for the given value(s):

$\frac{2 + 2|x - y|}{x + 3y}$, for $x = 11$ and $y = -6$: ☐

50. Evaluate each algebraic expression for the given value(s):

$\frac{2 + 4|x - y|}{x + 5y}$, for $x = 12$ and $y = 10$: ☐

Area Formulas

51. A rectangle's area can be calculated by the formula $A = LW$, where A stands for area, L for length and W for width.

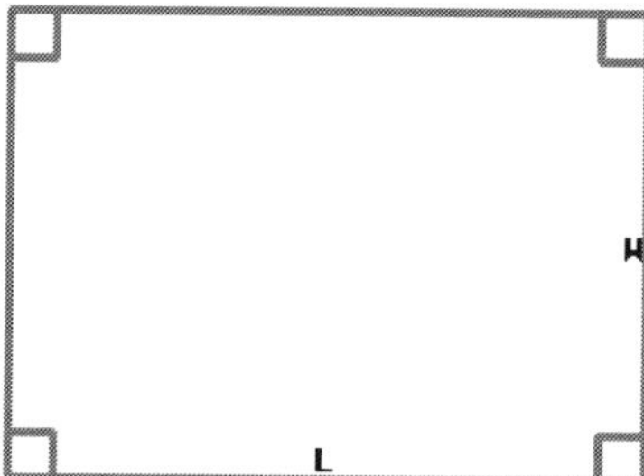

A rectangle's perimeter can be calculated by the formula $P = 2(L + W)$, where P stands for perimeter, L for length and W for width.

Use the formulas to calculate the rectangle's area and perimeter. Its length is 14 m and its width is 10 m.

Area = ______

Perimeter = ______

52. A rectangle's area can be calculated by the formula $A = LW$, where A stands for area, L for length and W for width.

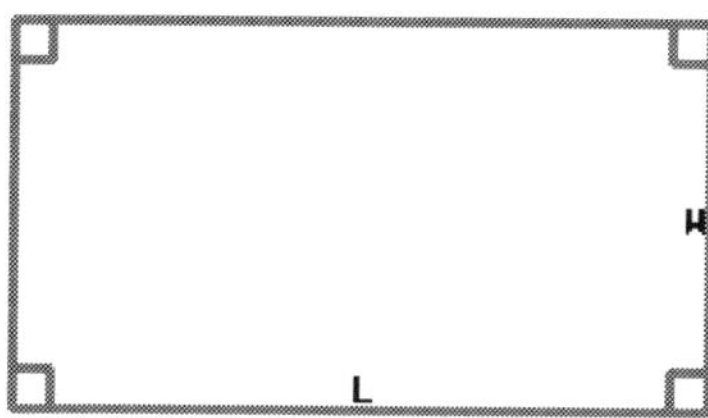

A rectangle's perimeter can be calculated by the formula $P = 2(L + W)$, where P stands for perimeter, L for length and W for width.

Use the formulas to calculate the rectangle's area and perimeter. Its length is 15 m and its width is 8 m.

Area = ______

Perimeter = ______

53. A triangle's area can be calculated by the formula $A = \frac{1}{2}bh$, where A stands for area, b for base and h for height.

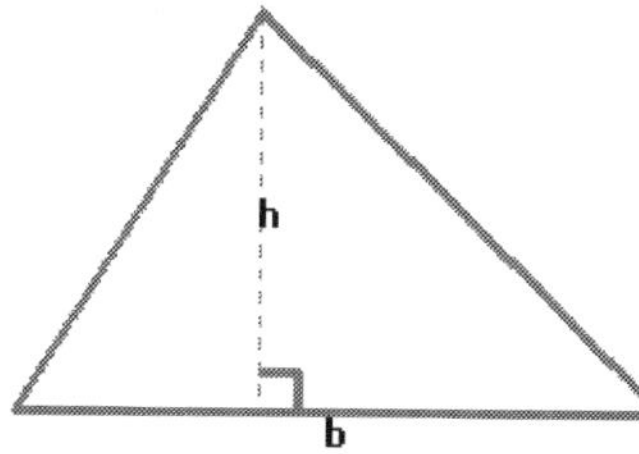

Find the triangle's area if $b = 20$ ft and $h = 12$ ft.

Area = ______

54. A triangle's area can be calculated by the formula $A = \frac{1}{2}bh$, where A stands for area, b for base and h for height.

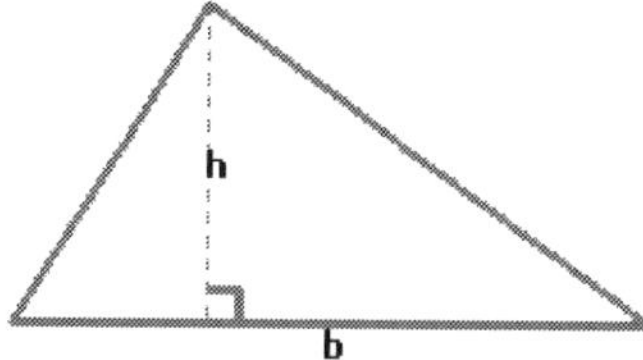

Find the triangle's area if $b = 21$ ft and $h = 10$ ft.

Area = ______

55. The formula

$$A = \frac{1}{2} r\, n\, s$$

gives the area of a regular polygon with side length s, number of sides n and, apothem r. (The *apothem* is the distance from the center of the polygon to one of its sides.)

What is the area of a regular pentagon with $s = 72$ in and $r = 94$ in?

The area is ______.

56. The formula

$$A = \frac{1}{2} r\, n\, s$$

gives the area of a regular polygon with side length s, number of sides n and, apothem r. (The *apothem* is the distance from the center of the polygon to one of its sides.)

What is the area of a regular 40-gon with $s = 84$ in and $r = 66$ in?

The area is ______.

57. A trapezoid's area can be calculated by the formula $A = \frac{1}{2}(b_1 + b_2)h$, where A stands for area, b_1 for the first base's length, b_2 for the second base's length, and h for height.

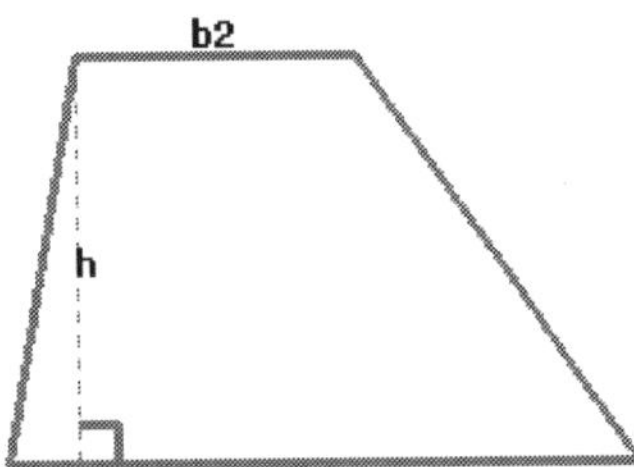

Use the formulas to calculate the trapezoid's area. Its first base's length is 18 m, its second base's length is 8 m and its height is 11 m.

Area = ______

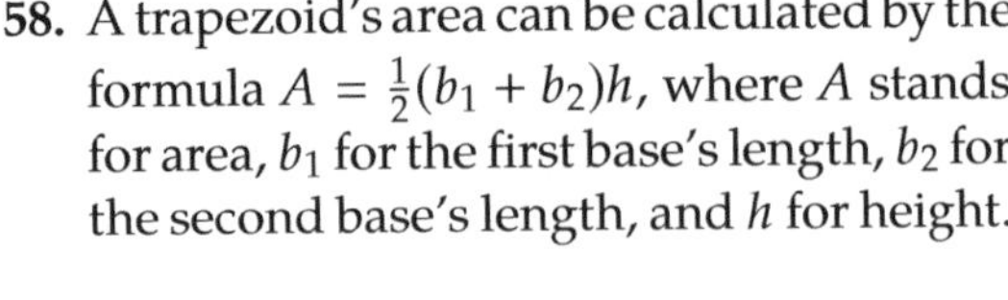

58. A trapezoid's area can be calculated by the formula $A = \frac{1}{2}(b_1 + b_2)h$, where A stands for area, b_1 for the first base's length, b_2 for the second base's length, and h for height.

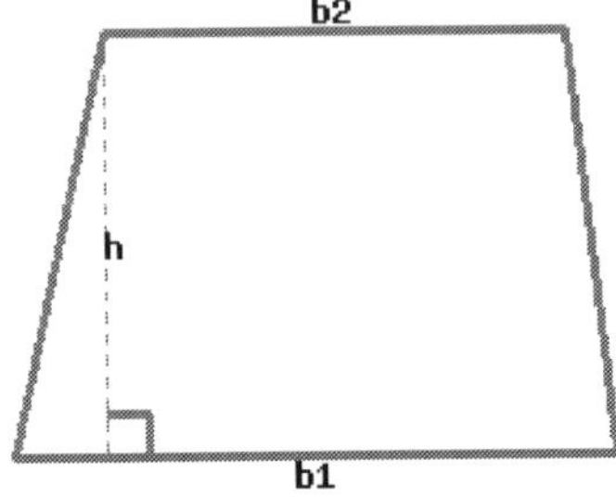

Use the formulas to calculate the trapezoid's area. Its first base's length is 13 m, its second base's length is 10 m and its height is 9 m.

Area = ______

59. A circle's circumference formula is $C = 2\pi r$, where C stands for circumference and r for radius. And a circle's area formula is $A = \pi r^2$, where A stands for area and r for radius.

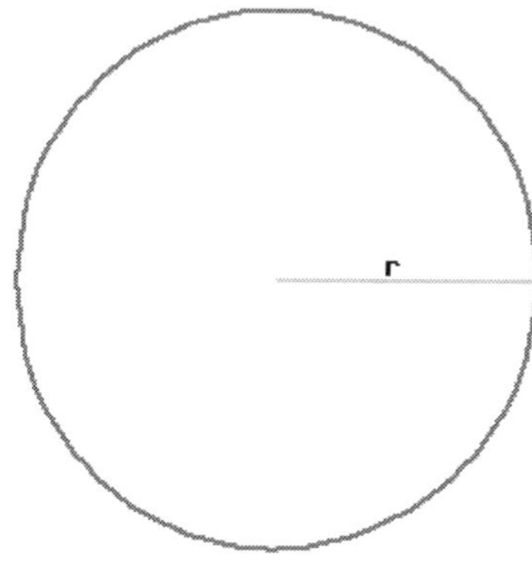

Use those two formulas to answer the following questions about a circle with radius 3 m.

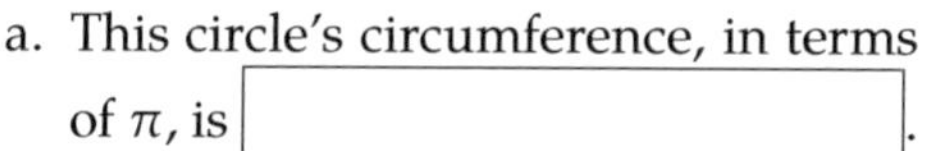

a. This circle's circumference, in terms of π, is ______.

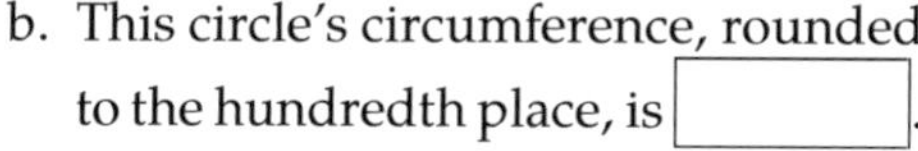

b. This circle's circumference, rounded to the hundredth place, is ______.

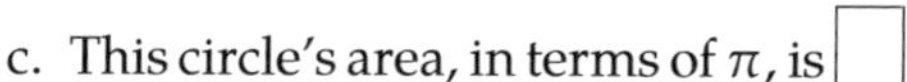

c. This circle's area, in terms of π, is ______.

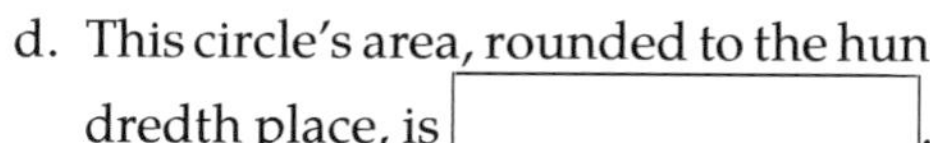

d. This circle's area, rounded to the hundredth place, is ______.

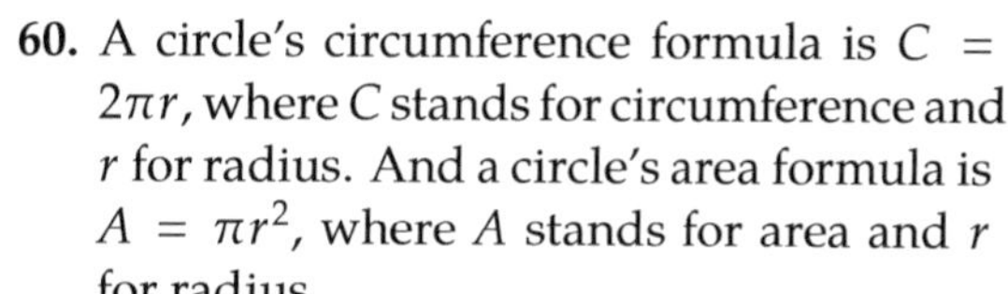

60. A circle's circumference formula is $C = 2\pi r$, where C stands for circumference and r for radius. And a circle's area formula is $A = \pi r^2$, where A stands for area and r for radius.

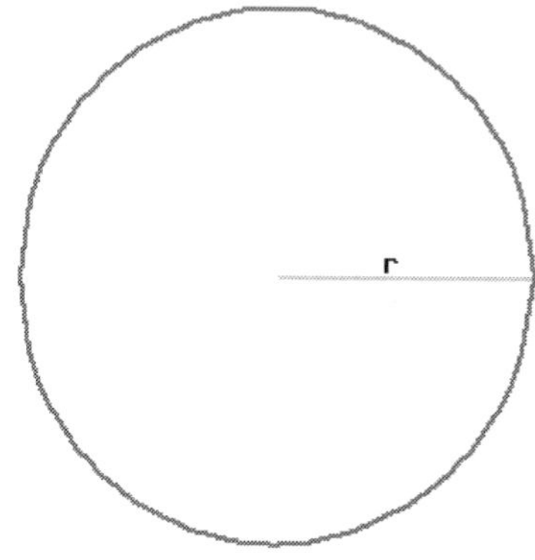

Use those two formulas to answer the following questions about a circle with radius 4 m.

a. This circle's circumference, in terms of π, is ______.

b. This circle's circumference, rounded to the hundredth place, is ______.

c. This circle's area, in terms of π, is ______.

d. This circle's area, rounded to the hundredth place, is ______.

Volume Formulas

61. The formula $V = \frac{1}{3} \cdot s^2 \cdot h$ gives the volume of a right square pyramid.

What is the volume of a right square pyramid with $s = 39$ in and $h = 92$ in?

The volume is ______.

62. The formula $V = \frac{1}{3} \cdot s^2 \cdot h$ gives the volume of a right square pyramid.

What is the volume of a right square pyramid with $s = 51$ in and $h = 58$ in?

The volume is ______.

63. The volume of a rectangular prism can be calculated by the formula $V = LWH$, where V stands for volume, L for the base's length, W for the base's width, and H for the prism's height.

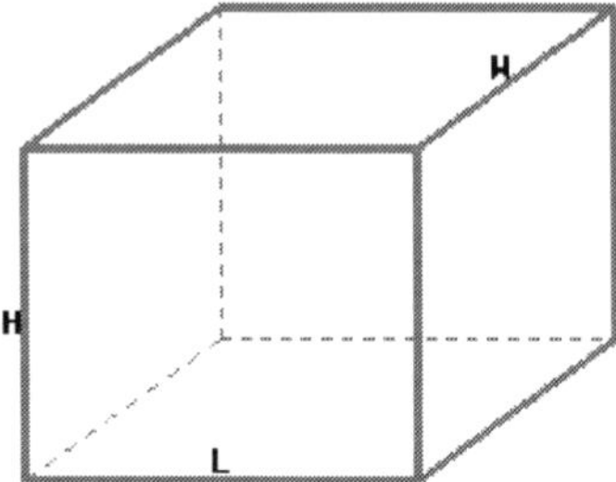

Use the formula to calculate a rectangular prism's volume, if its base's length is 13 in, its base's width is 6 in, and its height is 10 in.

Its volume is ______.

64. The volume of a rectangular prism can be calculated by the formula $V = LWH$, where V stands for volume, L for the base's length, W for the base's width, and H for the prism's height.

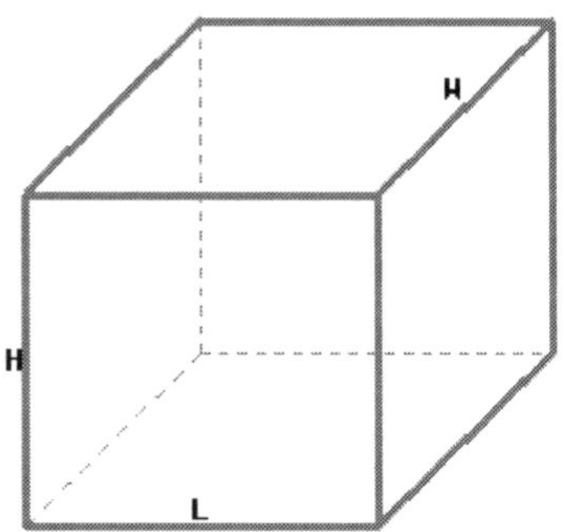

Use the formula to calculate a rectangular prism's volume, if its base's length is 14 in, its base's width is 9 in, and its height is 13 in.

Its volume is ______.

65. A cylinder's volume can be calculated by the formula $V = \pi r^2 h$, where V stands for volume, r for the base radius and h for the cylinder's height.

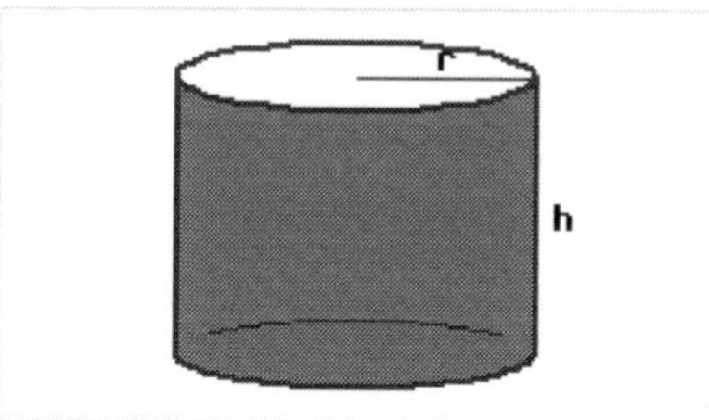

Use the formula to calculate the cylinder's volume if its base's radius is 6 m, and its height is 9 m.

a. This cylinder's volume, in terms of π, is ______.

b. This cylinder's volume, rounded to the hundredth place, is ______.

66. A cylinder's volume can be calculated by the formula $V = \pi r^2 h$, where V stands for volume, r for the base radius and h for the cylinder's height.

Use the formula to calculate the cylinder's volume if its base's radius is 6 m, and its height is 9 m.

a. This cylinder's volume, in terms of π, is ______.

b. This cylinder's volume, rounded to the hundredth place, is ______.

Evaluating Expressions in Context

67. The formula

$$y = \frac{1}{2} a t^2 + v_0 t + y_0$$

gives the vertical position of an object, at time t, thrown with an initial velocity v_0, from an initial position y_0 in a place where the acceleration of gravity is a. The acceleration of gravity on earth is $-9.8\,\frac{\text{m}}{\text{s}^2}$. It is negative, because we consider the upward direction as positive in this situation, and gravity pulls down.

What is the height of a baseball thrown with an initial velocity of $v_0 = 52\,\frac{\text{m}}{\text{s}}$, from an initial position of $y_0 = 92$ m, and at time $t = 12$ s?

After 12 s, the baseball was ______ high in the air.

68. The formula

$$y = \frac{1}{2} a t^2 + v_0 t + y_0$$

gives the vertical position of an object, at time t, thrown with an initial velocity v_0, from an initial position y_0 in a place where the acceleration of gravity is a. The acceleration of gravity on earth is $-9.8\,\frac{\text{m}}{\text{s}^2}$. It is negative, because we consider the upward direction as positive in this situation, and gravity pulls down.

What is the height of a baseball thrown with an initial velocity of $v_0 = 57\,\frac{\text{m}}{\text{s}}$, from an initial position of $y_0 = 75$ m, and at time $t = 5$ s?

After 5 s, the baseball was ______ high in the air.

69. To convert a temperature measured in degrees Fahrenheit to degrees Celsius, there is a formula:

$$C = \frac{5}{9}(F - 32)$$

where C represents the temperature in degrees Celsius and F represents the temperature in degrees Fahrenheit.

If a temperature is 23°F, what is that temperature measured in Celsius?

70. To convert a temperature measured in degrees Fahrenheit to degrees Celsius, there is a formula:

$$C = \frac{5}{9}(F - 32)$$

where C represents the temperature in degrees Celsius and F represents the temperature in degrees Fahrenheit.

If a temperature is 50°F, what is that temperature measured in Celsius?

71. A formula for converting meters into feet is

$$F = 3.28M$$

where M is a number of meters, and F is the corresponding number of feet. Use the formula to find the number of feet that corresponds to eleven meters.

[ ] feet corresponds to eleven meters.

72. A formula for converting hours into seconds is

$$S = 3600H$$

where H is a number of hours, and S is the corresponding number of seconds. Use the formula to find the number of seconds that corresponds to four hours.

[ ] seconds corresponds to four hours.

73. The percentage of births in the U.S. delivered via C-section can be given by the following formula for the years since 1996.

$$p = 0.8(y - 1996) + 21$$

In this formula y is a year after 1996 and p is the percentage of births delivered via C-section for that year. What percentage of births in the U.S. were delivered via C-section in the year 2009?

[ ] of births in the U.S. were delivered via C-section in the year 2009.

74. The percentage of births in the U.S. delivered via C-section can be given by the following formula for the years since 1996.

$$p = 0.8(y - 1996) + 21$$

In this formula y is a year after 1996 and p is the percentage of births delivered via C-section for that year. What percentage of births in the U.S. were delivered via C-section in the year 2011?

[ ] of births in the U.S. were delivered via C-section in the year 2011.

75. Target heart rate for moderate exercise is 50% to 70% of maximum heart rate. If we want to represent a certain percent of an individual's maximum heart rate, we'd use the formula

$$\text{rate} = p(220 - a)$$

where p is the percent, and a is age in years. Determine the target heart rate at 68% level for someone who is 13 years old. Round your answer to an integer.

The target heart rate at 68% level for someone who is 13 years old is [ ] beats per minute.

76. Target heart rate for moderate exercise is 50% to 70% of maximum heart rate. If we want to represent a certain percent of an individual's maximum heart rate, we'd use the formula

$$\text{rate} = p(220 - a)$$

where p is the percent, and a is age in years. Determine the target heart rate at 51% level for someone who is 53 years old. Round your answer to an integer.

The target heart rate at 51% level for someone who is 53 years old is [ ] beats per minute.

77. The diagonal length (D) of a rectangle with side lengths L and W is given by:

$$D = \sqrt{L^2 + W^2}$$

Determine the diagonal length of rectangles with $L = 8$ ft and $W = 6$ ft.

The diagonal length of rectangles with $L =$ 8 ft and $W = 6$ ft is ______.

78. The diagonal length (D) of a rectangle with side lengths L and W is given by:

$$D = \sqrt{L^2 + W^2}$$

Determine the diagonal length of rectangles with $L = 12$ ft and $W = 9$ ft.

The diagonal length of rectangles with $L =$ 12 ft and $W = 9$ ft is ______.

79. The height inside a camping tent when you are d feet from the edge of the tent is given by

$$h = -0.6|d - 5| + 6$$

where h stands for height in feet. Determine the height when you are:

a. 7.9 ft from the edge.

The height inside a camping tent when you 7.9 ft from the edge of the tent is ______

b. 1.1 ft from the edge.

The height inside a camping tent when you 1.1 ft from the edge of the tent is ______

80. The height inside a camping tent when you are d feet from the edge of the tent is given by

$$h = -1.8|d - 5.4| + 6$$

where h stands for height in feet. Determine the height when you are:

a. 7.1 ft from the edge.

The height inside a camping tent when you 7.1 ft from the edge of the tent is ______

b. 3.5 ft from the edge.

The height inside a camping tent when you 3.5 ft from the edge of the tent is ______

2.2 Equations and Inequalities as True/False Statements

This section introduces the concepts of algebraic **equations** and **inequalities**, and what it means for a number to be a **solution** to an equation or inequality.

2.2.1 Equations, Inequalities, and Solutions

An **equation** is two mathematical expressions with an equals sign between them. The two expressions can be relatively simple or more complicated:

A relatively simple equation:

$$x + 1 = 2$$

A more complicated equation:

$$\left(x^2 + y^2 - 1\right)^3 = x^2y^3$$

An **inequality** is quite similar, but the sign between the expressions is one of these: $<$, $\leq$, $>$, $\geq$, or $\neq$.

A relatively simple inequality:

$$x \geq 15$$

A more complicated inequality:

$$x^2 + y^2 < 1$$

A **linear equation** in one variable can be written in the form $ax + b = 0$, where a, b are real numbers, and $a \neq 0$. The variable could be any letter other than x. The variable cannot have other exponent than 1 ($x = x^1$), and the variable cannot be inside a root symbol (square root, cube root, etc.) or in a denominator.

The following are some linear equations in one variable:

$$4 - y = 5 \qquad 4 - z = 5z \qquad 0 = \frac{1}{2}p$$

$$3 - 2(q + 2) = 10 \qquad \sqrt{2} \cdot r + 3 = 10 \qquad \frac{s}{2} + 3 = 5$$

(Note that r is outside the square root symbol.) We will see in later sections that all equations above can be converted into the form $ax + b = 0$.

The following are some non-linear equations:

$1 + 2 = 3$	(There is no variable.)
$4 - 2y^2 = 5$	(The exponent of y is not 1.)
$\sqrt{2r} + 3 = 10$	(r is inside the square root.)
$\frac{2}{s} + 3 = 5$	(s is in a denominator.)

This chapter focuses on linear equations in *one* variable. We will study other types of equations in later chapters.

The simplest equations and inequalities have numbers and no variables. When this happens, the equation is either *true* or *false*.The following equations and inequalities are *true* statements:

$$2 = 2 \qquad -4 = -4 \qquad 2 > 1 \qquad -2 < -1 \qquad 3 \geq 3$$

The following equations and inequalities are *false* statements:

$$2 = 1 \qquad -4 = 4 \qquad 2 < 1 \qquad -2 \geq -1 \qquad 0 \neq 0$$

When equations and inequalities have variables, we can consider substituting values in for the variables. If replacing a variable with a number makes an equation or inequality *true*, then that number is called a **solution** to the equation.

Example 2.2.2 (A Solution). Consider the equation $y + 2 = 3$, which has only one variable, y. If we substitute in 1 for y and then simplify:

$$\begin{aligned} y + 2 &= 3 \\ 1 + 2 &\stackrel{?}{=} 3 \\ 3 &\stackrel{\checkmark}{=} 3 \end{aligned}$$

we get a true equation. So we say that 1 is a solution to $y+2 = 3$. Notice that we used a question mark at first because we are unsure if the equation is true or false until the end.

If replacing a variable with a number makes an equation or inequality false, then that number is not a solution.

Example 2.2.3 (Not a Solution). Consider the inequality $x + 4 > 5$, which has only one variable, x. If we substitute in 0 for x and then simplify:

$$\begin{aligned} x + 4 &> 5 \\ 0 + 4 &\stackrel{?}{>} 5 \\ 4 &\stackrel{\text{no}}{>} 5 \end{aligned}$$

we get a false equation. So we say that 0 is *not* a solution to $x + 4 > 5$.

2.2.2 Checking Possible Solutions

Given an equation or an inequality (with one variable), checking if some particular number is a solution is just a matter of replacing the value of the variable with the specified number and determining if the resulting equation/inequality is true or false. This may involve some amount of arithmetic simplification.

Example 2.2.4. Is 8 a solution to $x^2 - 5x = \sqrt{2x} + 20$?

To find out, substitute in 8 for x and see what happens.

$$\begin{aligned} x^2 - 5x &= \sqrt{2x} + 20 \\ 8^2 - 5(8) &\stackrel{?}{=} \sqrt{2(8)} + 20 \\ 64 - 5(8) &\stackrel{?}{=} \sqrt{16} + 20 \\ 64 - 40 &\stackrel{?}{=} 4 + 20 \\ 24 &\stackrel{\checkmark}{=} 24 \end{aligned}$$

So yes, 8 is a solution to $x^2 - 5x = \sqrt{2x} + 20$.

Example 2.2.5. Is -5 a solution to $\sqrt{169 - y^2} = y^2 - 2y$?

To find out, substitute in -5 for y and see what happens.

$$\begin{aligned} \sqrt{169 - y^2} &= y^2 - 2y \\ \sqrt{169 - (-5)^2} &\stackrel{?}{=} (-5)^2 - 2(-5) \\ \sqrt{169 - 25} &\stackrel{?}{=} 25 - 2(-5) \\ \sqrt{144} &\stackrel{?}{=} 25 - (-10) \\ 12 &\stackrel{\text{no}}{=} 35 \end{aligned}$$

So no, -5 is not a solution to $\sqrt{169 - y^2} = y^2 - 2y$.

But is -5 a solution to the *inequality* $\sqrt{169 - y^2} \leq y^2 - 2y$? Yes, because substituting -5 in for y would give you

$$12 \leq 35,$$

which is true.

Exercise 2.2.6. Is -3 a solution for x in the equation $2x - 3 = 5 - (4 + x)$? Evaluating the left and right sides gives:

$$\begin{array}{ccc} 2x - 3 & = & 5 - (4 + x) \\ \_\_\_\_ & \stackrel{?}{=} & \_\_\_\_ \end{array}$$

So -3 (□ is □ is not) a solution to $2x - 3 = 5 - (4 + x)$.

Exercise 2.2.7. Is $\frac{1}{3}$ a solution for t in the equation $2t = 4\left(t - \frac{1}{2}\right)$? Evaluating the left and right sides gives:

$$\begin{array}{ccc} 2t & = & 4\left(t-\frac{1}{2}\right) \\ \_\_\_ & \stackrel{?}{=} & \_\_\_ \end{array}$$

So $\frac{1}{3}$ (□ is □ is not) a solution to $2t = 4\left(t-\frac{1}{2}\right)$.

Exercise 2.2.8. Is -2 a solution for y in the inequality $y^2 + y - 5 \leq y - 1$? Evaluating the left and right sides gives:

$$\begin{array}{ccc} y^2 + y - 5 & \leq & y - 1 \\ \_\_\_ & \stackrel{?}{\leq} & \_\_\_ \end{array}$$

So -3 (□ is □ is not) a solution to $y^2 + y - 5 \leq y - 1$.

Exercise 2.2.9. Is 2 a solution to $\frac{z+3}{z-1} = \sqrt{18z}$? Evaluating the left and right sides gives:

$$\begin{array}{ccc} \frac{z+3}{z-1} & = & \sqrt{18z} \\ \_\_\_ & \stackrel{?}{=} & \_\_\_ \end{array}$$

So 2 (□ is □ is not) a solution to $\frac{z+3}{z-1} = \sqrt{18z}$.

Exercise 2.2.10. Is -3 a solution to $x^2 + x + 1 \leq \frac{3x+2}{x+2}$? Evaluating the left and right sides gives:

$$\begin{array}{ccc} x^2 + x + 1 & \leq & \frac{3x+2}{x+2} \\ \_\_\_ & \stackrel{?}{\leq} & \_\_\_ \end{array}$$

So -3 (□ is □ is not) a solution to $x^2 + x + 1 \leq \frac{3x+2}{x+2}$.

Example 2.2.11 (Cylinder Volume).

A cylinder's volume is related to its radius and its height by the equation

$$V = \pi r^2 h,$$

where V is the volume, r is the base's radius, and h is the height. If we know the volume is $96\pi\,\text{cm}^3$ and the radius is 4 cm, then this equation simplifies to

$$96\pi = 16\pi h$$

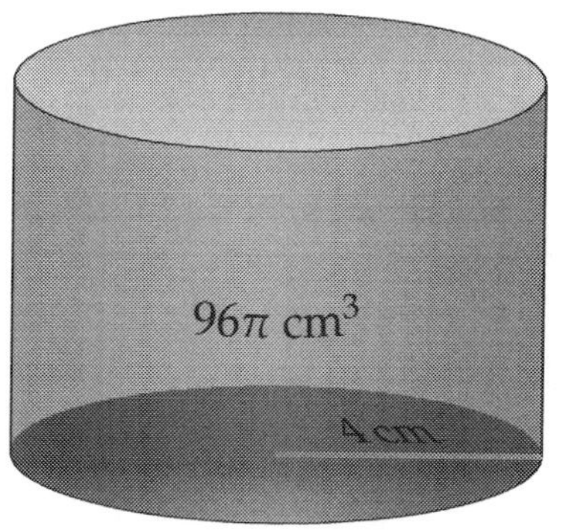

Figure 2.2.12: A cylinder

Is 4 cm the height of the cylinder? In other words, is 4 a solution to $96\pi = 16\pi h$? We will substitute h in the equation with 4 to check:

$$\begin{aligned} 96\pi &= 16\pi h \\ 96\pi &\stackrel{?}{=} 16\pi \cdot 4 \\ 96\pi &\stackrel{\text{no}}{=} 64\pi \end{aligned}$$

Since $96\pi = 64\pi$ is false, $h = 4$ does *not* satisfy the equation $96\pi = 16\pi h$.

Next, we will try $h = 6$:

$$\begin{aligned} 96\pi &= 16\pi h \\ 96\pi &\stackrel{?}{=} 16\pi \cdot 6 \\ 96\pi &\stackrel{\checkmark}{=} 96\pi \end{aligned}$$

When $h = 6$, the equation $96\pi = 16\pi h$ is true. This tells us that 6 *is* a solution for h in the equation $96\pi = 16\pi h$.

Remark 2.2.13. Note that we did not approximate π with 3.14 or any other approximation. We often leave π as π throughout our calculations. If we need to round, we do so as a final step.

Example 2.2.14. Ann has budgeted a maximum of \$300 for an appliance repair. The total cost of the repair can be modeled by $89+110(h-0.25)$, where \$89 is the initial cost and \$110 is the hourly labor charge after the first quarter hour. Is 2 a solution for h in the inequality $89+110(h-0.25) \leq 300$?

To determine if $h = 2$ satisfies the inequality $89 + 110(h - 0.25) \leq 300$, we will replace h with 2 and check if the statement is true or false:

$$89 + 110(h - 0.25) \leq 300$$

$$89 + 110(2 - 0.25) \overset{?}{\leq} 300$$
$$89 + 110(1.75) \overset{?}{\leq} 300$$
$$89 + 192.5 \overset{?}{\leq} 300$$
$$281.5 \overset{\checkmark}{\leq} 300$$

Thus 2 is a solution for h in the inequality $89 + 110(h - 0.25) \leq 300$. In context, this means that Ann would stay within her \$300 budget if 2 hours of labor were performed.

2.2.3 Exercises

Checking Solution for Equations

1. Is 7 a solution for x in the equation $x - 6 = -13$? (□ Yes □ No)

2. Is −5 a solution for x in the equation $x + 8 = 3$? (□ Yes □ No)

3. Is 2 a solution for y in the equation $4 - y = 3$? (□ Yes □ No)

4. Is −6 a solution for y in the equation $-8 - y = -2$? (□ Yes □ No)

5. Is 8 a solution for r in the equation $r - 9 = -2$? (□ Yes □ No)

6. Is 10 a solution for r in the equation $6r + 2 = 62$? (□ Yes □ No)

7. Is 7 a solution for t in the equation $-3t + 1 = 2t + 36$? (□ Yes □ No)

8. Is 6 a solution for t in the equation $6t - 6 = t + 24$? (□ Yes □ No)

9. Is −2 a solution for x in the equation $3(x + 7) = -4x$? (□ Yes □ No)

10. Is −7 a solution for x in the equation $7(x - 5) = 12x$? (□ Yes □ No)

11. Is 5 a solution for y in the equation $5(y + 1) = -6(y - 10)$? (□ Yes □ No)

12. Is 2 a solution for y in the equation $8(y - 9) = 11(y - 6)$? (□ Yes □ No)

13. Is $\frac{1}{9}$ a solution for x in the equation $-9x + 8 = 7$? (□ Yes □ No)

14. Is $\frac{1}{2}$ a solution for x in the equation $-6x + 5 = 1$? (□ Yes □ No)

15. Is $-\frac{4}{7}$ a solution for r in the equation $-\frac{3}{4}r + 4 = \frac{25}{7}$? (□ Yes □ No)

16. Is $-\frac{1}{7}$ a solution for t in the equation $\frac{5}{4}t + 10 = \frac{275}{28}$? (□ Yes □ No)

17. Is $-\frac{9}{5}$ a solution for t in the equation $\frac{4}{3}t - \frac{1}{5} = -\frac{1}{4}t - \frac{61}{20}$? (□ Yes □ No)

18. Is $-\frac{3}{5}$ a solution for x in the equation $-\frac{4}{5}x + 1 = -\frac{2}{3}x + \frac{27}{25}$? (□ Yes □ No)

Checking Solution for Inequalities

For the following exercises: Tell whether the following values are solutions to the given inequality.

19. $2x - 11 > -7$

a. $x = 7$ (□ is □ is not) a solution.

b. $x = 2$ (□ is □ is not) a solution.

c. $x = 0$ (□ is □ is not) a solution.

d. $x = -4$ (□ is □ is not) a solution.

20. $3x - 3 > 9$

a. $x = 0$ (□ is □ is not) a solution.

b. $x = 14$ (□ is □ is not) a solution.

c. $x = 4$ (□ is □ is not) a solution.

d. $x = -3$ (□ is □ is not) a solution.

21. $-3x + 9 \geq 3$

a. $x = 7$ (□ is □ is not) a solution.

b. $x = 2$ (□ is □ is not) a solution.

c. $x = 0$ (□ is □ is not) a solution.

d. $x = -5$ (□ is □ is not) a solution.

22. $3x - 15 \geq -3$

a. $x = -3$ (□ is □ is not) a solution.

b. $x = 14$ (□ is □ is not) a solution.

c. $x = 0$ (□ is □ is not) a solution.

d. $x = 4$ (□ is □ is not) a solution.

23. $4x - 13 \leq -9$

a. $x = 6$ (□ is □ is not) a solution.

b. $x = 0$ (□ is □ is not) a solution.

c. $x = -6$ (□ is □ is not) a solution.

d. $x = 1$ (□ is □ is not) a solution.

24. $-4x + 18 \leq 6$

a. $x = 0$ (□ is □ is not) a solution.

b. $x = 3$ (□ is □ is not) a solution.

c. $x = -4$ (□ is □ is not) a solution.

d. $x = 13$ (□ is □ is not) a solution.

Checking Solutions for Application Problems

25. A triangle's area is 171 square meters. Its height is 19 meters. Suppose we wanted to find how long is the triangle's base. A triangle's area formula is

$$A = \frac{1}{2}bh,$$

where A stands for area, b for base and h for height. If we let b be the triangle's base, in meters, we can solve this problem using the equation:

$$171 = \frac{1}{2}(b)(19).$$

Check whether 36 is a solution for b of this equation. (□ Yes □ No)

26. A triangle's area is 150 square meters. Its height is 15 meters. Suppose we wanted to find how long is the triangle's base. A triangle's area formula is

$$A = \frac{1}{2}bh,$$

where A stands for area, b for base and h for height. If we let b be the triangle's base, in meters, we can solve this problem using the equation:

$$150 = \frac{1}{2}(b)(15).$$

Check whether 20 is a solution for b of this equation. (□ Yes □ No)

27. A cylinder's volume is 160π cubic centimeters. Its height is 10 centimeters. Suppose we wanted to find how long is the cylinder's radius. A cylinder's volume formula is

$$V = \pi r^2 h,$$

where V stands for volume, r for radius and h for height. Let r represent the cylinder's radius, in centimeters. We can solve this problem using the equation:

$$160\pi = \pi r^2(10).$$

Check whether 16 is a solution for r of this equation. (□ Yes □ No)

28. A cylinder's volume is 704π cubic centimeters. Its height is 11 centimeters. Suppose we wanted to find how long is the cylinder's radius. A cylinder's volume formula is

$$V = \pi r^2 h,$$

where V stands for volume, r for radius and h for height. Let r represent the cylinder's radius, in centimeters. We can solve this problem using the equation:

$$704\pi = \pi r^2(11).$$

Check whether 8 is a solution for r of this equation. (□ Yes □ No)

29. A rectangular frame's perimeter is 7 feet. If its length is 2.2 feet, suppose we want to find how long is its width. A rectangle's perimeter formula is

$P = 2(l + w)$,

where P stands for perimeter, l for length and w for width. We can solve this problem using the equation:

$7 = 2(2.2 + w)$.

Check whether 4.8 is a solution for w of this equation. (□ Yes □ No)

30. A rectangular frame's perimeter is 6.8 feet. If its length is 2.3 feet, suppose we want to find how long is its width. A rectangle's perimeter formula is

$P = 2(l + w)$,

where P stands for perimeter, l for length and w for width. We can solve this problem using the equation:

$6.8 = 2(2.3 + w)$.

Check whether 4.5 is a solution for w of this equation. (□ Yes □ No)

31. When a plant was purchased, it was 2.6 inches tall. It grows 0.6 inches per day. How many days later will the plant be 11 inches tall?

Assume the plant will be 11 inches tall d days later. We can solve this problem using the equation:

$0.6d + 2.6 = 11$.

Check whether 14 is a solution for d of this equation. (□ Yes □ No)

32. When a plant was purchased, it was 1.9 inches tall. It grows 0.7 inches per day. How many days later will the plant be 14.5 inches tall?

Assume the plant will be 14.5 inches tall d days later. We can solve this problem using the equation:

$0.7d + 1.9 = 14.5$.

Check whether 18 is a solution for d of this equation. (□ Yes □ No)

33. A water tank has 210 gallons of water in it, and it is being drained at the rate of 16 gallons per minute. After how many minutes will there be 34 gallons of water left?

Assume the tank will have 34 gallons of water after m minutes. We can solve this problem using the equation:

$210 - 16m = 34$.

Check whether 11 is a solution for m of this equation. (□ Yes □ No)

34. A water tank has 349 gallons of water in it, and it is being drained at the rate of 17 gallons per minute. After how many minutes will there be 43 gallons of water left?

Assume the tank will have 43 gallons of water after m minutes. We can solve this problem using the equation:

$349 - 17m = 43$.

Check whether 18 is a solution for m of this equation. (□ Yes □ No)

35. A country's national debt was 140 million dollars in 2010. The debt increased at 70 million dollars per year. If this trend continues, when will the country's national debt increase to 910 million dollars?

Assume the country's national debt will become 910 million dollars y years after 2010. We can solve this problem using the equation:

$70y + 140 = 910.$

Check whether 11 is a solution for y of this equation. (This solution implies the country's national debt will become 910 million dollars in the year 2021.) (□ Yes □ No)

36. A country's national debt was 100 million dollars in 2010. The debt increased at 20 million dollars per year. If this trend continues, when will the country's national debt increase to 480 million dollars?

Assume the country's national debt will become 480 million dollars y years after 2010. We can solve this problem using the equation:

$20y + 100 = 480.$

Check whether 17 is a solution for y of this equation. (This solution implies the country's national debt will become 480 million dollars in the year 2027.) (□ Yes □ No)

37. A school district has a reserve fund worth 33.9 million dollars. It plans to spend 2.1 million dollars per year. After how many years, will there be 15 million dollars left?

Assume there will be 15 million dollars left after y years. We can solve this problem using the equation:

$33.9 - 2.1y = 15.$

Check whether 9 is a solution for y of this equation. (□ Yes □ No)

38. A school district has a reserve fund worth 26.4 million dollars. It plans to spend 2.2 million dollars per year. After how many years, will there be 11 million dollars left?

Assume there will be 11 million dollars left after y years. We can solve this problem using the equation:

$26.4 - 2.2y = 11.$

Check whether 7 is a solution for y of this equation. (□ Yes □ No)

2.3 Solving One-Step Equations

We have learned how to check whether a specific number is a solution to an equation or inequality. In this section, we will begin learning how to *find* the solution(s) to basic equations ourselves.

2.3.1 Imagine Filling in the Blanks

Let's start with a very simple situation — so simple, that you might have success entirely in your head without writing much down. It's not exactly the algebra we want you to learn, but the example may serve as a good warm-up.

Example 2.3.2. A number plus 2 is 6. What is that number?

You may be so familiar with basic arithmetic that you know the answer already. The *algebra* approach would be to start by translating "A number plus 2 is 6" into a math statement — in this case, an equation:

$$x + 2 = 6$$

where x is the number we are trying to find. In other words, what should be substituted in for x...

$$x + 2 = 6$$

... to make the equation true?

Now, how do you determine what x is? One valid option is to just *imagine* what number you could put in place of x that would result in a true equation. Would 0 work? No, that would mean $0+2 \stackrel{\text{no}}{=} 6$. Would 17 work? No, that would mean $17+2 \stackrel{\text{no}}{=} 6$. Would 4 work? Yes, because $4 + 2 = 6$ is a true equation.

So one solution to $x+2 = 6$ is 4. No other numbers are going to be solutions, because when you add 2 to something smaller or larger than 4, the result is going to be smaller or larger than 6.

This approach might work for you to solve *very basic equations*, but in general equations are going to be too complicated to solve in your head this way. So we move on to more systematic approaches.

2.3.2 The Basic Principle of Algebra

2.3.2.1 Opposite Operations

Let's revisit Example 2.3.2, thinking it through differently.

Example 2.3.3. If a number plus 2 is 6, what is the number?

One way to solve this riddle is to use the opposite operation. If a number *plus* 2 is 6, we should be able to *subtract* 2 from 6 and get that unknown number. So the unknown number is $6-2 = 4$.

Let's try this strategy with another riddle.

Example 2.3.4. If a number minus 2 is 6, what is the number? This time, we should be able to *add* 2 to 6 to get the unknown number. So the unknown number is $6 + 2 = 8$.

Does this strategy work with multiplication and division?

Example 2.3.5. If a number times 2 is 6, what is the number? This time, we should be able to *divide* 6 by 2 to get the unknown number. So the unknown number is $6 \div 2 = 3$.

Example 2.3.6. If a number divided by 2 equals 6, what is the number? This time, we can *multiply* 6 by 2, and the unknown number is $6 \cdot 2 = 12$.

These examples explore an important principle for solving an equation — applying an opposite arithmetic operation.

2.3.2.2 Balancing Equations Like a Scale

We can revisit Example 2.3.2 with yet another strategy.

If a number plus 2 is 6, what is the number?

As is common in algebra, we can use x to represent the unknown number. The question translates into the math equation

$$x + 2 = 6.$$

Try to envision the equals sign as the middle of a balanced scale. The left side has 2 one-pound objects and a block with unknown weight x lb. Together, the weight on the left is $x + 2$. The right side has 6 one-pound objects. Figure 2.3.7 shows the scale.

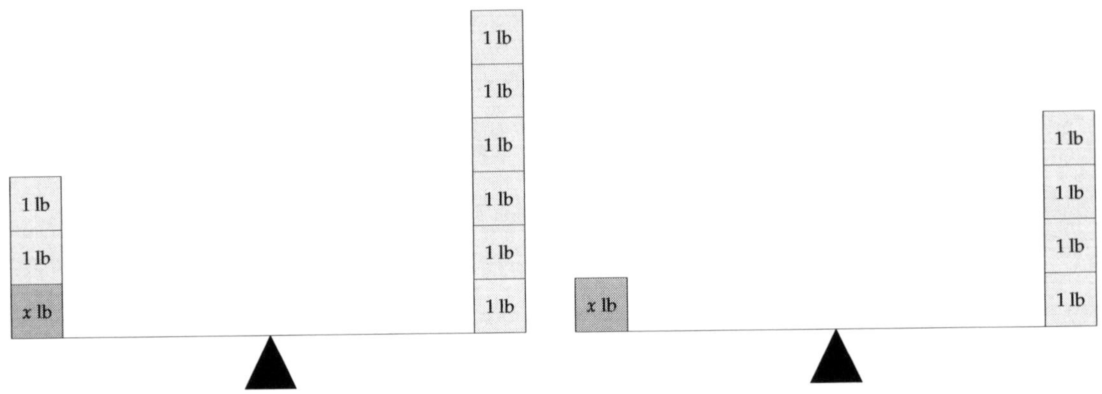

Figure 2.3.7: Scale to represent the equation $x + 2 = 6$

Figure 2.3.8: Scale to represent the solution to $x + 2 = 6$, after taking away 2 from each side

To find the weight of the unknown block, we can take away 2 one-pound blocks from *both* sides of the scale (to keep the scale balanced). Figure 2.3.8 shows the solution.

An equation is like a balanced scale, as the two sides of the equation are equal. In the same way that we can take away 2 lb from *both* sides of a balanced scale, we can subtract 2 on *both* sides of the equals sign. So instead of two pictures of balance scales, we can use algebra symbols and solve the equation $x + 2 = 6$ in the following manner:

$x + 2 = 6$	a balanced scale
$x + 2 - 2 = 6 - 2$	remove the same quanitty from each side
$x = 4$	still balanced; now it tells you the solution

It's important to note that each line shows what is called an **equivalent equation**. In other words, each equation shown is algebraically equivalent to the one above it and the one below it and will have exactly the same solution(s). The final equivalent equation $x = 4$ tells us that the **solution** to the equation is 4. The **solution set** to this equation is the set that lists every solution to the equation. For this example, the solution set is $\{4\}$.

We have learned we can add or subtract the same number on both sides of the equals sign, just like we can add or remove the same amount of weight on a balanced scale. Can we multiply and divide the same number on both sides of the equals sign?

Let's look at Example 2.3.5 again: If a number times 2 is 6, what is the number? Another balance scale can help visualize this.

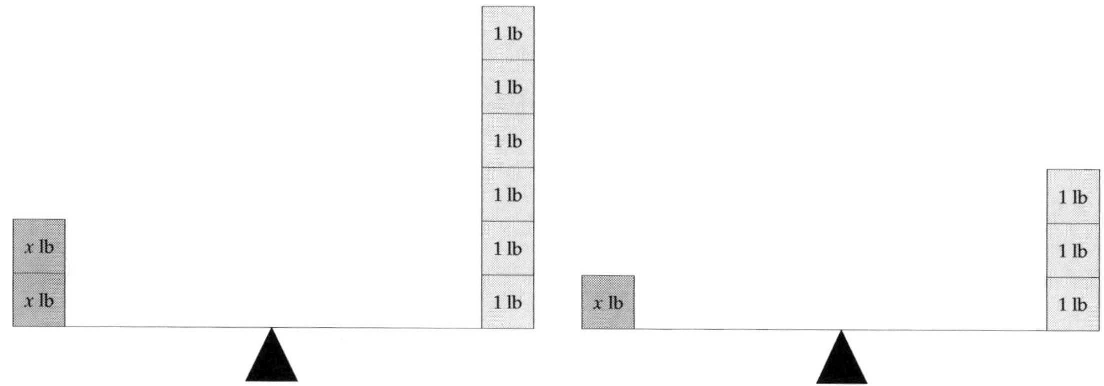

Figure 2.3.9: Scale to represent the equation $2x = 6$

Figure 2.3.10: Scale to represent the equation the solution to $2x = 6$, after cutting each side in half

Currently, the scale is balanced. If we cut the weight in half on both sides, the scale should still be balanced.

We can see from the scale that $x = 3$ is correct. Removing half of the weight on both sides of the scale is like dividing both sides of an equation by 2:

$$\begin{aligned} 2x &= 6 \\ \frac{2x}{2} &= \frac{6}{2} \\ x &= 3 \end{aligned}$$

The equivalent equation in this example is $x = 3$, which tells us that the solution to the equation is 3 and the solution set is $\{3\}$.

Remark 2.3.11. Note that when we divide each side of an equation by a number, we use the fraction line in place of the division symbol. The fact that $\frac{6}{2} = 6 \div 2$ is a reminder that the fraction line and division symbol have the same purpose. The division symbol is rarely used in later math courses.

Similarly, we could multiply both sides of an equation by 2, just like we can keep a scale balanced if we double the weight on each side. We will summarize these properties.

Fact 2.3.12 (Properties of Equivalent Equations). *If there is an equation $A = B$, we can do the following to obtain an equivalent equation:*

- *add the same number to each side: $A + c = B + c$*
- *subtract the same number from each side: $A - c = B - c$*
- *multiply each side of the equation by the same non-zero number: $A \cdot c = B \cdot c$*
- *divide each side of the equation by the same non-zero number: $\frac{A}{c} = \frac{B}{c}$*

With practice, you will learn when it is helpful to use each of these properties.

2.3.3 Solving One-Step Equations and Stating Solution Sets

Notice that when we solved equations in Subsection 2.3.2.2, the final equation looked like $x =$ number, where the variable x is separated from other numbers and stands alone on one side of the equals sign. The goal of solving any equation is to *isolate the variable* in this same manner.

Putting together both strategies (applying the opposite operation and balancing equations like a scale) that we just explored, we will summarize how to solve a one-step linear equation.

Apply Apply the opposite operation to both sides of the equation. If a number was added to the variable, subtract that number, and vice versa. If a number was multiplied by the variable, divide by that number, and vice versa.

Check Check the solution. This means, verify that what you think is the solution actually solves the equation. For one reason, it's human to have made a simple arithmetic mistake, and by checking you will protect yourself from this. For another reason, there are situations where solving an equation will tell you that certain numbers are *possible* solutions, but they do not actually solve the original equation. Checking solutions will catch these situations.

Summarize State the solution set, or in the case of application problems, summarize the result using a complete sentence and appropriate units.

List 2.3.13: Steps to Solving Simple (One-Step) Linear Equations

Let's look at a few examples.

Example 2.3.14. Solve for y in the equation $7 + y = 3$.

Solution.

To isolate y, we need to remove 7 from the left side. Since 7 is being *added* to y, we need to *subtract* 7 from each side of the equation.

$$\begin{aligned} 7 + y &= 3 \\ 7 + y - 7 &= 3 - 7 \\ y &= -4 \end{aligned}$$

We should always check the solution when we solve equations. For this problem, we will substitute y in the original equation with -4:

$$\begin{aligned} 7 + y &= 3 \\ 7 + (-4) &\stackrel{?}{=} 3 \\ 3 &\stackrel{\checkmark}{=} 3 \end{aligned}$$

The solution -4 is checked, so the solution set is $\{-4\}$.

Exercise 2.3.15. Solve for z in the equation $-7.3 + z = 5.1$.

The solution is [ ].

The solution set is [ ].

Solution. To remove the -7.3 from the left side, we need to *add* 7.3 to each side of the equation. (If we make the mistake of subtracting 7.3 from each side, we would have $-7.3+z-7.3 = 5.1-7.3$, which simplifies to $z - 14.3 = -2.2$, which would not isolate z.)

Here is how to solve this equation:

$$\begin{aligned} -7.3 + z &= 5.1 \\ -7.3 + z + 7.3 &= 5.1 + 7.3 \\ z &= 12.4 \end{aligned}$$

We will check the solution by substituting z in the original equation with 12.4:

$$\begin{aligned} -7.3 + z &= 5.1 \\ -7.3 + (12.4) &\stackrel{?}{=} 5.1 \\ 5.1 &\stackrel{\checkmark}{=} 5.1 \end{aligned}$$

The solution 12.4 is checked and the solution set is $\{12.4\}$.

Exercise 2.3.16. Solve for a in the equation $10 = -2a$.

The solution is ______.

The solution set is ______.

Solution. To isolate the variable a, we need to divide each side by -2 (because a is being *multiplied* by -2). One common mistake is to add 2 to each side. This would not isolate a, but would instead leave us with the expression $-2a + 2$ on the right-hand side.

Here is how to solve for a:

$$\begin{aligned} 10 &= -2a \\ \frac{10}{-2} &= \frac{-2a}{-2} \\ -5 &= a \end{aligned}$$

We will check the solution by substituting a in the original equation with -5:

$$\begin{aligned} 10 &= -2a \\ 10 &\stackrel{?}{=} -2(-5) \\ 10 &\stackrel{\checkmark}{=} 10 \end{aligned}$$

The solution -5 is checked and the solution set is $\{-5\}$.

Isolating on the Right Note that in solving the equation in Exercise 2.3.16 we found that $-5 = a$, which is equivalent to $a = -5$. We did not write $a = -5$ as an extra step though, as $-5 = a$ identified the solution.

Example 2.3.17. The formula for a circle's circumference is $C = \pi d$, where C stands for circumference, d stands for diameter, and π is a constant with the value of 3.1415926.... If a circle's circumference is 12π ft, find this circle's diameter and radius.

Solution. The circumference is given as 12π feet. Approximating π with 3.14, this means the circumference is approximately 37.68 ft. It is nice to have a rough understanding of how long the circumference is, but if we use 3.14 instead of π, we are using a slightly smaller number than π, and the result of any calculations we do would not be as accurate. This is why we will use the symbol π throughout solving this equation and round only at the end in the conclusion summary (if necessary).

We will substitute C in the formula with 12π and solve for d:

$$\begin{aligned} C &= \pi d \\ 12\pi &= \pi d \\ \frac{12\pi}{\pi} &= \frac{\pi d}{\pi} \\ 12 &= d \end{aligned}$$

So the circle's diameter is 12 ft. And since radius is half of diameter, the radius is 6 ft.

Example 2.3.18. Solve for b in $-b = 2$.

Solution. Note that the variable b has not been isolated yet as there is a negative sign in front of it. One way to solve for b is to recognize that multiplying on both sides by -1 would clear away that negative sign:

$$\begin{aligned} -b &= 2 \\ -1 \cdot (-b) &= -1 \cdot (2) \\ b &= -2 \end{aligned}$$

We removed the negative sign from $-b$ using the fact that $-1 \cdot (-b) = b$. A second way to remove the negative sign -1 from $-b$ is to divide both sides by -1. If you view the original $-b$ as $-1 \cdot b$, then this approach resembles the solution from Exercise 2.3.16.

$$\begin{aligned} -b &= 2 \\ -1 \cdot b &= 2 \\ \frac{-1 \cdot b}{-1} &= \frac{2}{-1} \\ b &= -2 \end{aligned}$$

A third way to remove the original negative sign, is to recognize that the opposite operation of negation is negation. So negating both sides will work out too:

$$\begin{aligned} -b &= 2 \\ -(-b) &= -2 \\ b &= -2 \end{aligned}$$

We will check the solution by substituting b in the original equation with -2:

$$\begin{aligned} -b &= 2 \\ -(-2) &\stackrel{?}{=} 2 \\ 2 &\stackrel{\checkmark}{=} 2 \end{aligned}$$

The solution -2 is checked and the solution set is $\{-2\}$.

2.3.4 Solving One-Step Equations Involving Fractions

When equations have fractions, solving them will make use of the same principles. You may need to use fraction arithmetic, and there may be special considerations that will make the calculations easier. So we have separated the following examples.

Example 2.3.19. Solve for g in $\frac{2}{3} + g = \frac{1}{2}$.

Solution.

In Section 3.2, we will learn a skill to avoid fraction operations entirely in equations like this one. For now, let's solve the equation by using subtraction to isolate g:

$$\begin{aligned} \frac{2}{3} + g &= \frac{1}{2} \\ \frac{2}{3} + g - \frac{2}{3} &= \frac{1}{2} - \frac{2}{3} \\ g &= \frac{3}{6} - \frac{4}{6} \\ g &= -\frac{1}{6} \end{aligned}$$

We will check the solution by substituting g in the original equation with $-\frac{1}{6}$:

$$\begin{aligned} \frac{2}{3} + g &= \frac{1}{2} \\ \frac{2}{3} + \left(-\frac{1}{6}\right) &\stackrel{?}{=} \frac{1}{2} \\ \frac{4}{6} + \left(-\frac{1}{6}\right) &\stackrel{?}{=} \frac{1}{2} \\ \frac{3}{6} &\stackrel{?}{=} \frac{1}{2} \\ \frac{1}{2} &\stackrel{\checkmark}{=} \frac{1}{2} \end{aligned}$$

The solution $-\frac{1}{6}$ is checked and the solution set is $\left\{-\frac{1}{6}\right\}$.

Exercise 2.3.20. Solve for q in the equation $q - \frac{3}{7} = \frac{3}{2}$.

The solution is ______.

The solution set is ______.

Solution. To remove the $\frac{3}{7}$ from the left side, we need to *add* $\frac{3}{7}$ to each side of the equation.

$$\begin{aligned} q - \frac{3}{7} &= \frac{3}{2} \\ q - \frac{3}{7} + \frac{3}{7} &= \frac{3}{2} + \frac{3}{7} \\ q &= \frac{27}{14} \end{aligned}$$

We will check the solution by substituting q in the original equation with $\frac{27}{14}$:

$$\begin{aligned} q - \frac{3}{7} &= \frac{3}{2} \\ \frac{27}{14} - \frac{3}{7} &\stackrel{?}{=} \frac{3}{2} \\ \frac{27}{14} - \frac{6}{14} &\stackrel{?}{=} \frac{3}{2} \\ \frac{21}{14} &\stackrel{?}{=} \frac{3}{2} \\ \frac{3}{2} &\stackrel{\checkmark}{=} \frac{3}{2} \end{aligned}$$

The solution $\frac{27}{14}$ is checked and the solution set is $\left\{\frac{27}{14}\right\}$.

Example 2.3.21. Solve for c in $\frac{c}{5} = 4$.

Solution.

Note that the fraction line here implies division, so our variable c is being divided by 5. The opposite operation is to *multiply* by 5:

$$\begin{aligned} \frac{c}{5} &= 4 \\ 5 \cdot \frac{c}{5} &= 5 \cdot 4 \\ c &= 20 \end{aligned}$$

We will check the solution by substituting c in the original equation with 20:

$$\begin{aligned} \frac{c}{5} &= 4 \\ \frac{20}{5} &\stackrel{?}{=} 4 \\ 4 &\stackrel{\checkmark}{=} 4 \end{aligned}$$

The solution 20 is checked and the solution set is $\{20\}$.

Example 2.3.22. Solve for d in $-\frac{1}{3}d = 6$.

Solution. It's true that in this example, the variable d is *multiplied* by $-\frac{1}{3}$. This means that *dividing* each side by $-\frac{1}{3}$ would be a valid strategy for solving this equation. However, dividing by a fraction could lead to human error, so consider this alternative strategy.

Another way to be rid of the $-\frac{1}{3}$ is to multiply by -3. Indeed, $-\frac{1}{3}d$ is the same as $\frac{d}{-3}$, and when we view the expression this way, d is being *divided* by -3. So multiplying by -3 would be the opposite operation.

$$\begin{aligned} -\frac{1}{3}d &= 6 \\ (-3)\cdot\left(-\frac{1}{3}d\right) &= (-3)\cdot 6 \\ d &= -18 \end{aligned}$$

If you choose to divide each side by $-\frac{1}{3}$, that will work out as well:

$$\begin{aligned} -\frac{1}{3}d &= 6 \\ \frac{-\frac{1}{3}d}{-\frac{1}{3}} &= \frac{6}{-\frac{1}{3}} \\ d &= \frac{6}{1}\cdot\frac{-3}{1} \\ d &= -18 \end{aligned}$$

This gives the same solution.

We will check the solution by substituting d in the original equation with -18:

$$\begin{aligned} -\frac{1}{3}d &= 6 \\ -\frac{1}{3}\cdot(-18) &\stackrel{?}{=} 6 \\ 6 &\stackrel{\checkmark}{=} 6 \end{aligned}$$

The solution -18 is checked and the solution set is $\{-18\}$.

Example 2.3.23. Solve for x in $\frac{3x}{4} = 10$.

Solution. The variable x appears to have *two* operations that apply to it: first multiplication by 3, and then division by 4. But note that

$$\frac{3x}{4} = \frac{3}{4}\cdot\frac{x}{1} = \frac{3}{4}x.$$

If we view the left side this way, we can get away with solving the equation in one step, by multiplying on each side by the reciprocal of $\frac{3}{4}$.

$$\begin{aligned}\frac{3x}{4} &= 10\\ \frac{3}{4}x &= 10\\ \frac{4}{3}\cdot\frac{3}{4}x &= \frac{4}{3}\cdot 10\\ x &= \frac{4}{3}\cdot\frac{10}{1}\\ x &= \frac{40}{3}\end{aligned}$$

We will check the solution by substituting x in the original equation with $\frac{40}{3}$:

$$\begin{aligned}\frac{3x}{4} &= 10\\ \frac{3\left(\frac{40}{3}\right)}{4} &\stackrel{?}{=} 10\\ \frac{40}{4} &\stackrel{?}{=} 10\\ 10 &\stackrel{\checkmark}{=} 10\end{aligned}$$

The solution $\frac{40}{3}$ is checked and the solution set is $\left\{\frac{40}{3}\right\}$.

Exercise 2.3.24. Solve for H in the equation $\frac{-7H}{12} = \frac{2}{3}$.

The solution is ______.

The solution set is ______.

Solution. The left side is effectively the same things as $-\frac{7}{12}H$, so multiplying by $-\frac{12}{7}$ will isolate H.

$$\begin{aligned}\frac{-7H}{12} &= \frac{2}{3}\\ -\frac{7}{12}H &= \frac{2}{3}\\ \left(-\frac{12}{7}\right)\cdot\left(-\frac{7}{12}H\right) &= \left(-\frac{12}{7}\right)\cdot\frac{2}{3}\\ H &= -\frac{\cancelto{4}{12}}{7}\cdot\frac{2}{\cancelto{1}{3}}\\ H &= -\frac{8}{7}\end{aligned}$$

We will check the solution by substituting H in the original equation with $-\frac{8}{7}$:

$$\begin{aligned}\frac{-7H}{12} &= \frac{2}{3}\\ \frac{-7\left(-\frac{8}{7}\right)}{12} &\stackrel{?}{=} \frac{2}{3}\\ \frac{8}{12} &\stackrel{?}{=} \frac{2}{3}\\ \frac{2}{3} &\stackrel{\checkmark}{=} \frac{2}{3}\end{aligned}$$

The solution $-\frac{8}{7}$ is checked and the solution set is $\{-\frac{8}{7}\}$.

2.3.5 Exercises

Solving One-Step Equations with Addition/Subtraction

For the following exercises: Solve the equation.

1. $t+7=12$

2. $t+3=12$

3. $x+10=1$

4. $x+6=1$

5. $-3=x+7$

6. $6=y+12$

7. $-9=y-8$

8. $-14=r-10$

9. $r+69=0$

10. $t+41=0$

11. $t-2=4$

12. $x-9=1$

13. $-13=x-6$

14. $-5=x-3$

15. $y-(-6)=8$

16. $y-(-8)=12$

17. $-5=r-(-5)$

18. $-6=r-(-2)$

19. $2+t=-4$

20. $1+t=-8$

21. $4=-4+x$

22. $-5=-7+x$

23. $x+\frac{9}{10}=\frac{9}{10}$

24. $y+\frac{7}{8}=\frac{7}{8}$

25. $-\frac{4}{5}+y=-\frac{9}{10}$

26. $-\frac{10}{11}+r=-\frac{15}{22}$

27. $\frac{4}{7}+C=-\frac{7}{4}$

28. $\frac{8}{3}+n=-\frac{7}{2}$

Solving One-Step Equations with Multiplication/Division

For the following exercises: Solve the equation.

29. $10t=50$

30. $6x=48$

31. $27=-9x$

32. $36=-4x$

33. $0=1a$

34. $0=-34c$

35. $\frac{1}{9}A=7$

36. $\frac{1}{6}C=10$

37. $\frac{3}{8}m = 12$

38. $\frac{6}{13}p = 18$

39. $\frac{4}{5}x = 2$

40. $\frac{7}{2}y = 8$

41. $\frac{4}{3} = -\frac{t}{7}$

42. $\frac{5}{6} = -\frac{a}{3}$

43. $2c = -7$

44. $8A = -4$

45. $-\frac{C}{15} = \frac{8}{5}$

46. $-\frac{m}{18} = \frac{2}{9}$

47. $-\frac{p}{40} = -\frac{7}{8}$

48. $-\frac{x}{16} = -\frac{9}{4}$

49. $-\frac{3}{10} = \frac{9y}{2}$

50. $-\frac{8}{7} = \frac{9t}{10}$

51. $\frac{x}{25} = \frac{4}{5}$

52. $\frac{x}{54} = \frac{7}{6}$

53. $\frac{2}{7} = \frac{x}{49}$

54. $\frac{9}{8} = \frac{x}{40}$

Comparisons

For the following exercises: Solve the equation.

55.

a. $4t = 16$

b. $4 + r = 16$

56.

a. $4t = 16$

b. $4 + r = 16$

57.

a. $50 = -5t$

b. $50 = -5 + x$

58.

a. $36 = -9x$

b. $36 = -9 + t$

59.

a. $-x = 12$

b. $-y = -12$

60.

a. $-y = 20$

b. $-x = -20$

61.

a. $-\frac{1}{5}c = 10$

b. $-\frac{1}{5}m = -10$

62.

a. $-\frac{1}{9}A = 7$

b. $-\frac{1}{9}b = -7$

63.

a. $12 = -\frac{4}{5}C$

b. $-12 = -\frac{4}{5}y$

64.

a. $4 = -\frac{2}{5}m$

b. $-4 = -\frac{2}{5}C$

65.

a. $2t = 10$

b. $16y = 52$

66.

a. $6t = 24$

b. $32x = 52$

67.

a. $20 = -10x$

b. $52 = -16r$

68.

a. $20 = -4x$

b. $65 = -40y$

Geometry Application Problems

69. A circle's circumference is 8π mm.

a. This circle's diameter is ________.

b. This circle's radius is ________.

70. A circle's circumference is 10π mm.

a. This circle's diameter is ________.

b. This circle's radius is ________.

71. A circle's circumference is 41 cm. Find the following values. Round your answer to at least 2 decimal places.

a. This circle's diameter is [ ].

b. This circle's radius is [ ].

72. A circle's circumference is 43 cm. Find the following values. Round your answer to at least 2 decimal places.

a. This circle's diameter is [ ].

b. This circle's radius is [ ].

73. A circle's area is 64π in^2.

a. This circle's radius is [ ].

b. This circle's diameter is [ ].

74. A circle's area is 81π in^2.

a. This circle's radius is [ ].

b. This circle's diameter is [ ].

75. A circle's area is 50 mm^2. Find the following values. Round your answer to 2 decimal places.

a. This circle's radius is [ ].

b. This circle's diameter is [ ].

76. A circle's area is 32 mm^2. Find the following values. Round your answer to 2 decimal places.

a. This circle's radius is [ ].

b. This circle's diameter is [ ].

77. A cylinder's base's radius is 3 m, and its volume is 27π m^3.

This cylinder's height is [ ].

78. A cylinder's base's radius is 9 m, and its volume is 324π m^3.

This cylinder's height is [ ].

79. A rectangle's area is 360 mm^2. Its height is 15 mm.

Its base is [ ].

80. A rectangle's area is 286 mm^2. Its height is 11 mm.

Its base is [ ].

81. A rectangular prism's volume is 11664 ft^3. The prism's base is a rectangle. The rectangle's length is 27 ft and the rectangle's width is 18 ft.

This prism's height is [ ].

82. A rectangular prism's volume is 3360 ft^3. The prism's base is a rectangle. The rectangle's length is 28 ft and the rectangle's width is 15 ft.

This prism's height is [ ].

83. A triangle's area is 170.5 m^2. Its base is 31 m.

Its height is [ ].

84. A triangle's area is 263.5 m^2. Its base is 31 m.

Its height is [ ].

2.4 Solving One-Step Inequalities

We have learned how to check whether a specific number is a solution to an equation or inequality. In this section, we will begin learning how to *find* the solution(s) to basic inequalities ourselves.

With one small complication, we can use very similar properties to Fact 2.3.12 when we solve inequalities (as opposed to equations).

Here are some numerical examples.

Add to both sides If $2 < 4$, then $2 + 1 \stackrel{\checkmark}{<} 4 + 1$.

Subtract from both sides If $2 < 4$, then $2 - 1 \stackrel{\checkmark}{<} 4 - 1$.

Multiply on both sides by a *positive* number If $2 < 4$, then $3 \cdot 2 \stackrel{\checkmark}{<} 3 \cdot 4$.

Divide on both sides by a *positive* number If $2 < 4$, then $\frac{2}{2} \stackrel{\checkmark}{<} \frac{4}{2}$.

However, something interesting happens when we multiply or divide by the same *negative* number on both sides of an inequality: the direction reverses! To understand why, consider Figure 2.4.2, where the numbers 2 and 4 are multiplied by the negative number -1.

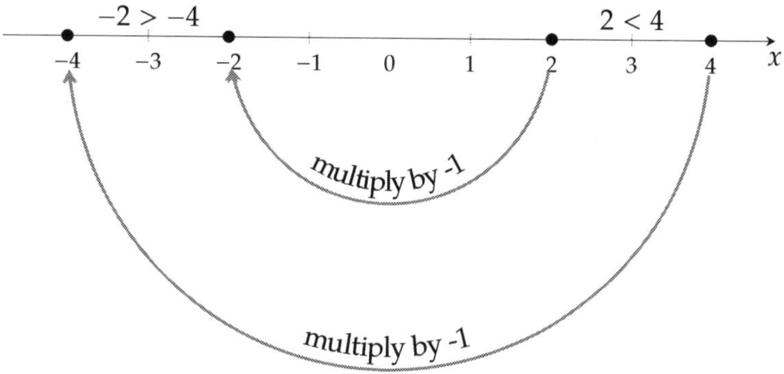

Figure 2.4.2: Multiplying two numbers by a negative number, and how their relationship changes

So even though $2 < 4$, if we multiply both sides by -1, we have $-2 \stackrel{\text{no}}{<} -4$. (The true inequality is $-2 > -4$.)

In general, we must apply the following property when solving an inequality.

Fact 2.4.3 (Changing the Direction of the Inequality Sign). *When we multiply or divide each side of an inequality by the same negative number, the inequality sign must change direction. Do not change the inequality sign when multiplying/dividing by a positive number, or when adding/subtracting by any number.*

Example 2.4.4. Solve the inequality $-2x \geq 12$. State the solution set graphically, using interval notation, and using set-builder notation. (Interval notation and set-builder notation are discussed in Section 1.7.

Solution. To solve this inequality, we will divide each side by -2:

$$\begin{aligned} -2x &\geq 12 \\ \frac{-2x}{-2} &\leq \frac{12}{-2} \qquad \text{Note the change in direction.} \\ x &\leq -6 \end{aligned}$$

Note that the inequality sign changed direction at the step where we divided each side of the inequality by a *negative* number.

When we solve a linear *equation*, there is usually exactly one solution. When we solve a linear *inequality*, there are usually infinitely many solutions. For this example, any number smaller than -6 or equal to -6 is a solution.

There are at least three ways to represent the solution set for the solution to an inequality: graphically, with set-builder notation, and with interval notation. Graphically, we represent the solution set as:

−10 −8 −6 −4 −2 0 2 4 6 8 10 x

Using interval notation, we write the solution set as $(-\infty, -6]$. Using set-builder notation, we write the solution set as $\{x \mid x \leq -6\}$.

As with equations, we should check solutions to catch both human mistakes as well as for possible extraneous solutions (numbers which were *possible* solutions according to algebra, but which actually do not solve the inequality).

Since there are infinitely many solutions, it's impossible to literally check them all. We found that all values of x for which $x \leq -6$ are solutions. One approach is to check that -6 satisfies the inequality, and also that one number less than -6 (any number, your choice) is a solution.

$$\begin{aligned} -2x &\geq 12 \\ -2(-6) &\overset{?}{\geq} 12 \\ 12 &\overset{\checkmark}{\geq} 12 \end{aligned} \qquad\qquad \begin{aligned} -2x &\geq 12 \\ -2(-7) &\overset{?}{\geq} 12 \\ 14 &\overset{\checkmark}{\geq} 12 \end{aligned}$$

Thus both -6 and -7 are solutions. It's important to note this doesn't directly verify that *all* solutions to this inequality check. But it is evidence that our solution is correct, and it's valuable in that making these two checks would likely help us catch an error if we had made one. Consult your instructor to see if you're expected to check your answer in this manner.

Example 2.4.5. Solve the inequality $t + 7 < 5$. State the solution set graphically, using interval notation, and using set-builder notation.

Solution. To solve this inequality, we will subtract 7 from each side. There is not much difference between this process and solving the *equation* $t + 7 = 5$, because we are not going to multiply or divide by negative numbers.

$$\begin{aligned} t + 7 &< 5 \\ t + 7 - 7 &< 5 - 7 \\ t &< -2 \end{aligned}$$

Note again that the direction of the inequality did not change, since we did not multiply or divide each side of the inequality by a negative number at any point.

Graphically, we represent this solution set as:

−8 −6 −4 −2 0 2 4 6 8 t

Using interval notation, we write the solution set as $(-\infty, -2)$. Using set-builder notation, we write the solution set as $\{t \mid t < -2\}$.

We should check that -2 is *not* a solution, but that some number less than -2 *is* a solution.

$$\begin{aligned} t + 7 &< 5 \\ -2 + 7 &\stackrel{?}{<} 5 \\ 5 &\stackrel{\text{no}}{<} 5 \end{aligned} \qquad \begin{aligned} t + 7 &< 5 \\ -10 + 7 &\stackrel{?}{<} 5 \\ -3 &\stackrel{\checkmark}{<} 5 \end{aligned}$$

So our solution is reasonably checked.

Exercise 2.4.6. Solve the inequality $x - 5 > -4$. State the solution set using interval notation and using set-builder notation.

In interval notation, the solution set is __________.

In set-builder notation, the solution set is __________.

Solution. To solve this inequality, we will add 5 to each side.

$$\begin{aligned} x - 5 &> -4 \\ x - 5 + 5 &> -4 + 5 \\ x &> 1 \end{aligned}$$

Note again that the direction of the inequality did not change, since we did not multiply or divide each side of the inequality by a negative number at any point.

Graphically, we represent this solution set as:

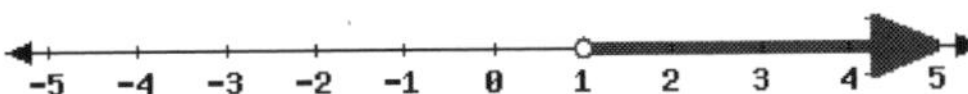

Using interval notation, we write the solution set as $(1, \infty)$. Using set-builder notation, we write the solution set as $\{x \mid x > 1\}$.

We should check that 1 is *not* a solution, but that some number greater than 1 *is* a solution.

$$\begin{aligned} x - 5 &> -4 & x - 5 &> -4 \\ 1 - 5 &\overset{?}{<} -4 & 10 - 5 &\overset{?}{<} -4 \\ -4 &\overset{\text{no}}{<} -4 & 5 &\overset{\checkmark}{<} -4 \end{aligned}$$

So our solution is reasonably checked.

Exercise 2.4.7. Solve the inequality $-\frac{1}{2}z \geq -1.74$. State the solution set using interval notation and using set-builder notation.

In interval notation, the solution set is ______.

In set-builder notation, the solution set is ______.

Solution. To solve this inequality, we will multiply by -2 to each side.

$$\begin{aligned} -\frac{1}{2}z &\geq -1.74 \\ (-2)\left(-\frac{1}{2}z\right) &\leq (-2)(-1.74) \\ z &\leq 3.48 \end{aligned}$$

In this exercise, we *did* multiply by a negative number and so the direction of the inequality sign changed.

Graphically, we represent this solution set as:

Using interval notation, we write the solution set as $(-\infty, 3.48]$. Using set-builder notation, we write the solution set as $\{z \mid z \leq 3.48\}$.

We should check that 3.48 *is* a solution, and also that some number less than 3.48 is a solution.

$$\begin{aligned} -\frac{1}{2}z &\geq -1.74 & -\frac{1}{2}z &\geq -1.74 \\ -\frac{1}{2}(3.48) &\stackrel{?}{\geq} -1.74 & -\frac{1}{2}(0) &\stackrel{?}{\geq} -1.74 \\ -1.74 &\stackrel{\checkmark}{\geq} -1.74 & 0 &\stackrel{\checkmark}{\geq} -1.74 \end{aligned}$$

So our solution is reasonably checked.

2.4.1 Exercises

Solving One-Step Inequalities using Addition/Subtraction

For the following exercises: Solve this inequality.

1. $x+2>9$

In set-builder notation, the solution set is ☐.

In interval notation, the solution set is ☐.

2. $x+3>7$

In set-builder notation, the solution set is ☐.

In interval notation, the solution set is ☐.

3. $x-3 \leq 6$

In set-builder notation, the solution set is ☐.

In interval notation, the solution set is ☐.

4. $x-4 \leq 9$

In set-builder notation, the solution set is ☐.

In interval notation, the solution set is ☐.

5. $4 \leq x+7$

In set-builder notation, the solution set is ☐.

In interval notation, the solution set is ☐.

6. $5 \leq x+10$

In set-builder notation, the solution set is ☐.

In interval notation, the solution set is ☐.

7. $5 > x-9$

In set-builder notation, the solution set is ☐.

In interval notation, the solution set is ☐.

8. $1 > x-7$

In set-builder notation, the solution set is ☐.

In interval notation, the solution set is ☐.

Solving One-Step Inequalities using Multiplication/Division

For the following exercises: Solve this inequality.

9. $2x \le 8$

In set-builder notation, the solution set is [ ].

In interval notation, the solution set is [ ].

10. $3x \le 9$

In set-builder notation, the solution set is [ ].

In interval notation, the solution set is [ ].

11. $4x > 5$

In set-builder notation, the solution set is [ ]. In interval notation, the solution set is [ ].

12. $7x > 6$

In set-builder notation, the solution set is [ ]. In interval notation, the solution set is [ ].

13. $-4x \ge 12$

In set-builder notation, the solution set is [ ].

In interval notation, the solution set is [ ].

14. $-5x \ge 10$

In set-builder notation, the solution set is [ ].

In interval notation, the solution set is [ ].

15. $20 \ge -5x$

In set-builder notation, the solution set is [ ].

In interval notation, the solution set is [ ].

16. $15 \ge -5x$

In set-builder notation, the solution set is [ ].

In interval notation, the solution set is [ ].

17. $2 < -x$

In set-builder notation, the solution set is [ ].

In interval notation, the solution set is [ ].

18. $3 < -x$

In set-builder notation, the solution set is [ ].

In interval notation, the solution set is [ ].

19. $-x \le 4$

In set-builder notation, the solution set is [ ].

In interval notation, the solution set is [ ].

20. $-x \le 5$

In set-builder notation, the solution set is [ ].

In interval notation, the solution set is [ ].

21. $\frac{6}{7}x > 6$

In set-builder notation, the solution set is [ ].

In interval notation, the solution set is [ ].

22. $\frac{7}{5}x > 28$

In set-builder notation, the solution set is [ ].

In interval notation, the solution set is [ ].

23. $-\frac{8}{5}x \leq 24$

In set-builder notation, the solution set is [ ].

In interval notation, the solution set is [ ].

24. $-\frac{9}{8}x \leq 36$

In set-builder notation, the solution set is [ ].

In interval notation, the solution set is [ ].

25. $-10 < \frac{10}{7}x$

In set-builder notation, the solution set is [ ].

In interval notation, the solution set is [ ].

26. $-3 < \frac{1}{2}x$

In set-builder notation, the solution set is [ ].

In interval notation, the solution set is [ ].

27. $-2 < -\frac{2}{7}x$

In set-builder notation, the solution set is [ ].

In interval notation, the solution set is [ ].

28. $-8 < -\frac{4}{5}x$

In set-builder notation, the solution set is [ ].

In interval notation, the solution set is [ ].

29. $3x > -12$

In set-builder notation, the solution set is [ ].

In interval notation, the solution set is [ ].

30. $4x > -12$

In set-builder notation, the solution set is [ ].

In interval notation, the solution set is [ ].

31. $-8 < -4x$

In set-builder notation, the solution set is [ ].

In interval notation, the solution set is [ ].

32. $-20 < -5x$

In set-builder notation, the solution set is [ ].

In interval notation, the solution set is [ ].

33. $\frac{7}{10} \geq \frac{x}{40}$

In set-builder notation, the solution set is ______.

In interval notation, the solution set is ___.

34. $\frac{9}{10} \geq \frac{x}{40}$

In set-builder notation, the solution set is ______.

In interval notation, the solution set is ___.

35. $-\frac{z}{8} < -\frac{9}{2}$

In set-builder notation, the solution set is ______.

In interval notation, the solution set is ___.

36. $-\frac{z}{12} < -\frac{7}{2}$

In set-builder notation, the solution set is ______.

In interval notation, the solution set is ___.

2.5 One-Step Equations With Percentages

With the skills we have learned for solving one-step linear equations, we can solve a variety of percent-related problems that arise in our everyday life.

In many situations when translating from English to math, the word "of" translates as multiplication. For example:

One third of thirty is ten.

$$\frac{1}{3} \cdot 30 = 10$$

It's also helpful to note that when we translate English into math, the word "is" (and many similar words related to "to be") translates to an equals sign.

Here is another example, this time involving a percentage. We know that "2 is 50% of 4," so we have:

2 is 50% of 4

$$2 = 0.5 \cdot 4$$

Example 2.5.2. Translate each statement involving percents below into an equation. Define any variables used. (Solving these equations is an exercise).

a. How much is 30% of $24.00?

b. $7.20 is what percent of $24.00?

c. $7.20 is 30% of how much money?

Solution. Each question can be translated from English into a math equation by reading it slowly and looking for the right signals.

a. The word "is" means about the same thing as the = sign. "How much" is a question phrase, and we can let x be the unknown amount (in dollars). The word "of" translates to multiplication, as discussed earlier. So we have:

$$\underbrace{x}_{\text{How much}} \underbrace{=}_{\text{is}} \underbrace{0.30}_{30\%} \underbrace{\cdot}_{\text{of}} \underbrace{24}_{\$24}$$

b. Let P be the unknown value. We have:

$$\underbrace{7.2}_{\$7.20} \underbrace{=}_{\text{is}} \underbrace{P}_{\text{what percent}} \underbrace{\cdot}_{\text{of}} \underbrace{24}_{\$24}$$

With this setup, P is going to be a decimal value (0.30) that you would translate into a percentage (30%).

c. Let x be the unknown amount (in dollars). We have:

$$\underbrace{7.2}_{\$7.20} \underbrace{=}_{\text{is}} \underbrace{0.30}_{30\%} \underbrace{\cdot}_{\text{of}} \underbrace{x}_{\text{how much money}}$$

Exercise 2.5.3. Solve each equation from Example 2.5.2.

a. How much is 30% of \$24.00?

$$x = 0.3 \cdot 24$$

x is ______.

b. \$7.20 is what percent of \$24.00?

$$7.2 = P \cdot 24$$

P is ______.

c. \$7.20 is 30% of how much money?

$$7.2 = 0.3 \cdot x$$

x is ______.

2.5.1 Modeling and Solving Percent Equations

An important skill for solving percent-related problems is to boil down a complicated word problem into a simple form like "2 is 50% of 4." Let's look at some further examples.

Remark 2.5.4. When solving problems that are *not* applications, we state the **solution set**, a mathematical object. This communicates the set of all solution(s) to that equation or inequality. However, when solving an equation or inequality that arises in an application problem, it makes more sense to summarize our result with a sentence, using the context of the application. This allows us to communicate the full result, including appropriate units.

Example 2.5.5.

As of Fall 2016, Portland Community College had 89,900 enrolled students. According to the chart below, how many black students were enrolled at PCC as of Fall 2016 term?

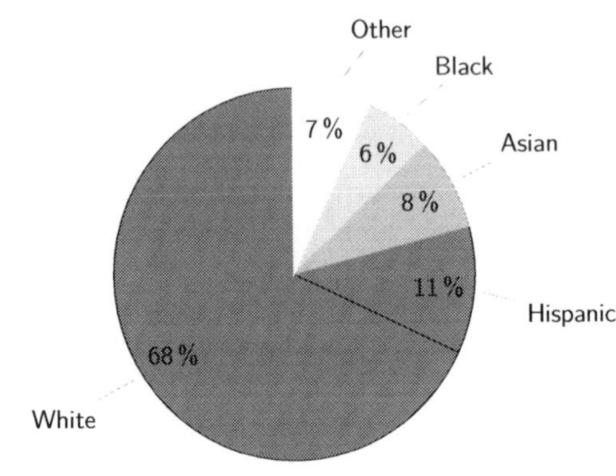

Figure 2.5.6: Racial breakdown of PCC students in Fall 2016

Solution. After reading this word problem and the chart, we can translate the problem into "what is 6% of 89,900?" Let x be the number of black students enrolled at PCC in Fall 2016. We can set up and solve the equation:

$$x = 0.06 \cdot 89900$$
$$x = 5394$$

(There was not much "solving" to do, since the variable we wanted to isolate was already isolated.)

As of Fall 2016, Portland Community College had 5394 black students. (Note: this is not likely to be perfectly accurate, because the information we started with (89,900 enrolled students and 6%) appear to be rounded.)

Example 2.5.7.

According to the bar graph, what percentage of the school's student population is freshman?

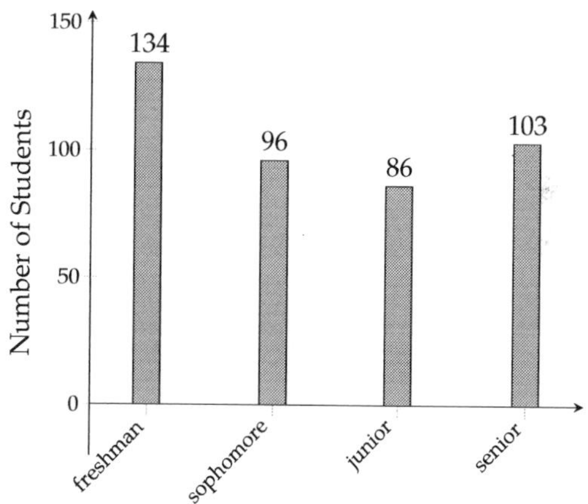

Figure 2.5.8: Number of students at a high school by class

Solution. The school's total number of students is:

$$134 + 96 + 86 + 103 = 419$$

Now, we can translate the main question:

"What percentage of the school's student population is freshman?"

into:

"What percent of 419 is 134?"

Using P to represent the unknown quantity, we write and solve the equation:

$$\overbrace{P}^{\text{what percent}} \overbrace{\cdot}^{\text{of}} \overbrace{419}^{419} \overbrace{=}^{\text{is}} \overbrace{134}^{134}$$

$$\frac{P \cdot 419}{419} = \frac{134}{419}$$

$$P = 0.319809\ldots$$

$$P \approx 31.98\%$$

Approximately 31.98% of the school's student population is freshman.

Example 2.5.9. Kim just received her monthly paycheck. Her gross pay (the amount before taxes and related things are deducted) was \$2,346.19, and her total tax and other deductions was \$350.21. The rest was deposited directly into her checking account. What percent of her gross pay went into her checking account?

Solution. Train yourself to read the word problem and not try to pick out numbers to substitute into formulas. You may find it helps to read the problem over to yourself three or more times before you attempt to solve it. There are *three* dollar amounts to discuss in this problem, and many students fall into a trap of using the wrong values in the wrong places. There is the gross pay, the amount that was deducted, and the amount that was deposited.

Even though there are three dollar amounts to think about, only two have been explicitly written down. We need to compute a subtraction to find the dollar amount that was deposited:

$$2346.19 - 350.21 = 1995.98$$

Now, we can translate the main question:

> "What percent of her gross pay went into her checking account?"

into:

> "What percent of 2346.19 is 1995.98?"

Using P to represent the unknown quantity, we write and solve the equation:

$$\overbrace{P}^{\text{what percent}} \overbrace{\cdot}^{\text{of}} \overbrace{2346.19}^{\$2346.19} \overbrace{=}^{\text{is}} \overbrace{1995.98}^{\$1995.98}$$

$$\frac{P \cdot 2346.19}{2346.19} = \frac{1995.98}{2346.19}$$

$$P = 0.85073\ldots$$

$$P \approx 85.07\%$$

Approximately 85.07% of her gross pay went into her checking account.

Example 2.5.10. Kandace sells cars for a living, and earns 28% of the dealership's sales profit as commission. In a certain month, she plans to earn \$2200 in commissions. How much total sales profit does she need to bring in for the dealership?

Solution. Be careful that you do not calculate 28% of \$2200. That might be what a student would do who fails to read the question repeatedly. If you have ever trained yourself to quickly find numbers in word problems and substitute them into formulas, you must *unlearn* this. The issue is that \$2200 is not the dealership's sales profit, and if you mistakenly multiply $0.28 \cdot 2200 = 616$, then \$616 makes no sense as an answer to this question. How could Kandace bring in only \$616 of sales profit, and be rewarded with \$2200 in commission?

We can translate the problem into "2200 is 28% of what?" Letting x be the sales profit for the dealership (in dollars), we can write and solve the equation:

$$\begin{aligned} 2200 &= 0.28 \cdot x \\ \frac{2200}{0.28} &= \frac{0.28x}{0.28} \\ 7857.142\ldots &= x \\ x &\approx 7857.14 \end{aligned}$$

To earn \$2200 in commission, Kandace needs to bring in approximately \$7857.14 of sales profit for the dealership.

In the following price increase/decrease problems, the key is to identify the original price, which represents 100%. For these problems, it is important to note that "a percent *of*" and "a percent *off*" are two very different things. To find 10% *of* \$50, simply multiply the percentage with the number: $0.10 \cdot \$50 = \$5$. So, \$5 is 10% of \$50. To find 10% *off* from an original value of \$50, consider how if we took 10% off, we would only need to pay 90% of the original value. This means to find 10% off, multiply the 90% with the number: $0.90 \cdot \$50 = \$45$. So, \$45 is 10% off from \$50.

Example 2.5.11. A shirt is on sale with 15% off. The current price is \$51.00. What was the original price?

Solution. If the shirt was 15% off, then what you pay for it is 100%−15%, or 85%, of its original price. So we can translate this problem into "51 is 85% of what?" It's very important to note that we are using 85%, not 15%. Let x represent the shirt's original price, in dollars. We can set up and solve the equation:

$$\begin{aligned} 51 &= 0.85 \cdot x \\ \frac{51}{0.85} &= \frac{0.85x}{0.85} \end{aligned}$$

$$60 = x$$

The shirt's original price was \$60.00.

Example 2.5.12.

According to e-Literate, the average cost of a new college textbook has been increasing. Find the percentage of increase from 2009 to 2013.

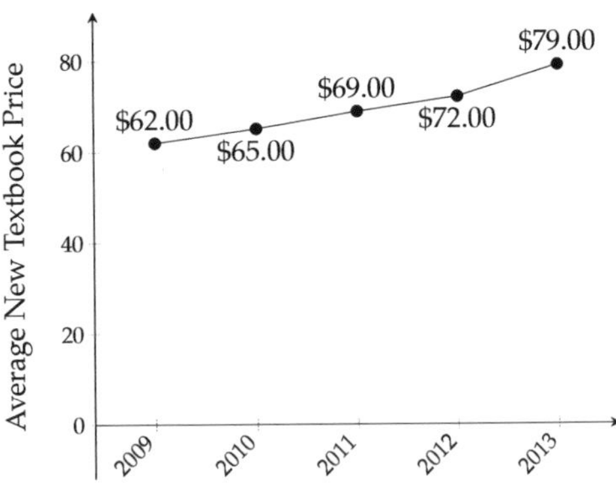

Figure 2.5.13: Average New Textbook Price from 2009 to 2013

Solution. The actual amount of increase from 2009 to 2013 was $79 - 62 = 17$, dollars. We need to answer the question "17 is what percent of 62?" Note that we are comparing the 17 to 62, not to 79. In these situations where one amount is the earlier amount, the earlier original amount is the one that represents 100%. Let x represent the percent of increase. We can set up and solve the equation:

$$\begin{aligned}17 &= x \cdot 62\\ 17 &= 62x\\ \frac{17}{62} &= \frac{62x}{62}\\ 0.274193\ldots &= x\\ x &\approx 27.42\%\end{aligned}$$

From 2009 to 2013, the average cost of a new textbook increased by approximately 27.42%.

Exercise 2.5.14. Last month, a full tank of gas for a car you drive cost you \$40.00. You hear on the news that gas prices have risen by 12%. What's the new cost for a full take of gas?

A full tank of gas will now cost ________.

Solution. After a 12% increase, the new cost would be 100% + 12% = 112% of the original price. We can translate this problem into "what is 112% of 40?". Let x represent the new price in dollars.

We can set up and solve the equation:

$$x = 1.12 \cdot 40$$
$$x = 44.8$$

A full tank now costs \$44.80.

Example 2.5.15. Enrollment at your neighborhood's elementary school two years ago was 417 children. After a 15% increase last year and a 15% decrease this year, what's the new enrollment?

Solution. It is tempting to think that increasing by 15% and then decreasing by 15% would bring the enrollment right back to where it started. But the 15% decrease applies to the enrollment *after* it had already increased. So that 15% decrease is going to translate to *more* students lost than were gained.

Using 100% as corresponding to the enrollment from two years ago, the enrollment last year was 100% + 15% = 115% of that. But then using 100% as corresponding to the enrollment from last year, the enrollment this year was 100% − 15% = 85% of that. So we can set up and solve the equation

$$\overbrace{x}^{\text{this year's enrollment}} \overbrace{=}^{\text{is}} \overbrace{0.85}^{85\%} \overbrace{\cdot}^{\text{of}} \overbrace{1.15}^{115\%} \overbrace{\cdot}^{\text{of}} \overbrace{417}^{\text{enrollment two years ago}}$$
$$x = 0.85 \cdot 1.15 \cdot 417$$
$$x = 407.6175$$

We would round and report that enrollment is now 408 students. (The percentage rise and fall of 15% were probably rounded in the first place, which is why we did not end up with a whole number of children.)

2.5.2 Exercises

Basic Percentage Problems

1. 6% of 580 is ______.

2. 3% of 670 is ______.

3. 80% of 770 is ______.

4. 50% of 870 is ______.

5. 290% of 970 is ______.

6. 750% of 180 is ______.

7. Fill in the blank with a percent:

147 is ______ of 490.

8. Fill in the blank with a percent:

72 is ______ of 180.

9. Fill in the blank with a percent:

136.8 is [ ] of 76.

10. Fill in the blank with a percent:

90 is [ ] of 45.

11. Fill in the blank with a percent. Round to the hundredth place, like 12.34%.

14 is approximately [ ] of 18.

12. Fill in the blank with a percent. Round to the hundredth place, like 12.34%.

16 is approximately [ ] of 74.

13. 42% of [ ] is 365.4.

14. 11% of [ ] is 106.7.

15. 7% of [ ] is 11.9.

16. 4% of [ ] is 10.8.

17. 580% of [ ] is 2146.

18. 400% of [ ] is 1880.

Percentage Application Problems

19. A town has 3100 registered residents. Among them, 33% were Democrats, 32% were Republicans. The rest were Independents. How many registered Independents live in this town?

There are [ ] registered Independent residents in this town.

20. A town has 3500 registered residents. Among them, 39% were Democrats, 21% were Republicans. The rest were Independents. How many registered Independents live in this town?

There are [ ] registered Independent residents in this town.

21. Kimball is paying a dinner bill of $43.00. Kimball plans to pay 16% in tips. How much in total (including bill and tip) will Kimball pay?

Kimball will pay [ ] in total (including bill and tip).

22. Holli is paying a dinner bill of $46.00. Holli plans to pay 12% in tips. How much in total (including bill and tip) will Holli pay?

Holli will pay [ ] in total (including bill and tip).

23. In the past few seasons' basketball games, Benjamin attempted 450 free throws, and made 405 of them. What percent of free throws did Benjamin make?

Fill in the blank with a percent.

Benjamin made [ ] of free throws in the past few seasons.

24. In the past few seasons' basketball games, Ronda attempted 310 free throws, and made 62 of them. What percent of free throws did Ronda make?

Fill in the blank with a percent.

Ronda made [ ] of free throws in the past few seasons.

25. A painting is on sale at $405.00. Its original price was $450.00. What percentage is this off its original price?

Fill in the blank with a percent.

The painting was [ ] off its original price.

26. A painting is on sale at $350.00. Its original price was $500.00. What percentage is this off its original price?

Fill in the blank with a percent.

The painting was [ ] off its original price.

27. The pie chart represents a collector's collection of signatures from various artists.

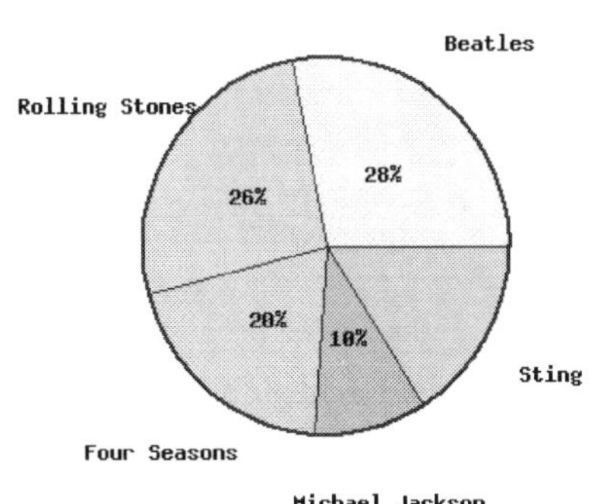

If the collector has a total of 950 signatures, there are [ ] signatures by Sting.

28. The pie chart represents a collector's collection of signatures from various artists.

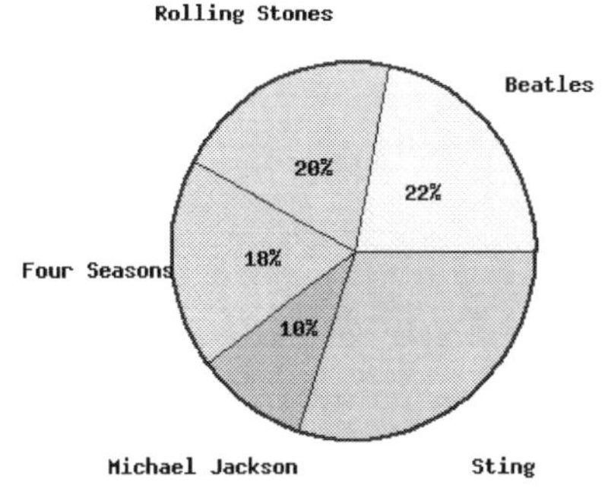

If the collector has a total of 1150 signatures, there are [ ] signatures by Sting.

29. Haley sells cars for a living. Each month, she earns $1,100.00 of base pay, plus a certain percentage of commission from her sales.

One month, Haley made $65,200.00 in sales, and earned a total of $4,842.48 in that month (including base pay and commission). What percent commission did Haley earn?

Fill in the blank with a percent.

Haley earned [ ] in commission.

30. Gustav sells cars for a living. Each month, he earns $1,100.00 of base pay, plus a certain percentage of commission from his sales.

One month, Gustav made $69,600.00 in sales, and earned a total of $3,271.52 in that month (including base pay and commission). What percent commission did Gustav earn?

Fill in the blank with a percent.

Gustav earned [ ] in commission.

31. A community college conducted a survey about the number of students riding each bus line available. The following bar graph is the result of the survey.

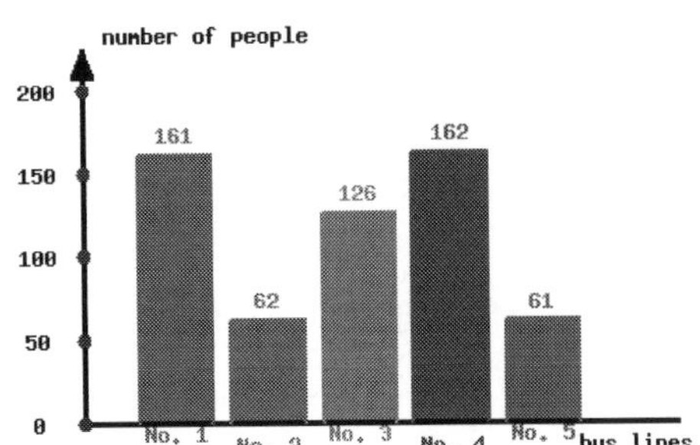

What percent of students ride Bus No. 1?

Fill in the blank with a percent. Round your percent to whole numbers, like 11%.

Approximately ______ of students ride Bus No. 1.

32. A community college conducted a survey about the number of students riding each bus line available. The following bar graph is the result of the survey.

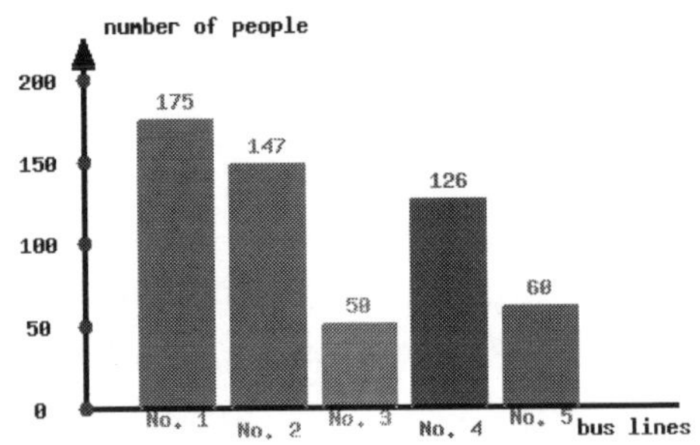

What percent of students ride Bus No. 1?

Fill in the blank with a percent. Round your percent to whole numbers, like 11%.

Approximately ______ of students ride Bus No. 1.

33. In the last election, 46% of a county's residents, or 5888 people, turned out to vote. How many residents live in this county?

This county has ______ residents.

34. In the last election, 32% of a county's residents, or 5536 people, turned out to vote. How many residents live in this county?

This county has ______ residents.

35. Alyson earned \$158.41 of interest from a mutual fund, which was 0.73% of his total investment. How much money did Alyson invest into this mutual fund?

Alyson invested ______ in this mutual fund.

36. Aleric earned \$96.57 of interest from a mutual fund, which was 0.37% of his total investment. How much money did Aleric invest into this mutual fund?

Aleric invested ______ in this mutual fund.

37. Corey paid a dinner and left 10%, or \$3.50, in tips. How much was the original bill (without counting the tip)?

The original bill (not including the tip) was ______.

38. Sydney paid a dinner and left 17%, or \$6.63, in tips. How much was the original bill (without counting the tip)?

The original bill (not including the tip) was ______.

39. The following is a nutrition fact label from a certain macaroni and cheese box.

Nutrition Facts	
Serving Size 1 cup	
Servings Per Container 2	
Amount Per Serving	
Calories 300	Calories from Fat 110
	% Daily Value
Total Fat 9 g	12%
Saturated Fat 2 g	15%
Trans Fat 2 g	
Cholesterol 30 mg	10%
Sodium 400 mg	20%
Total Carbohydrate 30 g	11%
Dietary Fiber 0 g	0%
Sugars 5 g	
Protein 5 g	
Vitamin A 2 mg	3%
Vitamin C 2 mg	2.5%
Calcium 2 mg	20%
Iron 3 mg	4%

The highlighted row means each serving of macaroni and cheese in this box contains 9 g of fat, which is 12% of an average person's daily intake of fat. What's the recommended daily intake of fat for an average person?

The recommended daily intake of fat for an average person is ______.

Use *g* for grams.

40. The following is a nutrition fact label from a certain macaroni and cheese box.

Nutrition Facts	
Serving Size 1 cup	
Servings Per Container 2	
Amount Per Serving	
Calories 300	Calories from Fat 110
	% Daily Value
Total Fat 15 g	20%
Saturated Fat 2 g	15%
Trans Fat 2 g	
Cholesterol 30 mg	10%
Sodium 400 mg	20%
Total Carbohydrate 30 g	11%
Dietary Fiber 0 g	0%
Sugars 5 g	
Protein 5 g	
Vitamin A 2 mg	3%
Vitamin C 2 mg	2.5%
Calcium 2 mg	20%
Iron 3 mg	4%

The highlighted row means each serving of macaroni and cheese in this box contains 15 g of fat, which is 20% of an average person's daily intake of fat. What's the recommended daily intake of fat for an average person?

The recommended daily intake of fat for an average person is ______.

Use *g* for grams.

Percent of Increase/Decrease Problems

41. The population of cats in a shelter decreased from 200 to 140. What is the percentage decrease of the shelter's cat population?

The percentage decrease is ______.

42. The population of cats in a shelter decreased from 20 to 17. What is the percentage decrease of the shelter's cat population?

The percentage decrease is ______.

43. The population of cats in a shelter increased from 22 to 41. What is the percentage increase of the shelter's cat population?

Fill in the blank with percent. Round your answer to 2 decimal places, like 12.34%.

The percentage increase is approximately ______.

44. The population of cats in a shelter increased from 30 to 44. What is the percentage increase of the shelter's cat population?

Fill in the blank with percent. Round your answer to 2 decimal places, like 12.34%.

The percentage increase is approximately ______.

45. Last year, a small town had 620 population. This year, the population decreased to 615. What is the percentage decrease of the town's population?

Fill in blank with a percent. Round your answer to two decimal places, like 1.23%.

The percentage decrease of the town's population was approximately ______.

46. Last year, a small town had 650 population. This year, the population decreased to 649. What is the percentage decrease of the town's population?

Fill in blank with a percent. Round your answer to two decimal places, like 1.23%.

The percentage decrease of the town's population was approximately ______.

47. Your salary used to be $41,000 per year.

You had to take a 4% pay cut. After the cut, your salary was ______ per year.

Then, you earned a 4% raise. After the raise, your salary was ______ per year.

48. Your salary used to be $33,000 per year.

You had to take a 4% pay cut. After the cut, your salary was ______ per year.

Then, you earned a 4% raise. After the raise, your salary was ______ per year.

49. This line graph shows a certain stock's price change over a few days.

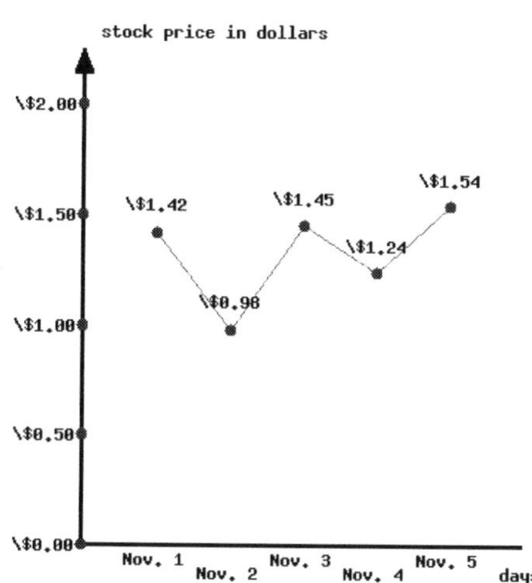

From Nov. 1 to Nov. 5, what is the stock price's percentage change?

Fill in the blank with a percent. Round your percent to two decimal places, like 12.34%.

From Nov. 1 to Nov. 5, the stock price's percentage change was approximately ☐.

50. This line graph shows a certain stock's price change over a few days.

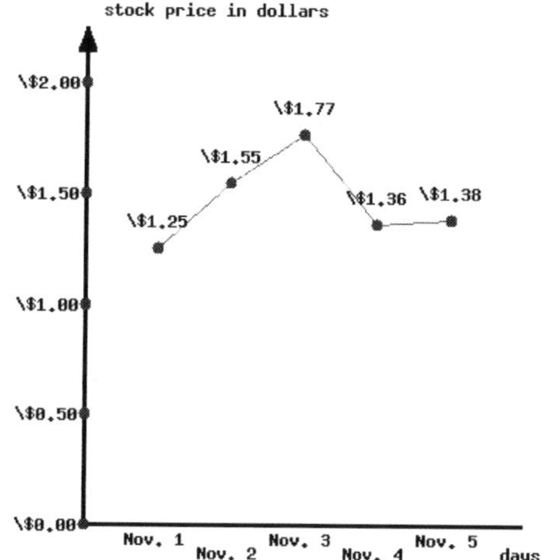

From Nov. 1 to Nov. 5, what is the stock price's percentage change?

Fill in the blank with a percent. Round your percent to two decimal places, like 12.34%.

From Nov. 1 to Nov. 5, the stock price's percentage change was approximately ☐.

51. A house was bought two years ago at the price of \$150,000. Each year, the house's value decreased by 2%. What's the house's value this year?

The house's value this year is ☐.

52. A house was bought two years ago at the price of \$420,000. Each year, the house's value decreased by 3%. What's the house's value this year?

The house's value this year is ☐.

53. After a 35% increase, a town has 405 people. What was the population before the increase?

Before the increase, the town's population was ☐.

54. After a 15% increase, a town has 460 people. What was the population before the increase?

Before the increase, the town's population was ☐.

2.6 Modeling with Equations and Inequalities

One purpose of learning math is to be able to model real-life situations and then use the model to ask and answer questions about the situation. In this lesson, we will cover the basics of modeling.

2.6.1 Setting Up Equations for Models

To set up an equation modeling a real world scenario, the first thing we need to do is identify what variable we will use. The variable we use will be determined by whatever is unknown in our problem statement. Once we've identified and defined our variable, we'll use the numerical information provided to set up our equation.

Example 2.6.2.

A savings account starts with \$500. Each month, an automatic deposit of \$150 is made. Write an equation that represents the number of months it will take for the balance to reach \$1,700.

To determine this equation, we'll start by making a table in order to identify a general pattern for the total amount in the account after m months:

Months Since Saving Started	Total Amount Saved (in Dollars)
0	500
1	$500 + 150 = 650$
2	$500 + 150(2) = 800$
3	$500 + 150(3) = 950$
4	$500 + 150(4) = 1100$
⋮	⋮
m	$500 + 150m$

Table 2.6.3: Amount in Savings Account

Using this pattern, we can determine that an equation showing the unknown number of months, m, when the total savings equals \$1700 would look like this:

$$500 + 150m = 1700$$

Remark 2.6.4. To determine the solution to the equation in Example 2.6.2, we can continue the pattern in Table 2.6.3:

We can see that the value of m that makes the equation true is 8 as $500 + 150(8) = 1700$. Thus it would take 8 months for an account starting with \$500 to reach \$1,700 if \$150 is saved each month.

Months Since Saving Started	Total Amount Saved (in Dollars)
5	$500 + 150(5) = 1250$
6	$500 + 150(6) = 1400$
7	$500 + 150(7) = 1550$
8	$500 + 150(8) = 1700$

Table 2.6.5: Amount in Savings Account

In the previous example, we were able to determine the solution by creating a table and using inputs that were integers. Often the solution will not be something we are able to arrive at this way. We will need to solve for it using algebra, as we'll see in later sections. For this section, we'll only focus on setting up the equation.

Example 2.6.6. A bathtub contains 2.5 ft^3 of water. More water is being poured in at a rate of 1.75 ft^3 per minute. Write an equation representing when the amount of water in the bathtub will reach 6.25 ft^3.

Solution.

Since this problem refers to *when* the amount of water will reach a certain amount, we immediately know that the unknown quantity is time. As the volume of water in the tub is measured in ft^3 per minute, we know that time needs to be measured in minutes. We'll define t to be the number of minutes that water is poured into the tub. To determine this equation, we'll start by making a table of values:

Minutes Water Has Been Poured	Total Amount of Water (in ft^3)
0	2.5
1	$2.5 + 1.75 = 4.25$
2	$2.5 + 1.75(2) = 6$
3	$2.5 + 1.75(3) = 7.75$
$\vdots$	$\vdots$
t	$2.5 + 1.75t$

Table 2.6.7: Amount of Water in the Bathtub

Using this pattern, we can determine that the equation representing when the amount will be 6.25 ft^3 is:

$$2.5 + 1.75t = 6.25$$

Example 2.6.8. Jakobi's annual salary as a nurse in Portland, Oregon, is \$73,290. His salary increased by 4% from last year. Write a linear equation modeling this scenario, where the unknown value is Jakobi's salary last year.

Solution. We need to know Jakobi's salary last year. So we'll introduce s, defined to be Jakobi's salary last year (in dollars). To determine how to set up this equation, we need to think about how he arrived at this year's salary. To get to this year's salary, his employer took last year's salary and added 4% to it. Conceptually, this means we have:

(last year's salary) + (4% of last year's salary) = (this year's salary)

We'll represent 4% of last year's salary with $0.04s$ since 0.04 is the decimal representation of 4%. This means that the equation we set up is:

$$s + 0.04s = 73290$$

Example 2.6.9. The price of a refrigerator after a 15% discount is \$612. Write a linear equation modeling this scenario, where the original price of the refrigerator (before the discount was applied) is the unknown quantity.

Solution. We'll let c be the original price of the refrigerator. To obtain the discounted price, we take the original price and subtract 15% of that amount. Conceptually, this looks like:

$$\text{(original price)} - \text{(15\% of the original price)} = \text{(discounted price)}$$

Since the amount of the discount is 15% of the original price, we'll represent this with $0.15c$. The equation we set up is then:

$$c - 0.15c = 612$$

Example 2.6.10. A cone-shaped paper cup needs to have a volume of $6.3\,\text{in}^3$. The radius of the cup is 1.3125 in. Write a linear equation modeling this scenario. (Hint: The volume for a cone is $\frac{1}{3}\pi r^2 h$.)

Solution. We know that we need to find the height of this cone. So we'll introduce h, defined to be the height of the cone in inches. We know that the units need to be inches because the radius is also given in inches and the volume is given in cubic inches. Using the provided values $r = 1.3125$ and $V = 6.3$ and the formula for a cone's volume, we set this equation up as:

$$\frac{1}{3}\pi(1.3125)^2 h = 6.3$$

2.6.2 Setting Up Inequalities for Models

In general, we'll model using inequalities when we want to determine a maximum or minimum value. To identify that an inequality is needed instead of an equality, we'll look for phrases like *at least, at most, at a minimum* or *at a maximum*.

Example 2.6.11. The car share company car2go has a one-time registration fee of \$5 and charges \$14.99 per hour for use of their vehicles. Rafael wants to use car2go and has a maximum budget of \$300. Write a linear inequality representing this scenario, where the unknown quantity is the number of hours he uses their vehicles.

Solution. We'll let h be the number of hours that Rafael uses car2go. We need the initial cost and the cost from the hourly charge to be less than or equal to \$300, which we set up as:

$$5 + 14.99h \leq 300$$

Example 2.6.12. When an oil tank is decommissioned, it is drained of its remaining oil and then re-filled with an inert material, such as sand. A cylindrical oil tank has a volume of 275 gal and is being filled with sand at a rate of 700 gal per hour. Write a linear inequality representing this scenario, where the time it takes for the tank to overflow with sand is the unknown quantity.

Solution. The unknown in this scenario is time, so we'll define t to be the number of hours that sand is poured into the tank. (Note that we chose hours based on the rate at which the sand is being poured.) We'll represent the amount of sand poured in as $700t$ as each hour an additional 700 gal are added. Given that we want to know when this amount exceeds 275 gal, we set this equation up as:

$$700t > 275$$

2.6.3 Translating Phrases into Mathematical Expressions and Equations/Inequalities

The following table shows how to translate common phrases into mathematical expressions:

English Phrases	Math Expressions
the sum of 2 and a number	$x + 2$ or $2 + x$
2 more than a number	$x + 2$ or $2 + x$
a number increased by 2	$x + 2$ or $2 + x$
a number and 2 together	$x + 2$ or $2 + x$
the difference between a number and 2	$x - 2$
the difference of 2 and a number	$2 - x$
2 less than a number	$x - 2$ (*not* $2 - x$)
a number decreased by 2	$x - 2$
2 decreased by a number	$2 - x$
2 subtracted from a number	$x - 2$
a number subtracted from 2	$2 - x$
the product of 2 and a number	$2x$
twice a number	$2x$
a number times 2	$x \cdot 2$ or $2x$
two thirds of a number	$\frac{2}{3}x$
25% of a number	$0.25x$
the quotient of a number and 2	$x/2$
the quotient of 2 and a number	$2/x$
the ratio of a number and 2	$x/2$
the ratio of 2 and a number	$2/x$

Table 2.6.13: Translating English Phrases into Math Expressions

We can extend this to setting up equations and inequalities. Let's look at some examples. The key is to break a complicated phrase or sentence into smaller parts, identifying key vocabulary such as "is," "of," "greater than," "at most," etc.

English Sentences	Math Equations and Inequalities
The sum of 2 and a number is 6.	$x + 2 = 6$
2 less than a number is at least 6.	$x - 2 \geq 6$
Twice a number is at most 6.	$2x \leq 6$
6 is the quotient of a number and 2.	$6 = \frac{x}{2}$
4 less than twice a number is greater than 10.	$2x - 4 > 10$
Twice the difference between 4 and a number is 10.	$2(4 - x) = 10$
The product of 2 and the sum of 3 and a number is less than 10.	$2(x + 3) < 10$
The product of 2 and a number, subtracted from 5, yields 8.	$5 - 2x = 8$
Two thirds of a number subtracted from 10 is 2.	$10 - \frac{2}{3}x = 2$
25% of the sum of 7 and a number is 2.	$0.25(x + 7) = 2$

Table 2.6.14: Translating English Sentences into Math Equations

2.6.4 Exercises

Modeling with Linear Equations and Inequalities

1. Dawn's annual salary as a radiography technician is \$46,125.00. Her salary increased by 2.5% from last year. What was her salary last year?

Assume her salary last year was s dollars. Write an equation to model this scenario. There is no need to solve it.

2. A bicycle for sale costs \$223.02, which includes 6.2% sales tax. What was the cost before sales tax?

Assume the bicycle's price before sales tax is p dollars. Write an equation to model this scenario. There is no need to solve it.

3. The price of a washing machine after 5% discount is \$218.50. What was the original price of the washing machine (before the discount was applied)?

Assume the washing machine's price before the discount is p dollars. Write an equation to model this scenario. There is no need to solve it.

4. The price of a restaurant bill, including an 17% gratuity charge, was \$105.30. What was the price of the bill before gratuity was added?

Assume the bill without gratuity is b dollars. Write an equation to model this scenario. There is no need to solve it.

5. In May 2016, the median rent price for a one-bedroom apartment in a city was reported to be \$1,507.50 per month. This was reported to be an increase of 0.5% over the previous month. Based on this reporting, what was the median price of a one-bedroom apartment in April 2016?

   Assume the median price of a one-bedroom apartment in April 2016 was p dollars. Write an equation to model this scenario. There is no need to solve it.

6. Sarah is driving an average of 31 miles per hour, and she is 31 miles away from home. After how many hours will she reach his home?

   Assume Sarah will reach home after h hours. Write an equation to model this scenario. There is no need to solve it.

7. Uhaul charges an initial fee of \$23.40 and then \$0.77 per mile to rent a 15-foot truck for a day. If the total bill is \$72.68, how many miles were driven?

   Assume m miles were driven. Write an equation to model this scenario. There is no need to solve it.

8. Ibuprofen for infants comes in a liquid form and contains 35 milligrams of ibuprofen for each 0.875 milliliters of liquid. If a child is to receive a dose of 40 milligrams of ibuprofen, how many milliliters of liquid should they be given?

   Assume l milliliters of liquid should be given. Write an equation to model this scenario. There is no need to solve it.

9. The property taxes on a 2500-square-foot house are \$3,325.00 per year. Assuming these taxes are proportional, what are the property taxes on a 2100-square-foot house?

   Assume property taxes on a 2100-square-foot house is t dollars. Write an equation to model this scenario. There is no need to solve it.

10. A cat litter box has a rectangular base that is 18 inches by 18 inches. What will the height of the cat litter be if 3.375 cubic feet of cat litter is poured? (Hint: $1\ \text{ft}^3 = 1728\ \text{in}^3$)

    Assume h inches will be the height of the cat litter if 3.375 cubic feet of cat litter is poured. Write an equation to model this scenario. There is no need to solve it.

11. A truck that hauls water is capable of carrying a maximum of 1900 lb. Water weighs $8.3454\frac{\text{lb}}{\text{gal}}$, and the plastic tank on the truck that holds water weighs 81 lb. What's the maximum number of gallons of water the truck can carry?

    Assume the truck can carry a maximum of g gallons of water. Write an equation to model this scenario. There is no need to solve it.

12. Kimball's maximum lung capacity is 6.9 liters. If his lungs are full and he exhales at a rate of 0.8 liters per second, when will he have 4.82 liters of air left in his lungs?

    Assume s seconds later, there would be 4.82 liters of air left in Kimball's lungs. Write an equation to model this scenario. There is no need to solve it.

13. A swimming pool is being filled with water from a garden hose at a rate of 8 gallons per minute. If the pool already contains 90 gallons of water and can hold 266 gallons, after how long will the pool overflow?

Assume m minutes later, the pool would overflow. Write an equation to model this scenario. There is no need to solve it.

14. An engineer is designing a cylindrical springform pan. The pan needs to be able to hold a volume of 247 cubic inches and have a diameter of 15 inches. What's the minimum height it can have? (Hint: The formula for the volume of a cylinder is $V = \pi r^2 h$).

Assume the pan's minimum height is h inches. Write an equation to model this scenario. There is no need to solve it.

Translating English Phrases into Math Expressions

15. Translate the following phrase into a math expression or equation (whichever is appropriate). Use x to represent the unknown number.

nine more than a number

16. Translate the following phrase into a math expression or equation (whichever is appropriate). Use r to represent the unknown number.

six less than a number

17. Translate the following phrase into a math expression or equation (whichever is appropriate). Use t to represent the unknown number.

the sum of a number and two

18. Translate the following phrase into a math expression or equation (whichever is appropriate). Use b to represent the unknown number.

the difference between a number and nine

19. Translate the following phrase into a math expression or equation (whichever is appropriate). Use A to represent the unknown number.

the difference between five and a number

20. Translate the following phrase into a math expression or equation (whichever is appropriate). Use B to represent the unknown number.

the difference between two and a number

21. Translate the following phrase into a math expression or equation (whichever is appropriate). Use m to represent the unknown number.

eight subtracted from a number

22. Translate the following phrase into a math expression or equation (whichever is appropriate). Use n to represent the unknown number.

five added to a number

23. Translate the following phrase into a math expression or equation (whichever is appropriate). Use q to represent the unknown number.

one decreased by a number

24. Translate the following phrase into a math expression or equation (whichever is appropriate). Use x to represent the unknown number.

eight increased by a number

25. Translate the following phrase into a math expression or equation (whichever is appropriate). Use r to represent the unknown number.

    a number decreased by five

26. Translate the following phrase into a math expression or equation (whichever is appropriate). Use t to represent the unknown number.

    a number increased by one

27. Translate the following phrase into a math expression or equation (whichever is appropriate). Use b to represent the unknown number.

    eight times a number, increased by five

28. Translate the following phrase into a math expression or equation (whichever is appropriate). Use c to represent the unknown number.

    five times a number, decreased by nine

29. Translate the following phrase into a math expression or equation (whichever is appropriate). Use B to represent the unknown number.

    one less than four times a number

30. Translate the following phrase into a math expression or equation (whichever is appropriate). Use m to represent the unknown number.

    seven less than eight times a number

31. Translate the following phrase into a math expression or equation (whichever is appropriate). Use n to represent the unknown number.

    four less than the quotient of two and a number

32. Translate the following phrase into a math expression or equation (whichever is appropriate). Use q to represent the unknown number.

    ten less than the quotient of six and a number

Translating English Sentences into Math Equations

33. Translate the following sentence into a math equation. Use x to represent the unknown number.

    Seven times a number is fifty-six.

34. Translate the following sentence into a math equation. Use r to represent the unknown number.

    Four times a number is twelve.

35. Translate the following sentence into a math equation. Use t to represent the unknown number.

    The difference between forty-eight and a number is thirty-two.

36. Translate the following sentence into a math equation. Use b to represent the unknown number.

    The difference between twenty-five and a number is twenty-one.

37. Translate the following sentence into a math equation. Use c to represent the unknown number.

    The quotient of a number and eleven is twelve elevenths.

38. Translate the following sentence into a math equation. Use B to represent the unknown number.

    The quotient of a number and thirty-nine is one thirty-ninth.

39. Translate the following sentence into a math equation. Use m to represent the unknown number.

The quotient of twenty-one and a number is seven fifths.

40. Translate the following sentence into a math equation. Use n to represent the unknown number.

The quotient of ten and a number is five thirds.

41. Translate the following sentence into a math equation. Use q to represent the unknown number.

The sum of seven times a number and nine is fifty-eight.

42. Translate the following sentence into a math equation. Use x to represent the unknown number.

The sum of five times a number and twenty-one is 236.

43. Translate the following sentence into a math equation. Use r to represent the unknown number.

One less than three times a number is twenty-six.

44. Translate the following sentence into a math equation. Use t to represent the unknown number.

Two less than seven times a number is 194.

45. Translate the following sentence into a math equation. Use b to represent the unknown number.

The product of five and a number, added to eight, is thirty-three.

46. Translate the following sentence into a math equation. Use c to represent the unknown number.

The product of two and a number, increased by four, is eighty-two.

47. Translate the following sentence into a math equation. Use B to represent the unknown number.

The product of seven and a number added to seven, is 231.

48. Translate the following sentence into a math equation. Use C to represent the unknown number.

The product of four and a number increased by three, is sixty-four.

49. Translate the following phrase into a math expression or equation (whichever is appropriate). Use n to represent the unknown number.

one half of a number

50. Translate the following phrase into a math expression or equation (whichever is appropriate). Use q to represent the unknown number.

one eighth of a number

51. Translate the following phrase into a math expression or equation (whichever is appropriate). Use x to represent the unknown number.

twelve twenty-sixths of a number

52. Translate the following phrase into a math expression or equation (whichever is appropriate). Use r to represent the unknown number.

two twenty-eighths of a number

53. Translate the following phrase into a math expression or equation (whichever is appropriate). Use t to represent the unknown number.

a number decreased by seven sixteenths of itself

54. Translate the following phrase into a math expression or equation (whichever is appropriate). Use b to represent the unknown number.

a number decreased by three twenty-fourths of itself

55. Translate the following phrase into a math expression or equation (whichever is appropriate). Use c to represent the unknown number.

A number decreased by three fourths is one fourth of that number.

56. Translate the following phrase into a math expression or equation (whichever is appropriate). Use B to represent the unknown number.

A number decreased by two sevenths is three fifths of that number.

57. Translate the following phrase into a math expression or equation (whichever is appropriate). Use C to represent the unknown number.

Seven less than the product of one ninth and a number gives three halves of that number.

58. Translate the following phrase into a math expression or equation (whichever is appropriate). Use n to represent the unknown number.

Seven more than the product of three sevenths and a number yields one fifth of that number.

2.7 Introduction to Exponent Rules

In this section, we're going to look at some rules or properties we use when simplifying expressions that involve multiplication and exponents.

2.7.1 Exponent Basics

Before we discuss any exponent rules, we need to quickly remind ourselves of some important concepts and vocabulary.

When working with expressions with exponents, we have the following vocabulary:

$$\text{base}^{\text{exponent}} = \text{power}$$

For example, when we calculate $8^2 = 64$, the **base** is 8, the **exponent** is 2, and the expression 8^2 is called the 2nd **power** of 8.

The other foundational concept is that if an exponent is a positive integer, the power can be rewritten as repeated multiplication of the base. For example, the 4th power of 3 can be written as 4 factors of 3 like so:

$$3^4 = 3 \cdot 3 \cdot 3 \cdot 3$$

2.7.2 Products and Exponents

Product Rule If we write out $3^5 \cdot 3^2$ without using exponents, we'd have:

$$3^5 \cdot 3^2 = (3 \cdot 3 \cdot 3 \cdot 3 \cdot 3) \cdot (3 \cdot 3)$$

If we then count how many 3s are being multiplied together, we find we have $5 + 2 = 7$, a total of seven 3s.

$$\begin{aligned} 3^5 \cdot 3^2 &= 3^{5+2} \\ &= 3^7 \end{aligned}$$

Example 2.7.2. Simplify $x^2 \cdot x^3$.

To simplify $x^2 \cdot x^3$, we write this out in its expanded form, as a product of x's, we have

$$\begin{aligned} x^2 \cdot x^3 &= (x \cdot x)(x \cdot x \cdot x) \\ &= x \cdot x \cdot x \cdot x \cdot x \\ &= x^5 \end{aligned}$$

Note that we obtained the exponent of 5 by adding 2 and 3.

This is our first rule, the **Product Rule**: when multiplying two expressions that have the same base, we can simplify the product by adding the exponents.

$$x^m \cdot x^n = x^{m+n} \tag{2.7.1}$$

Exercise 2.7.3. Use the properties of exponents to simplify the expression.

$t^{19} \cdot t^{13}$

Solution. We *add* the exponents as follows

$$\begin{aligned} t^{19} \cdot t^{13} &= t^{19+13} \\ &= t^{32} \end{aligned}$$

Power to a Power Rule The second rule is an extension of the first rule. If we write out $\left(3^5\right)^2$ without using exponents, we'd have 3^5 multiplied by itself:

$$\begin{aligned} \left(3^5\right)^2 &= \left(3^5\right) \cdot \left(3^5\right) \\ &= (3 \cdot 3 \cdot 3 \cdot 3 \cdot 3) \cdot (3 \cdot 3 \cdot 3 \cdot 3 \cdot 3) \end{aligned}$$

If we again count how many 3s are being multiplied, we have a total of two groups each with five 3s. So we'd have $2 \cdot 5 = 10$ instances of a 3.

$$\begin{aligned} \left(3^5\right)^2 &= 3^{2 \cdot 5} \\ &= 3^{10} \end{aligned}$$

Example 2.7.4. Simplify $\left(x^2\right)^3$.

To simplify $\left(x^2\right)^3$, we write this out in its expanded form, as a product of x's, we have

$$\begin{aligned} \left(x^2\right)^3 &= \left(x^2\right) \cdot \left(x^2\right) \cdot \left(x^2\right) \\ &= (x \cdot x) \cdot (x \cdot x) \cdot (x \cdot x) \\ &= x^6 \end{aligned}$$

Note that we obtained the exponent of 6 by multiplying 2 and 3.

We have our second rule, the **Power to a Power Rule**: when a base is raised to an exponent and that expression is raised to another exponent, we multiply the exponents.

$$\left(x^m\right)^n = x^{m \cdot n}$$

Exercise 2.7.5. Use the properties of exponents to simplify the expression.

$(x^2)^4$

Solution. We *multiply* the exponents as follows

$$\begin{aligned}(x^2)^4 &= x^{2\cdot 4} \\ &= x^8\end{aligned}$$

Product to a Power Rule The third exponent rule deals with having multiplication inside a set of parentheses and an exponent outside the parentheses. If we write out $(3t)^5$ without using an exponent, we'd have $3t$ multiplied by itself five times:

$$(3t)^5 = (3t)(3t)(3t)(3t)(3t)$$

Keeping in mind that there is multiplication between every 3 and t and multiplication between all of the parentheses, we can reorder and regroup the factors:

$$\begin{aligned}(3t)^5 &= (3\cdot t)\cdot(3\cdot t)\cdot(3\cdot t)\cdot(3\cdot t)\cdot(3\cdot t) \\ &= (3\cdot 3\cdot 3\cdot 3\cdot 3)\cdot(t\cdot t\cdot t\cdot t\cdot t) \\ &= 3^5 t^5\end{aligned}$$

We essentially applied the outer exponent to each factor inside the parentheses.

Example 2.7.6. Simplify $(xy)^5$.

To simplify $(xy)^5$, we write this out in its expanded form, as a product of x's and y's, we have

$$\begin{aligned}(xy)^5 &= (x\cdot y)\cdot(x\cdot y)\cdot(x\cdot y)\cdot(x\cdot y)\cdot(x\cdot y) \\ &= (x\cdot x\cdot x\cdot x\cdot x)\cdot(y\cdot y\cdot y\cdot y\cdot y) \\ &= x^5 y^5\end{aligned}$$

Note that the exponent on xy can simply be applied to both x and y.

This is our third rule, the **Product to a Power Rule**: when a product is raised to an exponent, we can apply the exponent to each factor in the product.

$$(x\cdot y)^n = x^n \cdot y^n$$

Exercise 2.7.7. Use the properties of exponents to simplify the expression.

$(5r)^2$

Solution. We *multiply* the exponents and apply the rule $(ab)^m = a^m \cdot b^m$ as follows

$$\begin{aligned}(5r)^2 &= (5)^2 r^2 \\ &= 25r^2\end{aligned}$$

If a and b are real numbers, and n and m are positive integers, then we have the following rules:

Product Rule $a^n \cdot a^m = a^{n+m}$

Power to a Power Rule $(a^n)^m = a^{n \cdot m}$

Product to a Power Rule $(ab)^n = a^n \cdot b^n$

List 2.7.8: Summary of the Rules of Exponents for Multiplication

Many examples we'll come across will make use of more than one exponent rule. In deciding which exponent rule to work with first, it's important to remember that the order of operations still applies.

Example 2.7.9. Simplify the following expressions.

1. $(3^7 r^5)^4$
2. $(t^3)^2 \cdot (t^4)^5$

Solution.

1. Since we cannot simplify anything inside the parentheses, we'll begin simplifying this expression using the Product to a Power Rule. We'll apply the outer exponent of 4 to each factor inside the parentheses. Then we'll use the Power to a Power Rule to finish out simplification process:

$$\begin{aligned}(3^7 r^5)^4 &= (3^7)^4 \cdot (r^5)^4 \\ &= 3^{7 \cdot 4} \cdot r^{5 \cdot 4} \\ &= 3^{28} r^{20}\end{aligned}$$

2. According to the order of operations, we should first simplify any exponents before carrying out any multiplication. Therefore, we'll begin simplifying this by applying the Power to a Power Rule and then finish using the Product Rule:

$$\begin{aligned}(t^3)^2 \cdot (t^4)^5 &= t^{3 \cdot 2} \cdot t^{4 \cdot 5} \\ &= t^6 \cdot t^{20} \\ &= t^{6+20} \\ &= t^{26}\end{aligned}$$

Remark 2.7.10. We cannot simplify an expression like $x^2 y^3$ using the Product Rule, as the factors x^2 and y^3 do not have the same base.

2.7.3 Exercises

Calculate Exponentional Expressions

For the following exercises: Evaluate the following expressions that have integer exponents:

1.

a. $3^2 =$ ☐

b. $4^3 =$ ☐

c. $(-5)^2 =$ ☐

d. $(-4)^3 =$ ☐

2.

a. $3^2 =$ ☐

b. $2^3 =$ ☐

c. $(-4)^2 =$ ☐

d. $(-5)^3 =$ ☐

3.

a. $(-5)^2 =$ ☐

b. $-2^2 =$ ☐

4.

a. $(-5)^2 =$ ☐

b. $-6^2 =$ ☐

5.

a. $(-2)^3 =$ ☐

b. $-1^3 =$ ☐

6.

a. $(-1)^3 =$ ☐

b. $-4^3 =$ ☐

Use the Rules of Exponents

For the following exercises: Use the properties of exponents to simplify the expression.

7. $9 \cdot 9^6$

8. $2 \cdot 2^2$

9. $3^8 \cdot 3^5$

10. $4^5 \cdot 4^8$

11. $x^9 \cdot x^3$

12. $r^{11} \cdot r^{15}$

13. $t^{13} \cdot t^9 \cdot t^2$

14. $r^{15} \cdot r^2 \cdot r^{10}$

15. $(17^8)^9$

16. $(19^4)^4$

17. $(r^2)^{12}$

18. $(x^3)^8$

19. $(3r)^2$

20. $(5t)^3$

21. $(4yt)^3$

22. $(2ty)^3$

23. $(2x^9)^4$

24. $(4t^{10})^3$

25. $(3x^{19}) \cdot (-7x^5)$

26. $(-5y^2) \cdot (8y^{18})$

27. $\left(-\frac{y^4}{8}\right) \cdot \left(-\frac{y^{11}}{7}\right)$

28. $\left(\frac{y^6}{5}\right) \cdot \left(\frac{y^4}{8}\right)$

29. $(5y^8) \cdot (4y^{17}) \cdot (-5y^8)$

30. $(4r^{10}) \cdot (8r^{10}) \cdot (-r^3)$

31. a.

$(-10y^5)^2 =$ ☐

b.

$-(10y^5)^2 =$ ☐

32. a.

$(-2m^2)^6 =$ ☐

b.

$-(2m^2)^6 =$ ☐

33. $(-7y^{10})^2$

34. $(-3x^{12})^2$

2.8 Simplifying Expressions

We know that if we have two apples and add three more, then our result is the same as if we'd had three apples and added two more. In this section, we'll formally define and extend these basic properties we know about numbers to variable expressions.

2.8.1 Identities and Inverses

We will start with some definitions. The number 0 is called the **additive identity**. If the sum of two numbers is the additive identity, 0, these two numbers are called **additive inverses**. For example, 2 is the additive inverse of -2, and the additive inverse of -2 is 2.

Similarly, the number 1 is called the **multiplicative identity**. If the product of two numbers is the multiplicative identity, 1, these two numbers are called **multiplicative inverses**. For example, 2 is the multiplicative inverse of $\frac{1}{2}$, and the multiplicative inverse of $-\frac{2}{3}$ is $-\frac{3}{2}$. The multiplicative inverse is also called **reciprocal**.

2.8.2 Introduction to Algebraic Properties

Commutative Property When we compute the area of a rectangle, we generally multiply the length by the width. Does the result change if we multiply the width by the length?

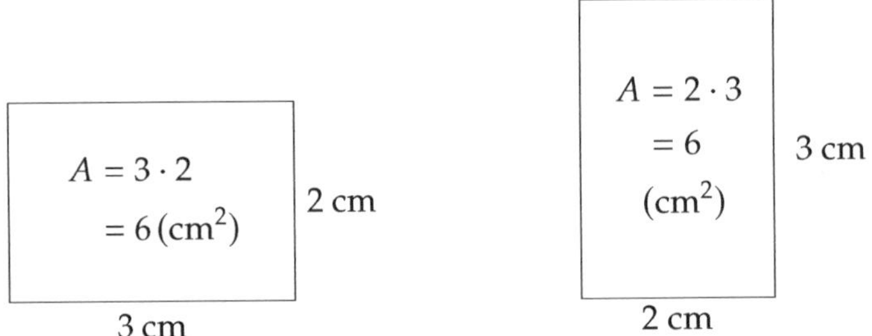

Figure 2.8.2: Horizontal and Vertical Rectangles

We can see $3 \cdot 2 = 2 \cdot 3$. If we denote the length of a rectangle with L and the width with W, this implies $LW = WL$. This is referred to as the **commutative property of multiplication**. The commutative property also applies to addition, as in $1+2 = 2+1$, where it is called the **commutative property of addition**. However, there is no commutative property of subtraction or division, as $2 - 1 \neq 1 - 2$, and $\frac{4}{2} \neq \frac{2}{4}$.

Associative Property Let's extend that example to a rectangular prism with length $L = 4$ cm, width $W = 3$ cm, and height $H = 2$ cm. To compute the volume of this solid, we multiply the length, width and height, which we write as LWH.

In the following figure, on the left side, we multiply the length and width first, and then multiply the height; on the right side, we multiply the width and height first, and then multiply the length. Let's compare the products.

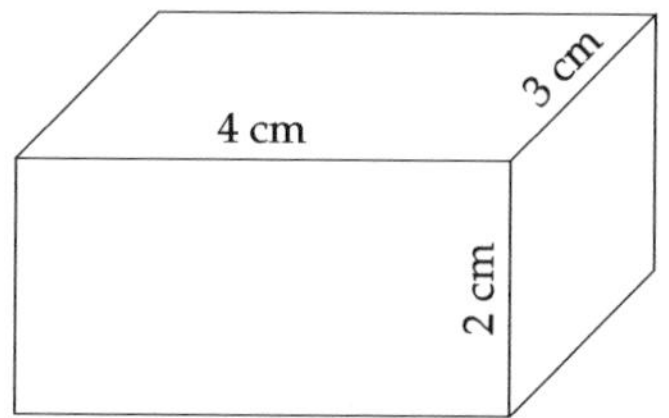

Figure 2.8.3: $(4 \cdot 3) \cdot 2 = 24$

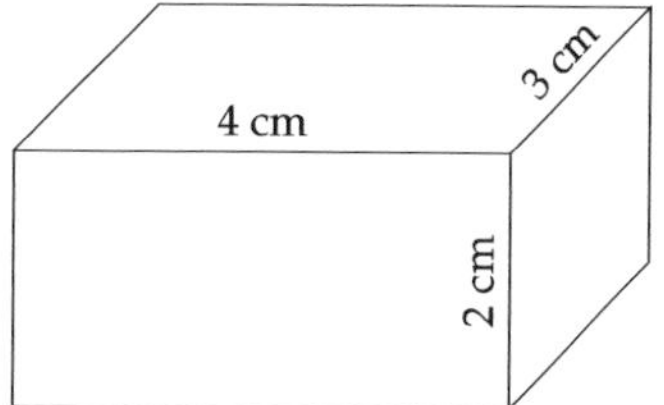

Figure 2.8.4: $4 \cdot (3 \cdot 2) = 24$

We can see $(LW)H = L(WH)$. This is known as the **associative property of multiplication**. The associative property also applies to addition, as in $(1 + 2) + 3 = 1 + (2 + 3)$, which is called the **associative property of addition**. However, there is no associative property of subtraction, as $(3 - 2) - 1 \neq 3 - (2 - 1)$.

Distributive Property The final property we'll explore is called the **distributive property**, which involves both multiplication and addition. To conceptualize this property, let's consider what happens if we buy 3 boxes that each contain one apple and one pear. This will have the same total cost as if we'd bought 3 apples and 3 pears. We write this algebraically:

$$3(a + p) = 3a + 3p.$$

Visually, we can see that it's just a means of re-grouping: 3(+) = 3() + 3().

2.8.3 Summary of Algebraic Properties

Let a, b, and c represent real numbers, variables, or algebraic expressions. Then the following properties hold:

Commutative Property of Multiplication: $a \cdot b = b \cdot a$

Associative Property of Multiplication: $a \cdot (b \cdot c) = (a \cdot b) \cdot c$

Commutative Property of Addition: $a + b = b + a$

Associative Property of Addition: $a + (b + c) = (a + b) + c$

Distributive Property: $a(b + c) = ab + ac$

List 2.8.5: Algebraic Properties

Let's practice these properties in the following exercises.

Exercise 2.8.6.

a. Use the commutative property of multiplication to write an equivalent expression to $53m$.

b. Use the associative property of multiplication to write an equivalent expression to $3(5n)$.

c. Use the commutative property of addition to write an equivalent expression to $q + 84$.

d. Use the associative property of addition to write an equivalent expression to $x + (20 + c)$.

e. Use the distributive property to write an equivalent expression to $3(r + 7)$ that has no grouping symbols.

Solution.

a. To use the commutative property of multiplication, we change the order in which two factors are multiplied:

$$\begin{aligned} &53m \\ &= m \cdot 53. \end{aligned}$$

b. To use the associative property of multiplication, we leave factors written in their original order, but change the grouping symbols so that a different multiplication has higher priority:

$$\begin{aligned} &3(5n) \\ &= (3 \cdot 5)n. \end{aligned}$$

You may further simplify by carrying out the multiplication between the two numbers:

$$\begin{aligned} &3(5n) \\ &= (3 \cdot 5)n \\ &= 15n \end{aligned}$$

c. To use the commutative property of addition, we change the order in which two terms are added:

$$\begin{aligned} &q + 84 \\ &= 84 + q. \end{aligned}$$

d. To use the associative property of addition, we leave terms written in their original order, but change the grouping symbols so that a different addition has higher priority:

$$\begin{aligned} &x + (20 + c) \\ &= (x + 20) + c. \end{aligned}$$

e. To use the distributive property, we multiply the number outside the parentheses, 3, with each term inside the parentheses:

$$\begin{aligned} &3(r+7) \\ &= 3 \cdot r + 3 \cdot 7 \\ &= 3r + 21 \end{aligned}$$

2.8.4 Applying the Commutative, Associative, and Distributive Properties

Like Terms One of the main ways that we will use the commutative, associative, and distributive properties is to simplify expressions. In order to do this, we need to recognize **like terms**. Two terms are considered **like terms** if they contain the same variable raised to the same exponent. Here's a table comparing like terms and non-like terms:

Like Terms	Non-Like Terms
$7x, 2x$	$7x, 7y$
$-5xy, \frac{1}{3}xy$	$-5x^2, -4x$
$4y^3, y^3$	$3x, 4$
$3, -10$	$2xy, 5y$

Combining Like Terms We combine like terms when we take $2a + 3a$ and write the result as $5a$. The formal process actually involves invoking the distributive property, as we obtain:

$$\begin{aligned} 2a + 3a &= (2+3)a \\ &= 5a \end{aligned}$$

In practice, it's helpful to think of this as having 2 of an object and then an additional 3 of that same object. In total, we then have 5 of that object.

Example 2.8.7. Where possible, simplify the following expressions by combining like terms.

a. $6c + 12c - 5c$

b. $-5q^2 - 3q^2$

c. $-3x - 5y + 4x$

d. $2x - 3y + 4z$

Solution.

a. All three terms are like terms, so they may combined. We combine them two at a time:

$$\begin{aligned} 6c + 12c - 5c &= 18c - 5c \\ &= 13c \end{aligned}$$

b. The two terms $-5q^2$ and $-3q^2$ are like terms, so we may combine them:

$$-5q^2 - 3q^2 = -8q^2$$

c. The two terms $-3x$ and $4x$ are like terms, while the other term is different. Using the associative and commutative properties of addition in the first step allows us to place the two like terms next to each other, and then combine them:

$$\begin{aligned} -3x - 5y + 4x &= -3x + 4x + (-5y) \\ &= x - 5y \end{aligned}$$

d. The expression $2x - 3y + 4z$ cannot be simplified as there are no like terms.

Remark 2.8.8. The expression x represents $1x$ and the expression $-x$ represents $-1x$, but we don't write either the "1" or the "−1" as each is implied. However, it's helpful when combining like terms to remember that $x = 1x$ and $-x = -1x$.

Adding Expressions When we add an expression like $4x - 5$ to an expression like $3x - 7$, we write them as follows:

$$(4x - 5) + (3x - 7)$$

In order to remove the given sets of parentheses and apply the commutative property of addition, we will rewrite the subtraction operation as "adding the opposite":

$$4x + (-5) + 3x + (-7)$$

At this point we can apply the commutative property of addition and then combine like terms. Here's how the entire problem will look:

$$\begin{aligned} (4x - 5) + (3x - 7) &= 4x + (-5) + 3x + (-7) \\ &= 4x + 3x + (-5) + (-7) \\ &= 7x + (-12) \\ &= 7x - 12 \end{aligned}$$

Remark 2.8.9. Once we get more comfortable simplifying such expressions, we will simply write the following:

$$(4x - 5) + (3x - 7) = 7x - 12$$

Example 2.8.10. Use the associative, commutative, and distributive properties to simplify the following expressions as much as possible.

a. $(2x + 3) + (4x + 5)$

b. $(-5x + 3) + (4x - 7)$

Solution.

a. We will remove parentheses, and then combine like terms::

$$\begin{aligned}(2x+3)+(4x+5) &= 2x+3+4x+5\\ &= 2x+4x+3+5\\ &= 6x+8\end{aligned}$$

b. We will remove parentheses, and then combine like terms::

$$\begin{aligned}(-5x+3)+(4x-7) &= -5x+3+4x+(-7)\\ &= 7x+(-4)\\ &= 7x-4\end{aligned}$$

Applying the Distributive Property with Negative Coefficients Applying the distributive property in an expression such as $2(3x+4)$ is fairly straightforward, in that this becomes $2(3x)+2(4)$ which we then simplify to $6x+8$. Applying the distributive property is a little trickier when subtraction or a negative constant is involved, for example, with the expression $2(3x-4)$. Recalling that subtraction is defined as "adding the opposite," we can change the subtraction of positive 4 to the addition of negative 4:

$$2(3x+(-4))$$

Now when we distribute, we obtain:

$$2(3x)+2(-4)$$

As a final step, we see that this simplifies to:

$$6x-8$$

Remark 2.8.11. We can also extend the distributive property to one involving subtraction, which states that $a(b-c)=ab-ac$. With this property, we would simplify $2(3x-4)$ more efficiently:

$$\begin{aligned}2(3x-4) &= 2(3x)-2(4)\\ &= 6x-8\end{aligned}$$

In general, we will use this approach.

Example 2.8.12. Apply the distributive property to each expression and simplify it as much as possible.

a. $-3(5x+7)$

b. $2(-4x-1)$

Solution.

a. We will distribute -3 to the $5x$ and 7:

$$\begin{aligned}-3(5x+7) &= -3(5x)+(-3)(7)\\ &= -15x-21\end{aligned}$$

b. We will distribute 2 to the $-4x$ and -1:

$$\begin{aligned}2(-4x-1) &= 2(-4x)-2(1)\\ &= -8x-2\end{aligned}$$

Exercise 2.8.13. Use the distributive property to write an equivalent expression to $-7(y+2)$ that has no grouping symbols.

Solution. To use the distributive property, we multiply the number outside the parentheses, -7, with each term inside the parentheses:

$$\begin{aligned}-7(y+2) &= -7\cdot y-7(2)\\ &= -7y-14\end{aligned}$$

Subtracting Expressions To subtract one expression from another expression, such as $(5x+9)-(3x+2)$, we will again rely on the fact that subtraction is defined as "adding the opposite." To add the *opposite* of an expression, we will technically distribute a constant factor of -1 and simplify from there:

$$\begin{aligned}(5x+9)-(3x+2) &= (5x+9)+(-1)(3x+2)\\ &= 5x+9+(-1)(3x)+(-1)(2)\\ &= 5x+9+(-3x)+(-2)\\ &= 2x+7\end{aligned}$$

Remark 2.8.14. The above example demonstrates *how* we apply the distributive property in order to subtract two expressions. But in practice, it can be pretty cumbersome. A shorter (and often clearer) approach is to instead subtract every term in the expression we are subtracting, which is shown like this:

$$\begin{aligned}(5x+9)-(3x+2) &= 5x+9-3x-2\\ &= 2x+7\end{aligned}$$

In general, we'll use this approach.

Example 2.8.15. Use the associative, commutative, and distributive properties to simplify the following expressions as much as possible.

a. $(-6x + 4) - (3x - 7)$

b. $(-2x - 5) - (-4x - 6)$

Solution.

a. We will remove parentheses using the distributive property, and then combine like terms:

$$\begin{aligned}(-6x + 4) - (3x - 7) &= -6x + 4 - 3x - (-7)\\ &= -6x + 4 - 3x + 7\\ &= -9x + 11\end{aligned}$$

b. We will remove parentheses using the distributive property, and then combine like terms:

$$\begin{aligned}(-2x - 5) - (-4x - 6) &= -2x - 5 - (-4x) - (-6)\\ &= -2x - 5 + 4x + 6\\ &= 2x + 1\end{aligned}$$

2.8.5 The Role of the Order of Operations in Applying the Commutative, Associative, and Distributive Properties

When simplifying an expression such as $3 + 4(5x + 7)$, we need to respect the order of operations. Since the terms inside the parentheses are not like terms, there is nothing to simplify there. The next highest priority operation is multiplying the 4 by $(5x + 7)$. This must be done *before* anything happens with the adding of that 3. We cannot say $3 + 4(5x + 7) = 7(5x + 7)$, because that would mean we treated the addition as having higher priority than the multiplication.

So to simplify $3 + 4(5x + 7)$, we will first examine the multiplication of 4 with $(5x + 7)$, and here we may apply the distributive property. After that, we will use the commutative and associative properties:

$$\begin{aligned}3 + 4(5x + 7) &= 3 + 4(5x) + 4(7)\\ &= 3 + 20x + 28\\ &= 20x + 3 + 28\\ &= 20x + 31\end{aligned}$$

Example 2.8.16. Simplify the following expressions using the commutative, associative, and distributive properties.

a. $4 - (3x - 9)$

b. $5x + 9(-2x + 3)$

c. $5(x - 9) + 4(x + 4)$

Solution.

a. We will remove parentheses using the distributive property, and then combine like terms:

$$\begin{aligned} 4-(3x-9) &= 4-3x-(-9) \\ &= 4-3x+9 \\ &= -3x+13 \end{aligned}$$

b. We will remove parentheses using the distributive property, and then combine like terms:

$$\begin{aligned} 5x+9(-2x+3) &= 5x+9(-2x)+9(3) \\ &= 5x-18x+27 \\ &= -13x+27 \end{aligned}$$

c. We will remove parentheses using the distributive property, and then combine like terms:

$$\begin{aligned} 5(x-9)+4(x+4) &= 5x-45+4x+16 \\ &= 9x-29 \end{aligned}$$

Exercise 2.8.17. Use the distributive property to simplify $2-10(-4-8t)$ completely.

Solution. We first use distributive property to get rid of parentheses, and then combine like terms:

$$\begin{aligned} 2-10(-4-8t) &= 2+(-10)(-4-8t) \\ &= 2+(-10)(-4)+(-10)(-8t) \\ &= 2+40+80t \\ &= 42+80t \\ &= 80t+42 \end{aligned}$$

Note that either of the last two expressions are acceptable final answers.

2.8.6 Rules of Exponents and Simplifying

In Section 2.7, we introduced three exponent rules. We continue to use these rules when simplifying expressions. Sometimes though, students incorrectly apply "rules" of exponents where they have misremembered the actual rule. Let's summarize what we can and cannot do.

When we add/subtract two expressions, we can only combine *like* terms. For example:

- $3x - x = 2x$
- $t^2 + t^2 = 2t^2$
- $q^2 + q$ cannot be combined.

However, we can multiply two expressions regardless of whether or not they are like terms. For example:

- $x \cdot x = x^2$
- $t^2 \cdot t^3 = t^5$
- $(q^2)(q) = q^3$

Consider:

- When we combine like terms that have a variable, the exponent doesn't change, as in $x^2 + x^2 = 2x^2$.
- When we multiply powers of a variable that use the same variable, the exponent *will* change, as in $(x^2)(x^2) = x^4$.
- We *cannot* combine "unlike terms," as something like $x^2 + x$ is as simplified as it can be.
- We *can* multiply powers with different exponents, as in $(x^2)(x) = x^3$.

The next few examples test your understanding of these concepts.

Example 2.8.18. Simplify the following expressions using the rules of exponents and the distributive property.

a. $3x^2 + 2x + x^2$

b. $(3x^2)(2x)(x^2)$

c. $2x(3x + 4)$

d. $x^3 - 3x^2(5x - 2)$

Solution.

a. We will combine like terms $3x^2$ and x^2:

$$3x^2 + 2x + x^2 = 4x^2 + 2x$$

b. We will apply the Product Rule:

$$(3x^2)(2x)(x^2) = 6x^5$$

c. To simplify $2x(3x + 4)$, we want to first distribute $2x$, and then we can apply the Product Rule:

$$\begin{aligned} 2x(3x + 4) &= 2x(3x) + 2x(4) \\ &= 6x^2 + 8x \end{aligned}$$

d. We will use the distributive property first, apply the Product Rule, and combine like terms:

$$\begin{aligned} x^3 - 3x^2(5x - 2) &= x^3 - 3x^2(5x) - (-3x^2)(2) \\ &= x^3 - 15x^3 + 6x^2 \\ &= -14x^3 + 6x^2 \end{aligned}$$

2.8.7 Exercises

These exercises involve the concepts of like terms and the commutative, associative, and distributive properties.

1. The additive inverse of -3 is [ ]

2. The additive inverse of -1 is [ ]

3. The multiplicative inverse of 2 is [ ]

4. The multiplicative inverse of 4 is [ ]

5. Use the associative property of addition to write an equivalent expression to $n + (12 + a)$.

6. Use the associative property of addition to write an equivalent expression to $p + (54 + q)$.

7. Use the associative property of addition to write an equivalent expression to $12 + (19 + x)$.

8. Use the associative property of addition to write an equivalent expression to $5 + (8 + y)$.

9. Use the associative property of multiplication to write an equivalent expression to $10(9t)$.

10. Use the associative property of multiplication to write an equivalent expression to $6(3a)$.

11. Use the commutative property of addition to write an equivalent expression to $c + 21$.

12. Use the commutative property of addition to write an equivalent expression to $n + 4$.

13. Use the commutative property of addition to write an equivalent expression to $5c + 73$.

14. Use the commutative property of addition to write an equivalent expression to $9n + 16$.

15. Use the commutative property of addition to write an equivalent expression to $4(p + 82)$.

16. Use the commutative property of addition to write an equivalent expression to $8(x + 47)$.

17. Use the commutative property of multiplication to write an equivalent expression to $12y$.

18. Use the commutative property of multiplication to write an equivalent expression to $77t$.

19. Use the commutative property of multiplication to write an equivalent expression to $42 + 10a$.

20. Use the commutative property of multiplication to write an equivalent expression to $8 + 5c$.

21. Use the commutative property of multiplication to write an equivalent expression to $6(m + 79)$.

22. Use the commutative property of multiplication to write an equivalent expression to $4(b + 21)$.

23. Use the distributive property to write an equivalent expression to $2(n + 7)$ that has no grouping symbols.

24. Use the distributive property to write an equivalent expression to $8(p + 2)$ that has no grouping symbols.

25. Use the distributive property to write an equivalent expression to $-7(x - 1)$ that has no grouping symbols.

26. Use the distributive property to write an equivalent expression to $-2(y + 8)$ that has no grouping symbols.

27. Use the distributive property to write an equivalent expression to $-(t+3)$ that has no grouping symbols.

28. Use the distributive property to write an equivalent expression to $-(a-4)$ that has no grouping symbols.

29. Use the distributive property to simplify $10+3(4+3c)$ completely.

30. Use the distributive property to simplify $3+2(10+6x)$ completely.

31. Use the distributive property to simplify $10-4(-8+t)$ completely.

32. Use the distributive property to simplify $10-7(-1-9n)$ completely.

33. Use the distributive property to simplify $6-(6-7p)$ completely.

34. Use the distributive property to simplify $3-(-6+9x)$ completely.

35. Use the distributive property to simplify $9-(3y+3)$ completely.

36. Use the distributive property to simplify $6-(-3t-10)$ completely.

37. Use the distributive property to simplify $\frac{3}{2}(-10+a)$ completely.

38. Use the distributive property to simplify $\frac{9}{2}(-10+3c)$ completely.

39. Use the distributive property to simplify $\frac{2}{3}(-10+\frac{3}{4}c)$ completely.

40. Use the distributive property to simplify $\frac{4}{5}(-10+\frac{3}{2}y)$ completely.

41. The expression $n+m+y$ would be ambiguous if we did not have a left-to-right reading convention. Use grouping symbols to emphasize the order that these additions should be carried out. Use the associative property of addition to write an equivalent (but different) algebraic expression.

42. The expression $p+b+q$ would be ambiguous if we did not have a left-to-right reading convention. Use grouping symbols to emphasize the order that these additions should be carried out. Use the associative property of addition to write an equivalent (but different) algebraic expression.

43. A student has (correctly) simplified an algebraic expression in the following steps. Between each pair of steps, identify the algebraic property that justifies moving from one step to the next.

$2(x+9)+5x$

(□ commutative property of addition □ commutative property of multiplication □ associative property of addition □ associative property of multiplication □ distributive property)

$=(2x+18)+5x$

(□ commutative property of addition □ commutative property of multiplication □ associative property of addition □ associative property of multiplication □ distributive property)

$=(18+2x)+5x$

(□ commutative property of addition □ commutative property of multiplication □ associative property of addition □ associative property of multiplication □ distributive property)

$=18+(2x+5x)$

(□ commutative property of addition □ commutative property of multiplication □ associative property of addition □ associative property of multiplication □ distributive property)

$=18+(2+5)x$

$=18+7x$

(□ commutative property of addition □ commutative property of multiplication □ associative

property of addition □ associative property of multiplication □ distributive property)
$= 7x + 18$

44. A student has (correctly) simplified an algebraic expression in the following steps. Between each pair of steps, identify the algebraic property that justifies moving from one step to the next.

$7(y + 4) + 3y$
(□ commutative property of addition □ commutative property of multiplication □ associative property of addition □ associative property of multiplication □ distributive property)
$= (7y + 28) + 3y$
(□ commutative property of addition □ commutative property of multiplication □ associative property of addition □ associative property of multiplication □ distributive property)
$= (28 + 7y) + 3y$
(□ commutative property of addition □ commutative property of multiplication □ associative property of addition □ associative property of multiplication □ distributive property)
$= 28 + (7y + 3y)$
(□ commutative property of addition □ commutative property of multiplication □ associative property of addition □ associative property of multiplication □ distributive property)
$= 28 + (7 + 3)\, y$
$= 28 + 10y$
(□ commutative property of addition □ commutative property of multiplication □ associative property of addition □ associative property of multiplication □ distributive property)
$= 10y + 28$

45. The number of students enrolled in math courses at Portland Community College has grown over the years.
The formulas

$$M = 0.37x + 4.1 \qquad W = 0.54x + 5.6$$

describe the numbers (of thousands) of men and women enrolled in math courses at PCC x years after 2005. Give a simplified formula for the total number T of thousands of students at PCC taking math classes x years after 2005. Be sure to give the entire formula, starting with T=.

46. The number of students enrolled in math courses at Portland Community College has grown over the years.
The formulas

$$M = 0.4x + 3.1 \qquad W = 0.36x + 4.8$$

describe the numbers (of thousands) of men and women enrolled in math courses at PCC x years after 2005. Give a simplified formula for the total number T of thousands of students at PCC taking math classes x years after 2005. Be sure to give the entire formula, starting with T=.

These exercises involve the rules of exponents.

For the following exercises: Find the product of the *mo*nomial and the *bi*nomial.

47. $-x\,(x + 8) =$ ______ **48.** $x\,(x - 4) =$ ______

49. $8x\,(6x - 9) =$ ______ **50.** $9x\,(-9x + 9) =$ ______

51. $8x^2(x-4) =$ ☐

52. $-10x^2(x+10) =$ ☐

53. $-7x^2(7x^2+6x) =$ ☐

54. $4x^2(5x^2-10x) =$ ☐

These exercises involve rules of exponents and combining like terms.

55. Simplify the following expressions if possible.

1) $a^4 + a^4 =$ ☐

2) $(a^4)(a^4) =$ ☐

3) $a^4 + a =$ ☐

4) $(a^4)(a) =$ ☐

56. Simplify the following expressions if possible.

1) $c^3 + c^3 =$ ☐

2) $(c^3)(c^3) =$ ☐

3) $c^3 + c^2 =$ ☐

4) $(c^3)(c^2) =$ ☐

57. Simplify the following expressions if possible.

1) $3r^2 + 2r^2 =$ ☐

2) $(3r^2)(3r^2) =$ ☐

3) $3r^2 - r =$ ☐

4) $(3r^2)(-r) =$ ☐

58. Simplify the following expressions if possible.

1) $-3p^4 - 2p^4 =$ ☐

2) $(-3p^4)(-3p^4) =$ ☐

3) $-3p^4 + 2p^2 =$ ☐

4) $(-3p^4)(2p^2) =$ ☐

59. Simplify the following expressions if possible.

1) $m^2 + 4m^3 + 2m^2 =$ ☐

2) $(m^2)(4m^3)(2m^2) =$ ☐

60. Simplify the following expressions if possible.

1) $-3p^3 - 4p^4 + 2p^3 =$ ☐

2) $(-3p^3)(-4p^4)(2p^3) =$ ☐

61. Simplify the following expression.

$3x^5(-3x^3)^2 =$ ☐

62. Simplify the following expression.

$-4y^3(-5y^5)^2 =$ ☐

63. Simplify the following expression.

$-2t^2y^5(2t^5y^2)^5 =$ ☐

64. Simplify the following expression.

$-2a^4m^2(4a^3m^4)^2 =$ ☐

65. Simplify the following expression.

$(-3c^5)(2c^3) - (3c^4)(4c^4) =$ ☐

66. Simplify the following expression.

$(-3y^4)(4y^5) - (y^5)(-5y^4) =$ ☐

67. Simplify the following expression.

$(5m^5)(4m^4)^2-(m^5)(5m^4)=$ ☐

68. Simplify the following expression.

$(-2m^5)(3m^4)^2-(5m^4)(-5m^5)=$ ☐

69. Simplify the following expression.

$(p^5)(2p^4)^3+(-3p)^2(5p^{15})=$ ☐

70. Simplify the following expression.

$(2q^5)(-q^2)^2+(2q)^4(-3q^5)=$ ☐

71. Simplify the following expression.

$(-3y^4)^3a^{12}-2(y^4a^4)^3=$ ☐

72. Simplify the following expression.

$(3t^3)^3q^9-3(t^3q^3)^3=$ ☐

These exercises involve the distributive property and rules of exponents.

73. Use the distributive property to write an equivalent expression to $-5a(2a-1)$ that has no grouping symbols.

74. Use the distributive property to write an equivalent expression to $-8c(9c+8)$ that has no grouping symbols.

75. Use the distributive property to write an equivalent expression to $-8q^3(q+1)$ that has no grouping symbols.

76. Use the distributive property to write an equivalent expression to $-8m^2(m+9)$ that has no grouping symbols.

77. Use the distributive property to simplify $4+3m(10+6m)$ completely.

78. Use the distributive property to simplify $10+6p(8+6p)$ completely.

79. Use the distributive property to simplify $7q-2q(-1-2q^2)$ completely.

80. Use the distributive property to simplify $3y-7y(-7-y^4)$ completely.

81. Use the distributive property to simplify $9t^3-3t^3(8-t^3)$ completely.

82. Use the distributive property to simplify $6a^3-9a^3(2-a^4)$ completely.

83. Fully simplify $-4(-2x-7)+6(-9x-1)$.

84. Fully simplify $-5(-8x+1)-7(-4x-3)$.

85. Fully simplify $6(-4x-5)+8(7x-5)$.

86. Fully simplify $7(-2x+9)-9(-2x-8)$.

2.9 Chapter Review

2.9.1 Variables and Evaluating Expressions Review

Evaluating Expressions When we evaluate an expression's value, we substitute each variable with its given value.

Example Evaluate the value of $\frac{5}{9}(F-32)$ if $F=212$.

$$\begin{aligned}\frac{5}{9}(F-32)&=\frac{5}{9}(212-32)\\&=\frac{5}{9}(180)\\&=100\end{aligned}$$

Substituting a Negative Number When we substitute a variable with a negative number, it's important to use parentheses around the number.

Example Evaluate the following expressions if $x=-3$.

a.

$$\begin{aligned}x^2&=(-3)^2\\&=9\end{aligned}$$

b.

$$\begin{aligned}x^3&=(-3)^3\\&=(-3)(-3)(-3)\\&=-27\end{aligned}$$

c.

$$\begin{aligned}-x^2&=-(-3)^2\\&=-9\end{aligned}$$

d.

$$\begin{aligned}-x^3&=-(-3)^3\\&=-(-27)\\&=27\end{aligned}$$

2.9.2 Equations and Inequalities as True/False Statements Review

Checking Possible Solutions Given an equation or an inequality (with one variable), checking if some particular number is a solution is just a matter of replacing the value of the variable with the specified number and determining if the resulting equation/inequality is true or false. This may involve some amount of arithmetic simplification.

Example Is −5 a solution to $2(x+3)-2=4-x$?

To find out, substitute in −5 for x and see what happens.

$$\begin{aligned} 2(x+3)-2 &= 4-x \\ 2((-5)+3)-2 &\stackrel{?}{=} 4-(-5) \\ 2(-2)-2 &\stackrel{?}{=} 9 \\ -4-2 &\stackrel{?}{=} 9 \\ -6 &\stackrel{?}{=} 9 \end{aligned}$$

So no, −5 is not a solution to $2(x+3)-2=4-x$.

2.9.3 Solving One-Step Equations Review

Solving One-Step Equations When we solve linear equations, we use Properties of Equivalent Equations and follow .

Example Solve for g in $\frac{1}{2}=\frac{2}{3}+g$.

We will subtract $\frac{2}{3}$ on both sides of the equation:

$$\begin{aligned} \frac{1}{2} &= \frac{2}{3}+g \\ \frac{1}{2}-\frac{2}{3} &= \frac{2}{3}+g-\frac{2}{3} \\ \frac{3}{6}-\frac{4}{6} &= g \\ -\frac{1}{6} &= g \end{aligned}$$

We will check the solution by substituting g in the original equation with $-\frac{1}{6}$:

$$\begin{aligned} \frac{1}{2} &= \frac{2}{3}+g \\ \frac{1}{2} &\stackrel{?}{=} \frac{2}{3}+\left(-\frac{1}{6}\right) \\ \frac{1}{2} &\stackrel{?}{=} \frac{4}{6}+\left(-\frac{1}{6}\right) \\ \frac{1}{2} &\stackrel{?}{=} \frac{3}{6} \\ \frac{1}{2} &\stackrel{\checkmark}{=} \frac{1}{2} \end{aligned}$$

The solution $-\frac{1}{6}$ is checked and the solution set is $\left\{-\frac{1}{6}\right\}$.

2.9.4 Solving One-Step Equations and Inequalities Review

Solving One-Step Inequalities When we solve linear inequalities, we also use Properties of Equivalent Equations with one small complication: When we multiply or divide by the same *negative* number on both sides of an inequality, the direction reverses!

Example Solve the inequality $-2x \geq 12$. State the solution set with both interval notation and set-builder notation.

To solve this inequality, we will divide each side by -2:

$$\begin{aligned} -2x &\geq 12 \\ \frac{-2x}{-2} &\leq \frac{12}{-2} \qquad \text{Note the change in direction.} \\ x &\leq -6 \end{aligned}$$

- The inequality's solution set in interval notation is $(-\infty, -6]$.
- The inequality's solution set in set-builder notation is $\{x \mid x \leq -6\}$.

2.9.5 Solving One-Step Equations Involving Percentages Review

Solving One-Step Equations Involving Percentages An important skill for solving percent-related problems is to boil down a complicated word problem into a simple form like "2 is 50% of 4."

Example What percent of 2346.19 is 1995.98?

Using P to represent the unknown quantity, we write and solve the equation:

$$\begin{aligned} \overbrace{P}^{\text{what percent}} \overbrace{\cdot}^{\text{of}} \overbrace{2346.19}^{\$2346.19} &\overbrace{=}^{\text{is}} \overbrace{1995.98}^{\$1995.98} \\ \frac{P \cdot 2346.19}{2346.19} &= \frac{1995.98}{2346.19} \\ P &= 0.85073\ldots \\ P &\approx 85.07\% \end{aligned}$$

In summary, 2346.19 is approximately 85.07% of 1995.98.

2.9.6 Modeling with Equations and Inequalities Review

Modeling with Equations and Inequalities To set up an equation modeling a real world scenario, the first thing we need to do is identifying what variable we will use. The variable we use will be determined by whatever is unknown in our problem statement. Once we've identified and defined our variable, we'll use the numerical information provided in the equation to set up our equation.

Examples A bathtub contains 2.5 ft^3 of water. More water is being poured in at a rate of 1.75 ft^3 per minute. When will the amount of water in the bathtub reach 6.25 ft^3?

Since the question being asked in this problem starts with "when," we immediately know that the unknown is time. As the volume of water in the tub is measured in ft^3 per minute, we know that time needs to be measured in minutes. We'll defined t to be the number of minutes that water is poured into the tub. Since each minute there are 1.75 ft^3 of water added, we will add the expression $1.75t$ to 2.5 to obtain the total amount of water. Thus the equation we set up is:

$$2.5 + 1.75t = 6.25$$

2.9.7 Exponent Rules Review (Multiplication Only)

Rules of Exponents Let x, and y represent real numbers, variables, or algebraic expressions, and let m and n represent positive integers. Then the following properties hold:

Product of Powers $x^m \cdot x^n = x^{m+n}$

Power to Power $(x^m)^n = x^{m \cdot n}$

Product to Power $(xy)^n = x^n \cdot y^n$

Examples Simplify the following expressions using the rules of exponents.

$$-2t^3 \cdot 4t^5 = -8t^8$$
$$5\left(v^4\right)^2 = 5v^8$$
$$-(3u)^2 = -9u^4$$
$$(-3u)^2 = 9u^4$$

2.9.8 Simplifying Expressions Review

Algebraic Properties Let a,, b, and c represent real numbers, variables, or algebraic expressions. Then the following properties hold:

Commutative Property of Multiplication:

$$a \cdot b = b \cdot a \tag{2.9.1}$$

Associative Property of Multiplication: $a \cdot (b \cdot c) = (a \cdot b) \cdot c$

Commutative Property of Addition: $a + b = b + a$

Associative Property of Addition: $a + (b + c) = (a + b) + c$

Distributive Property:

$$a(b + c) = ab + ac \tag{2.9.2}$$

Examples Use the associative, commutative, and distributive properties to simplify the following expression as much as possible:

$$5x + 9(-2x + 3)$$

We will remove parentheses by the distributive property, and then combine like terms:

$$\begin{aligned} &= 5x + 9(-2x + 3) \\ &= 5x + 9(-2x) + 9(3) \\ &= 5x - 18x + 27 \\ &= -13x + 27 \end{aligned}$$

2.9.9 Exercises

1. A trapezoid's area can be calculated by the formula $A = \frac{1}{2}(b_1 + b_2)h$, where A stands for area, b_1 for the first base's length, b_2 for the second base's length, and h for height.

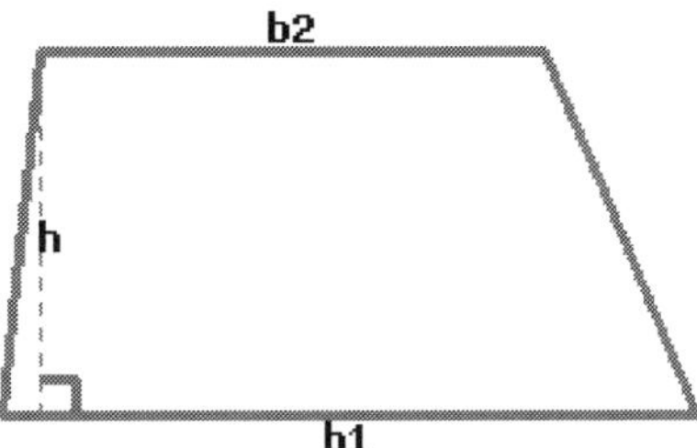

Use the formulas to calculate the trapezoid's area. Its first base's length is 18 m, its second base's length is 13 m and its height is 9 m.

Area = [ ]

2. To convert a temperature measured in degrees Fahrenheit to degrees Celsius, there is a formula:

$$C = \frac{5}{9}(F - 32)$$

where C represents the temperature in degrees Celsius and F represents the temperature in degrees Fahrenheit.
If a temperature is 122°F, what is that temperature measured in Celsius?

3. Evaluate the expression x^2:

a. When $x = 2$, $x^2 =$ ☐

b. When $x = -3$, $x^2 =$ ☐

4. Evaluate the expression x^3:

a. When $x = 4$, $x^3 =$ ☐

b. When $x = -4$, $x^3 =$ ☐

5. Is -2 a solution for x in the equation $3x + 5 = -4 - (-1 + x)$? Evaluating the left and right sides gives:

$$\begin{array}{ccc} 3x + 5 & = & -4 - (-1 + x) \\ \_\_\_ & \stackrel{?}{=} & \_\_\_ \end{array}$$

So -2 (☐ is ☐ is not) a solution to $3x + 5 = -4 - (-1 + x)$.

6. Is -1 a solution for x in the inequality $-5x^2 - x \leq 4x + 9$? Evaluating the left and right sides gives:

$$\begin{array}{ccc} -5x^2 - x & \leq & 4x + 9 \\ \_\_\_ & \stackrel{?}{\leq} & \_\_\_ \end{array}$$

So -1 (☐ is ☐ is not) a solution to $-5x^2 - x \leq 4x + 9$.

7. Solve the equation.

$$r + 7 = 4$$

8. Solve the equation.

$$-4 = -7 + r$$

9. Solve the equation.

$$t + \frac{1}{10} = \frac{9}{10}$$

10. Solve the equation.

$$60 = -5t$$

11. Solve the equation.

$$\frac{4}{9}q = 12$$

12. The pie chart represents a collector's collection of signatures from various artists.

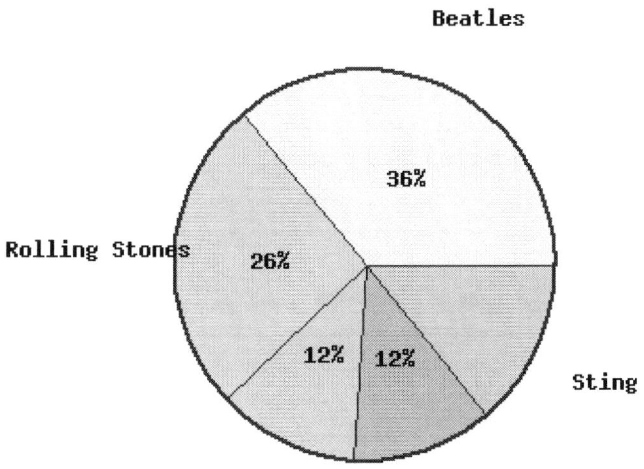

If the collector has a total of 450 signatures, there are [ ] signatures by Sting.

13. A community college conducted a survey about the number of students riding each bus line available. The following bar graph is the result of the survey.

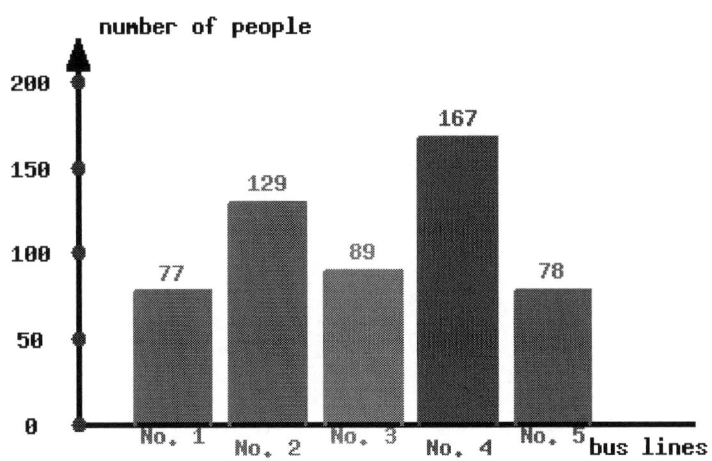

What percent of students ride Bus No. 1?
Fill in the blank with a percent. Round your percent to whole numbers, like 11%.
Approximately [ ] of students ride Bus No. 1.

14. The following is a nutrition fact label from a certain macaroni and cheese box.

Nutrition Facts	
Serving Size 1 cup	
Servings Per Container 2	
Amount Per Serving	
Calories 300	Calories from Fat 110
	% Daily Value
Total Fat 7.2 g	12%
Saturated Fat 2 g	15%
Trans Fat 2 g	
Cholesterol 30 mg	10%
Sodium 400 mg	20%
Total Carbohydrate 30 g	11%
Dietary Fiber 0 g	0%
Sugars 5 g	
Protein 5 g	
Vitamin A 2 mg	3%
Vitamin C 2 mg	2.5%
Calcium 2 mg	20%
Iron 3 mg	4%

The highlighted row means each serving of macaroni and cheese in this box contains 7.2 g of fat, which is 12% of an average person's daily intake of fat. What's the recommended daily intake of fat for an average person?

The recommended daily intake of fat for an average person is ______.
Use *g* for grams.

15. Evan used to make 14 dollars per hour. After he earned his Bachelor's degree, his pay rate increased to 52 dollars per hour. What is the percentage increase in Evan's salary?

The percentage increase is ______.

16. After a 60% increase, a town has 800 people. What was the population before the increase?

Before the increase, the town's population was ______.

17. A bicycle for sale costs $221.76, which includes 5.6% sales tax. What was the cost before sales tax?
Assume the bicycle's price before sales tax is p dollars. Write an equation to model this scenario. There is no need to solve it.

18. The property taxes on a 2400-square-foot house are $4,176.00 per year. Assuming these taxes are proportional, what are the property taxes on a 1700-square-foot house?
Assume property taxes on a 1700-square-foot house is t dollars. Write an equation to model this scenario. There is no need to solve it.

19. A swimming pool is being filled with water from a garden hose at a rate of 8 gallons per minute. If the pool already contains 90 gallons of water and can hold 298 gallons, after how long will the pool overflow?
Assume m minutes later, the pool would overflow. Write an equation to model this scenario. There is no need to solve it.

20. Use the distributive property to simplify $3 - 9(10 - 8q)$ completely.

21. Use the properties of exponents to simplify the expression.
$(8t^3) \cdot (-2t^{17})$

22. Use the properties of exponents to simplify the expression.
a.
$(-7r^3)^2 =$ [ ]
b.
$-(7r^3)^2 =$ [ ]

23. Find the product of the *mo*nomial and the *bi*nomial.
$-3y^2(-3y^2 + 4y) =$ [ ]

CHAPTER 3

Linear Equations and Inequalities

3.1 Solving Multistep Linear Equations and Inequalities

We have learned how to solve one-step equations and inequalities. In this section, we will learn how to solve multistep equations and inequalities.

3.1.1 Solving Two-Step Equations

Example 3.1.2. A water tank can hold 140 gallons of water, but it has only 5 gallons of water. A tap was turned on, pouring 15 gallons of water into the tank every minute. After how many minutes will the tank be full?

Let's find a pattern first.

Minutes since Tap Was Turned on	Amount of Water in the Tank (in Gallons)
0	5
1	$15 \cdot 1 + 5 = 20$
2	$15 \cdot 2 + 5 = 35$
3	$15 \cdot 3 + 5 = 50$
4	$15 \cdot 4 + 5 = 65$
$\vdots$	$\vdots$
m	$15m + 5$

Table 3.1.3: Amount of Water in the Tank

We can see the tap can pour $15m$ gallons of water into the tank in m minutes. The tank had 5 gallons of water in the beginning, so the amount of water in the tank can be modeled by $15m + 5$, where m is the number of minutes since the tap was turned on. To find when the tank will be full (with 140 gallons of water), we can write the equation

$$15m + 5 = 140$$

First, we need to isolate the variable term, $15m$, in the equation. In other words, we need to remove 5 from the left side of the equals sign. We can do this by subtracting 5 from both sides of the equation. Once the variable term is isolated, we can eliminate the coefficient and solve for m.

CC BY

The full process appears as:

$$15m + 5 = 140$$
$$15m + 5 - 5 = 140 - 5$$
$$15m = 135$$
$$\frac{15m}{15} = \frac{135}{15}$$
$$m = 9$$

Next, we need to substitute m with 9 in the equation $15m+5 = 140$ to check the solution:

$$15m + 5 = 140$$
$$15(9) + 5 \stackrel{?}{=} 140$$
$$135 + 5 \stackrel{?}{=} 140$$
$$140 \stackrel{\checkmark}{=} 140$$

The solution 9 is checked.

In summary, the tank will be full after 9 minutes.

In solving the two-step equation in Example 3.1.2, we first isolated the variable expression $15m$ and then eliminated the coefficient of 15 by dividing each side of the equation by 15. These two steps will be at the heart of our approach to solving linear equations. For more complicated equations, we may need to simplify some of the expressions first. Below is a general approach to solving linear equations that we will use as we solve more and more complicated equations.

Simplify Simplify the expressions on each side of the equation by distributing and combining like terms.

Isolate Use addition or subtraction to separate the variable terms and constant terms (numbers) so that they are on different sides of the equation.

Eliminate Use multiplication or division to eliminate the variable term's coefficient.

Check Check the solution.

Summarize State the solution set or (in the case of an application problem) summarize the result in a complete sentence using appropriate units.

List 3.1.4: Steps to Solving Linear Equations

Let's look at some more examples.

Example 3.1.5. Solve for y in the equation $7 - 3y = -8$.

Solution.

To solve, we will first separate the variable terms and constant terms into different sides of the equation. Then we will eliminate the variable term's coefficient:

$$\begin{aligned} 7-3y &= -8 \\ 7-3y-7 &= -8-7 \\ -3y &= -15 \\ \frac{-3y}{-3} &= \frac{-15}{-3} \\ y &= 5 \end{aligned}$$

To check our solution, we will replace y with 5 in the original equation:

$$\begin{aligned} 7-3y &= -8 \\ 7-3(5) &\stackrel{?}{=} -8 \\ 7-15 &\stackrel{?}{=} -8 \\ -8 &\stackrel{\checkmark}{=} -8 \end{aligned}$$

Therefore the solution to the equation $7-3y=-8$ is 5 and the solution set is $\{5\}$.

3.1.2 Solving Multistep Linear Equations

Example 3.1.6. Shane has saved \$2,500.00 in his savings account and is going to start saving \$550.00 per month. Tammy has saved \$4,600.00 in her savings account and is going to start saving \$250.00 per month. If this situation continues, how many months later would Shane catch up with Tammy in savings?

Shane saves \$550.00 per month, so he can save $550m$ dollars in m months. Counting \$2,500.00 already in his account, the amount of money in his account is $550m+2500$ dollars. Similarly, the amount of money in Tammy's account is $250m+4600$ dollars. To find when those two accounts will have the same amount of money, we write the equation

$$550m+2500=250m+4600$$

Here is the full process:

$$\begin{aligned}
550m + 2500 &= 250m + 4600\\
550m + 2500 - 2500 &= 250m + 4600 - 2500\\
550m &= 250m + 2100\\
550m - 250m &= 250m + 2100 - 250m\\
300m &= 2100\\
\frac{300m}{300} &= \frac{2100}{300}\\
m &= 7
\end{aligned}$$

Checking the solution 7 in the equation $550m + 2500 = 250m + 4600$, we get:

$$\begin{aligned}
550m + 2500 &= 250m + 4600\\
550(7) + 2500 &\stackrel{?}{=} 250(7) + 4600\\
3850 + 2500 &\stackrel{?}{=} 1750 + 4600\\
6350 &\stackrel{\checkmark}{=} 6350
\end{aligned}$$

In summary, Shane will catch up with Tammy in the savings account 7 months later.

Let's look at a few more examples.

Example 3.1.7. Solve for x in $5 - 2x = 5x - 9$.

Solution.

$$\begin{aligned}
5 - 2x &= 5x - 9\\
5 - 2x - 5 &= 5m - 9 - 5\\
-2x &= 5x - 14\\
-2x - 5x &= 5x - 14 - 5x\\
-7x &= -14\\
\frac{-7x}{-7} &= \frac{-14}{-7}\\
x &= 2
\end{aligned}$$

Checking the solution 2 in the equation $5 - 2x = 5x - 9$, we get:

$$\begin{aligned}
5 - 2x &= 5x - 9\\
5 - 2(2) &\stackrel{?}{=} 5(2) - 9\\
5 - 4 &\stackrel{?}{=} 10 - 9\\
1 &\stackrel{\checkmark}{=} 1
\end{aligned}$$

Therefore the solution is 2 and the solution set is $\{2\}$.

Remark 3.1.8.

In Example 3.1.7, we could have moved variable terms to the right side of the equals sign, and number terms to the left side. We chose not to. There's no reason we *couldn't* have moved variable terms to the right side though. Let's compare:

$$\begin{aligned} 5-2x &= 5x-9 \\ 5-2x+9 &= 5x-9+9 \\ 14-2x &= 5x \\ 14-2x+2x &= 5x+2x \\ 14 &= 7x \\ \frac{14}{7} &= \frac{7x}{7} \\ 2 &= x \end{aligned}$$

Lastly, we could save a step by moving variable terms and number terms in one step:

$$\begin{aligned} 5-2x &= 5x-9 \\ 5-2x+2x+9 & \quad 5x-9+2x+9 \\ 14 &= 7x \\ \frac{14}{7} &= \frac{7x}{7} \\ 2 &= x \end{aligned}$$

This textbook will move variable terms and number terms separately throughout this chapter. Check with your instructor for their expectations.

Exercise 3.1.9. Solve the equation.

$9b + 10 = b + 82$

Solution. The first step is to subtract terms in order to separate the variable and non-variable terms.

$$\begin{aligned} 9b+10 &= b+82 \\ 9b+10-\mathbf{b}-\mathbf{10} &= b+82-\mathbf{b}-\mathbf{10} \\ 8b &= 72 \\ \frac{8b}{8} &= \frac{72}{8} \\ b &= 9 \end{aligned}$$

The solution to this equation is 9. To stress that this is a value assigned to b, some report $b = 9$. We can also say that the solution set is $\{9\}$, or that $b \in \{9\}$. If we substitute 9 in for b in the original equation $9b+10 = b+82$, the equation will be true. Please check this on your own; it is an important habit.

The next example requires combining like terms.

Example 3.1.10. Solve for n in $n - 9 + 3n = n - 3n$.

Solution.

To start solving this equation, we'll need to combine like terms. After this, we can put all terms containing n on one side of the equation and finish solving for n.

$$\begin{aligned} n - 9 + 3n &= n - 3n \\ 4n - 9 &= -2n \\ 4n - 9 - 4n &= -2n - 4n \\ -9 &= -6n \\ \frac{-9}{-6} &= \frac{-6n}{-6} \\ n &= \frac{3}{2} \end{aligned}$$

Checking the solution $\frac{3}{2}$ in the equation $n - 9 + 3n = n - 3n$, we get:

$$\begin{aligned} n - 9 + 3n &= n - 3n \\ \frac{3}{2} - 9 + 3\left(\frac{3}{2}\right) &\stackrel{?}{=} \frac{3}{2} - 3\left(\frac{3}{2}\right) \\ \frac{3}{2} - 9 + \frac{9}{2} &\stackrel{?}{=} \frac{3}{2} - \frac{9}{2} \\ \frac{12}{2} - 9 &\stackrel{?}{=} -\frac{6}{2} \\ 6 - 9 &\stackrel{?}{=} -3 \\ -3 &\stackrel{?}{=} -3 \end{aligned}$$

The solution to the equation $n - 9 + 3n = n - 3n$ is $\frac{3}{2}$ and the solution set is $\left\{\frac{3}{2}\right\}$.

Exercise 3.1.11. Solve the equation.

$$-6 + 4 = -5A - A - 62$$

Solution. The first step is simply to combine like terms.

$$\begin{aligned} -6 + 4 &= -5A - A - 62 \\ -2 &= -6A - 62 \\ -2 + \mathbf{62} &= -6A - 62 + \mathbf{62} \\ 60 &= -6A \\ \frac{60}{-6} &= \frac{-6A}{-6} \\ -10 &= A \\ A &= -10 \end{aligned}$$

The solution to this equation is -10. To stress that this is a value assigned to A, some report $A = -10$. We can also say that the solution set is $\{-10\}$, or that $A \in \{-10\}$. If we substitute -10 in for A in the original equation $-6 + 4 = -5A - A - 62$, the equation will be true. Please check this on your own; it is an important habit.

Example 3.1.12. Virginia is designing a rectangular garden. The garden's length must be 4 meters less than three times the width, and its perimeter must be 40 meters. Find the garden's length and width.

Reminder: A rectangle's perimeter formula is $P = 2(L + W)$, where P stands for perimeter, L stands for length and W stands for width.

Assume the rectangle's width is W meters. We can then represent the length as $3W - 4$ meters since we are told that it is 4 meters less than three times the width. It's given that the perimeter is 40 meters. Substituting those values into the formula, we have:

$$\begin{aligned} P &= 2(L + W) \\ 40 &= 2(3W - 4 + W) \\ 40 &= 2(4W - 4) && \text{Like terms were combined.} \end{aligned}$$

The next step to solve this equation is to remove the parentheses by distribution.

$$\begin{aligned} 40 &= 2(4W - 4) \\ 40 &= 8W - 8 \\ 40 + 8 &= 8W - 8 + 8 \\ 48 &= 8W \\ \frac{48}{8} &= \frac{8W}{8} \\ 6 &= W. \end{aligned}$$

To check this result, we'll want to replace 6 in the equation $40 = 2(4W - 4)$:

$$\begin{aligned} 40 &= 2(4W - 4) \\ 40 &\stackrel{?}{=} 2(4(6) - 4) \\ 40 &\stackrel{?}{=} 2(20) \\ 40 &\stackrel{\checkmark}{=} 40. \end{aligned}$$

To determine the length, recall that this was represented by $3W - 4$, which is:

$$\begin{aligned} 3W - 4 &= 3(6) - 4 \\ &= 14. \end{aligned}$$

Thus the rectangle's width is 6 meters and the length is 14 meters.

Exercise 3.1.13. A rectangle's perimeter is 48 m. Its width is 8 m. Use an equation to solve for the rectangle's length.

Its length is ______.

Solution. When we deal with a geometric figure, it's always a good idea to sketch it to help us think. Let the length be x meters.

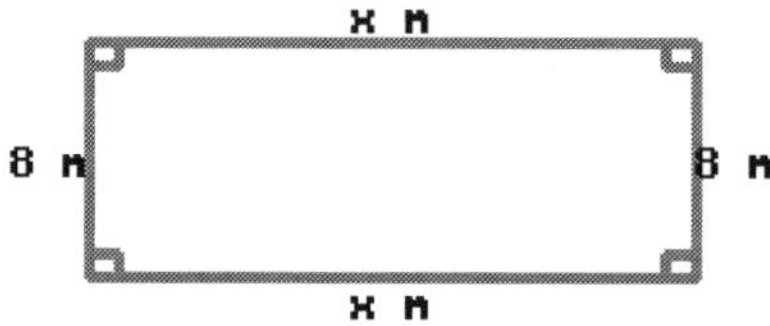

The perimeter is given as 48 m. Adding up the rectangle's 4 sides gives the perimeter. The equation is:

$$\begin{aligned}
x+x+8+8 &= 48\\
2x+16 &= 48\\
2x+16-\mathbf{16} &= 48-\mathbf{16}\\
2x &= 32\\
\frac{2x}{2} &= \frac{32}{2}\\
x &= 16
\end{aligned}$$

So the rectangle's length is 16 m. Don't forget the unit m.

We should be careful when we distribute a negative sign into the parentheses, like in the next example.

Example 3.1.14. Solve for a in $4-(3-a)=-2-2(2a+1)$.

Solution. To solve this equation, we will simplify each side of the equation, manipulate it so that all variable terms are on one side and all constant terms are on the other, and then solve for a:

$$\begin{aligned}
4-(3-a) &= -2-2(2a+1)\\
4-3+a &= -2-4a-2\\
1+a &= -4-4a\\
1+a+4a &= -4-4a+4a\\
1+5a &= -4\\
1+5a-1 &= -4-1\\
5a &= -5\\
\frac{5a}{5} &= \frac{-5}{5}\\
a &= -1
\end{aligned}$$

Checking the solution -1 in the original equation, we get:

$$\begin{aligned}
4-(3-a) &= -2-2(2a+1)\\
4-(3-(-1)) &\stackrel{?}{=} -2-2(2(-1)+1)\\
4-(4) &\stackrel{?}{=} -2-2(-1)\\
0 &\stackrel{\checkmark}{=} 0
\end{aligned}$$

Therefore the solution to the equation is -1 and the solution set is $\{-1\}$.

3.1.3 Differentiating among Simplifying Expressions, Evaluating Expressions and Solving Equations

Let's look at the following similar, yet different examples.

Example 3.1.15. Simplify the expression $10 - 3(x + 2)$.

Solution.

$$\begin{aligned} 10 - 3(x + 2) &= 10 - 3x - 6 \\ &= -3x + 4 \end{aligned}$$

An equivalent result is $4 - 3x$. Note that our final result is an *expression*.

Example 3.1.16. Evaluate the expression $10 - 3(x + 2)$ when $x = 2$ and when $x = 3$.

Solution. We will substitute $x = 2$ into the expression:

$$\begin{aligned} 10 - 3(x + 2) &= 10 - 3(2 + 2) \\ &= 10 - 3(4) \\ &= 10 - 12 \\ &= -2 \end{aligned}$$

When $x = 2$, $10 - 3(x + 2) = -2$.

Similarly, we will substitute $x = 3$ into the expression:

$$\begin{aligned} 10 - 3(x + 2) &= 10 - 3(3 + 2) \\ &= 10 - 3(5) \\ &= 10 - 15 = -5 \end{aligned}$$

When $x = 3$, $10 - 3(x + 2) = -5$.

Note that the final results here are *values of the original expression*.

Example 3.1.17. Solve the equation $10 - 3(x + 2) = x - 16$.

Solution.

$$
\begin{aligned}
10 - 3(x + 2) &= x - 16 \\
10 - 3x - 6 &= x - 16 \\
-3x + 4 &= x - 16 \\
-3x + 4 - 4 &= x - 16 - 4 \\
-3x &= x - 20 \\
-3x - x &= x - 20 - x \\
-4x &= -20 \\
\frac{-4x}{-4} &= \frac{-20}{-4} \\
x &= 5
\end{aligned}
$$

To check whether $x = 5$ is the correct solution of the equation, we substitute 5 for x into the equation, and we have:

$$
\begin{aligned}
10 - 3(x + 2) &= x - 16 \\
10 - 3(5 + 2) &\stackrel{?}{=} 5 - 16 \\
10 - 3(7) &\stackrel{?}{=} -11 \\
10 - 21 &\stackrel{?}{=} -11 \\
-11 &\stackrel{\checkmark}{=} -11
\end{aligned}
$$

We have checked that $x = 5$ is a solution of the equation $10 - 3(x + 2) = x - 16$.

Note that the final results here are *solutions* to the equations.

Let's summarize the differences among simplifying expressions, evaluating expressions and solving equations:

- An expression like $10 - 3(x + 2)$ can be simplified to $-3x + 4$ (as in Example 3.1.15), but we cannot solve for x in an expression.
- As x takes different values, an expression has different values. In Example 3.1.16, when $x = 2$, $10 - 3(x + 2) = -2$; but when $x = 3$, $10 - 3(x + 2) = -5$.
- An equation connects two expressions with an equals sign. In Example 3.1.17, $10 - 3(x + 2) = x - 16$ has the expression $10 - 3(x + 2)$ on the left side of equals sign, and the expression $x - 16$ on the right side.
- When we solve the equation $10 - 3(x + 2) = x - 16$, we are looking for a number which makes those two expressions have the same value. In Example 3.1.17, we found the solution to be $x = 5$, which makes both $10 - 3(x + 2) = -11$ and $x - 16 = -11$, as shown in the checking part.

3.1.4 Solving Multistep Inequalities

When solving a linear inequality, we follow the same steps in List 3.1.4. The only difference in our steps to solving is that when we multiply or divide by a negative number on both sides of an inequality, the direction of the inequality symbol must switch. We will look at some examples.

Simplify Simplify the expressions on each side of the inequality by distributing and combining like terms.

Isolate Use addition or subtraction to isolate the variable terms and constant terms (numbers) so that they are on different sides of the inequality symbol.

Eliminate Use multiplication or division to eliminate the variable term's coefficient. If each side of the inequality is multiplied or divided by a negative number, switch the direction of the inequality symbol.

Check When specified, verify the infinite solution set by checking multiple solutions.

Summarize State the solution set or (in the case of an application problem) summarize the result in a complete sentence using appropriate units.

List 3.1.18: Steps to Solve Linear Inequalities

Example 3.1.19. Solve for t in the inequality $-3t + 5 \geq 11$. Write the solution set in both set-builder notation and interval notation.

Solution.

$$\begin{aligned} -3t + 5 &\geq 11 \\ -3t + 5 - 5 &\geq 11 - 5 \\ -3t &\geq 6 \\ \frac{-3t}{-3} &\leq \frac{6}{-3} \\ t &\leq -2 \end{aligned}$$

Note that when we divided both sides of the inequality by -3, we had to switch the direction of the inequality symbol.

The solution set in set-builder notation is $\{t \mid t \leq -2\}$.

The solution set in interval notation is $(-\infty, -2]$.

Remark 3.1.20. Since the inequality solved in Example 3.1.19 has infinite solutions, it's difficult to check. We found that all values of t for which $t \leq -2$ are solutions, so one approach is to check if -2 is a solution and additionally if one other number less than -2 is a solution.

Here, we'll check that −2 satisfies this inequality:

$$
\begin{aligned}
-3t + 5 &\geq 11 \\
-3(-2) + 5 &\stackrel{?}{\geq} 11 \\
6 + 5 &\stackrel{?}{\geq} 11 \\
11 &\stackrel{\checkmark}{\geq} 11
\end{aligned}
$$

Next, we can check another number smaller than −2, such as −5:

$$
\begin{aligned}
-3t + 5 &\geq 11 \\
-3(-5) + 5 &\stackrel{?}{\geq} 11 \\
15 + 5 &\stackrel{?}{\geq} 11 \\
20 &\stackrel{\checkmark}{\geq} 11
\end{aligned}
$$

Thus both −2 and −5 are solutions. It's important to note that this doesn't directly verify that *all* solutions to this inequality check. It's valuable though in that it would likely help us catch an error if we had made one. Consult your instructor to see if you're expected to check your answer in this manner.

Example 3.1.21. Solve for z in the inequality $(6z + 5) - (2z - 3) < -12$. Write the solution set in both set-builder notation and interval notation.

Solution.

$$
\begin{aligned}
(6z + 5) - (2z - 3) &< -12 \\
6z + 5 - 2z + 3 &< -12 \\
4z + 8 &< -12 \\
4z + 8 - 8 &< -12 - 8 \\
4z &< -20 \\
\frac{4z}{4} &< \frac{-20}{4} \\
z &< -5
\end{aligned}
$$

Note that we divided both sides of the inequality by 4 and since this is a positive number we *did not* need to switch the direction of the inequality symbol.

The solution set in set-builder notation is $\{z \mid t < -5\}$.

The solution set in interval notation is $(-\infty, -5)$.

Example 3.1.22. Solve for x in $-2 - 2(2x + 1) > 4 - (3 - x)$. Write the solution set in both set-builder notation and interval notation.

Solution.

$$
\begin{aligned}
-2 - 2(2x + 1) &> 4 - (3 - x) \\
-2 - 4x - 2 &> 4 - 3 + x
\end{aligned}
$$

$$\begin{aligned}
-4x - 4 &> x + 1 \\
-4x - 4 - x &> x + 1 - x \\
-5x - 4 &> 1 \\
-5x - 4 + 4 &> 1 + 4 \\
-5x &> 5 \\
\frac{-5x}{-5} &< \frac{5}{-5} \\
x &< -1
\end{aligned}$$

Note that when we divided both sides of the inequality by -5, we had to switch the direction of the inequality symbol.

The solution set in set-builder notation is $\{x \mid x < -1\}$.

The solution set in interval notation is $(-\infty, -1)$.

Example 3.1.23. When a stopwatch started, the pressure inside a gas container was 4.2 atm (standard atmospheric pressure). As the container was heated, the pressure increased by 0.7 atm per minute. The maximum pressure the container can handle was 21.7 atm. Heating must be stopped once the pressure reaches 21.7 atm. In what time interval was the container safe?

The pressure increases by 0.7 atm per minute, so it increases by $0.7m$ after m minutes. Counting in the original pressure of 4.2 atm, pressure in the container can be modeled by $0.7m+4.2$, where m is the number of minutes since the stop watch started.

The container is safe when the pressure is 21.7 atm or lower. We can write and solve this inequality:

$$\begin{aligned}
0.7m + 4.2 &\le 21.7 \\
0.7m + 4.2 - 4.2 &\le 21.7 - 4.2 \\
0.7m &\le 17.5 \\
\frac{0.7m}{0.7} &\le \frac{17.5}{0.7} \\
m &\le 25
\end{aligned}$$

In summary, the container was safe as long as $m \le 25$. Assuming that m also must be greater than or equal to zero, this means $0 \le m \le 25$. We can write this as the time interval as $[0, 25]$. Thus the container was safe between 0 minutes and 25 minutes.

3.1.5 Exercises

Solving Two-Step Equations

For the following exercises: Solve the equation.

1. $8m + 1 = 25$

2. $5p + 5 = 40$

3. $2q - 3 = 17$

4. $8y - 2 = -18$

5. $-9 = 5r + 6$

6. $-12 = 2a + 4$

7. $18 = 7b - 3$

8. $-37 = 4A - 1$

9. $-2C + 9 = 21$

10. $-5m + 6 = 16$

11. $-8p - 3 = 29$

12. $-2q - 10 = -20$

13. $7 = -y + 2$

14. $15 = -r + 6$

15. $9a + 18 = 0$

16. $6b + 42 = 0$

Application Problems for Solving Two-Step Equations

17. A gym charges members $30 for a registration fee, and then $22 per month. You became a member some time ago, and now you have paid a total of $426 to the gym. How many months have passed since you joined the gym?

[ ] months have passed since you joined the gym.

18. Your cell phone company charges a $23 monthly fee, plus $0.18 per minute of talk time. One month your cell phone bill was $84.20. How many minutes did you spend talking on the phone that month?

You spent [ ] talking on the phone that month.

19. A school purchased a batch of T-shirts from a company. The company charged $9 per T-shirt, and gave the school a $75 rebate. If the school had a net expense of $3,795 from the purchase, how many T-shirts did the school buy?

The school purchased [ ] T-shirts.

20. Rita hired a face-painter for a birthday party. The painter charged a flat fee of $95, and then charged $2.50 per person. In the end, Rita paid a total of $122.50. How many people used the face-painter's service?

[ ] people used the face-painter's service.

21. A certain country has 788 million acres of forest. Every year, the country loses 9.85 million acres of forest mainly due to deforestation for farming purposes. If this situation continues at this pace, how many years later will the country have only 334.9 million acres of forest left? (Use an equation to solve this problem.)

After [ ] years, this country would have 334.9 million acres of forest left.

22. Priscilla has $71 in her piggy bank. She plans to purchase some Pokemon cards, which costs $1.85 each. She plans to save $56.20 to purchase another toy. At most how many Pokemon cards can he purchase?

Write an equation to solve this problem.

Priscilla can purchase at most [ ] Pokemon cards.

Solving Equations with Variable Terms on Both Sides

For the following exercises: Solve the equation.

23. $6r + 4 = r + 54$

24. $9a + 8 = a + 64$

25. $-5b + 3 = -b - 13$

26. $-8A + 8 = -A - 34$

27. $8 - 10B = 8B + 44$

28. $5 - 5m = 8m + 70$

29. $2p + 9 = 4p + 8$

30. $7q + 7 = 3q + 6$

31.

a. $9y + 8 = 4y + 28$

b. $4a + 8 = 9a - 42$

32.

a. $9r + 7 = 3r + 61$

b. $3q + 7 = 9q - 5$

Application Problems for Solving Equations with Variable Terms on Both Sides

For the following exercises: Use a linear equation to solve the word problem.

33. Two trees are 6.5 feet and 15.5 feet tall. The shorter tree grows 2.5 feet per year; the taller tree grows 2 feet per year. How many years later would the shorter tree catch up with the taller tree?

It would take the shorter tree [ ] years to catch up with the taller tree.

34. Massage Heaven and Massage You are competitors. Massage Heaven has 5300 registered customers, and it gets approximately 600 newly registered customers every month. Massage You has 7200 registered customers, and it gets approximately 500 newly registered customers every month. How many months would it take Massage Heaven to catch up with Massage You in the number of registered customers?

These two companies would have approximately the same number of registered customers [ ] months later.

35. Two truck rental companies have different rates. V-Haul has a base charge of \$65.00, plus \$0.75 per mile. W-Haul has a base charge of \$56.20, plus \$0.85 per mile. For how many miles would these two companies charge the same amount?

If a driver drives [ ] miles, those two companies would charge the same amount of money.

36. Massage Heaven and Massage You are competitors. Massage Heaven has 6500 registered customers, but it is losing approximately 200 registered customers every month. Massage You has 1500 registered customers, and it gets approximately 300 newly registered customers every month. How many months would it take Massage Heaven to catch up with Massage You in the number of registered customers?

These two companies would have approximately the same number of registered customers [ ] months later.

37. Diane has \$85.00 in her piggy bank, and she spends \$3.00 every day.

Alejandro has \$13.50 in his piggy bank, and he saves \$2.50 every day.

If they continue to spend and save money this way, how many days later would they have the same amount of money in their piggy banks?

[ ] days later, Diane and Alejandro will have the same amount of money in their piggy banks.

38. Sherial has \$90.00 in her piggy bank, and she spends \$1.50 every day.

Rebecca has \$26.00 in her piggy bank, and she saves \$2.50 every day.

If they continue to spend and save money this way, how many days later would they have the same amount of money in their piggy banks?

[ ] days later, Sherial and Rebecca will have the same amount of money in their piggy banks.

For the following exercises: Solve the equation.

39. $6q + 10q + 4 = 100$

40. $3y + 5y + 4 = 28$

41. $9r + 9 + 6 = 105$

42. $6a + 4 + 6 = 52$

43. $-9 + 7 = -7b - b - 50$

44. $-3 + 2 = -10A - A - 67$

45. $3r + 6 - 8r = 16$

46. $2r + 10 - 8r = 40$

47. $-8t + 4 + t = 39$

48. $-6t + 8 + t = -42$

49. $-18 = -2y - 3 - y$

50. $2 = -8r - 7 - r$

51. $5 - a - a = -10 + 29$

52. $2 - b - b = -5 + (-9)$

53. $4 - 7A - 10 = -6$

54. $3 - 4B - 7 = -4$

55. $-2 + 5 = 10m - 6 - 7m + 2 - 4m$

56. $-3 + (-7) = 7n - 9 - 2n + 2 - 6n$

57. $q - 4 - 6q = -6 - 3q + 10$

58. $x - 10 - 10x = -4 - 4x + 4$

59. $-8r + 3r = 10 - 4r - 15$

60. $-7a + 4a = 8 - 4a - 16$

61. $9b + 3 = -5b + 3 - 4b$

62. $6A + 7 = -3A + 7 - 4A$

Application Problems for Solving Linear Equations with Like Terms

63. A 162-meter rope is cut into two segments. The longer segment is 14 meters longer than the shorter segment. Write and solve a linear equation to find the length of each segment. Include units.

The segments are [ ] and [ ] long.

64. In a doctor's office, the receptionist's annual salary is \$149,000 less than that of the doctor. Together, the doctor and the receptionist make \$225,000 per year. Find each person's annual income.

The receptionist's annual income is [ ].

The doctor's annual income is [ ].

65. Emily and Brent went picking strawberries. Emily picked 122 fewer strawberries than Brent did. Together, they picked 250 strawberries. How many strawberries did Brent pick?

Brent picked [ ] strawberries.

66. Connor and Bobbi collect stamps. Bobbi collected 13 fewer than three times the number of Connor's stamps. Altogether, they collected 767 stamps. How many stamps did Connor and Bobbi collect?

Connor collected [ ] stamps.

Bobbi collected [ ] stamps.

67. Penelope and Charity sold girl scout cookies. Penelope's sales were $21 more than four times of Charity's. Altogether, their sales were $671. How much did each girl sell?

Penelope's sales were [ ].

Charity's sales were [ ].

68. A hockey team played a total of 98 games last season. The number of games they won was 14 more than three times of the number of games they lost.

Write and solve an equation to answer the following questions.

The team lost [ ] games.

The team won [ ] games.

Solving Linear Equations with Parentheses

For the following exercises: Solve the equation.

69. $2(a+2)=16$

70. $8(b+9)=152$

71. $5(A-7)=-55$

72. $2(B-4)=2$

73. $56=-8(m+2)$

74. $-60=-5(n+6)$

75. $38=-2(q-10)$

76. $77=-7(x-5)$

77. $-(r-9)=13$

78. $-(t-3)=-7$

79. $-4=-(7-b)$

80. $-12=-(3-A)$

81. $10(6B-10)=320$

82. $7(10m-10)=70$

83. $52=-2(4-3n)$

84. $60=-5(8-2q)$

85.

a. $6+(x+8)=7$

b. $6-(x+8)=7$

86.

a. $3+(r+5)=10$

b. $3-(r+5)=10$

87. $5 + 10(t + 10) = 35$

88. $3 + 8(b + 10) = 155$

89. $1 - 10(A + 9) = -119$

90. $4 - 7(B + 9) = -38$

91. $131 = 5 - 7(m - 9)$

92. $8 = 2 - 2(n - 9)$

93. $4 - 8(q - 9) = 68$

94. $3 - 10(x - 9) = 143$

95. $14 = 6 - (2 - r)$

96. $9 = 9 - (4 - t)$

97. $2 - (b + 6) = -2$

98. $1 - (c + 8) = 1$

99. $4(B + 10) - 10(B - 1) = 20$

100. $3(m + 5) - 8(m - 6) = 38$

101. $5 + 10(n - 5) = -4 - (1 - 2n)$

102. $4 + 7(q - 3) = 9 - (6 - 3q)$

103. $7(x - 7) - x = 49 - 4(2 + 3x)$

104. $10(r - 1) - r = 14 - 4(6 + 3r)$

105. $7(-14t + 6) = 14(-3 - 8t)$

106. $3(-14b + 10) = 6(-10 - 8b)$

107. $10 + 6(3 - 5c) = -5(c - 4) + 8$

108. $29 + 4(5 - 3B) = -5(B - 8) + 9$

Application Problems for Solving Linear Equations with Parentheses

109. A rectangle's perimeter is 76 cm. Its base is 27 cm.

Its height is ______.

110. A rectangle's perimeter is 58 m. Its width is 12 m. Use an equation to solve for the rectangle's length.

Its length is ______.

111. A rectangle's perimeter is 114 in. Its length is 7 in longer than its width. Use an equation to find the rectangle's length and width.

Its width is ______.

Its length is ______.

112. A rectangle's perimeter is 120 cm. Its length is 2 times as long as its width. Use an equation to find the rectangle's length and width.

It's width is ______.

Its length is ______.

113. A rectangle's perimeter is 100 ft. Its length is 5 ft shorter than four times of its width. Use an equation to find the rectangle's length and width.

Its width is ______.

Its length is ______.

114. A rectangle's perimeter is 180 ft. Its length is 2 ft longer than three times of its width. Use an equation to find the rectangle's length and width.

Its width is ______.

Its length is ______.

Comparisons

115. Solve the equation.

a. $-b + 6 = 6$

b. $-y + 6 = -6$

c. $-C - 6 = 6$

d. $-c - 6 = -6$

116. Solve the equation.

a. $-c + 6 = 6$

b. $-m + 6 = -6$

c. $-y - 6 = 6$

d. $-r - 6 = -6$

117.

a. Solve the following linear equation:

$$r - 2 = 4$$

b. Evaluate the following expression when $r = 6$:

$r - 2 =$ ______

118.

a. Solve the following linear equation:

$$r - 8 = 2$$

b. Evaluate the following expression when $r = 10$:

$r - 8 =$ ______

119.

a. Solve the following linear equation:

$$4(t - 3) - 3 = 9$$

b. Evaluate the following expression when $t = 6$:

$4(t - 3) - 3 =$ ______

c. Simplify the following expression:

$4(t - 3) - 3 =$ ______

120.

a. Solve the following linear equation:

$$3(t + 5) - 9 = 24$$

b. Evaluate the following expression when $t = 6$:

$3(t + 5) - 9 =$ ______

c. Simplify the following expression:

$3(t + 5) - 9 =$ ______

121. Choose True or False for the following questions about the difference between expressions and equations.

a. $-10x - 10 = -10x - 10$ is an equation. (□ True □ False)

b. We can check whether $x = 1$ is a solution of $-10x-10 = -10x-10$. (□ True □ False)

c. $-10x - 10$ is an equation. (□ True □ False)

d. We can check whether $x = 1$ is a solution of $-10x - 10$. (□ True □ False)

e. We can evaluate $-10x - 10 = -10x - 10$ when $x = 1$ (□ True □ False)

f. $-10x - 10$ is an expression. (□ True □ False)

g. We can evaluate $-10x - 10$ when $x = 1$ (□ True □ False)

h. $-10x - 10 = -10x - 10$ is an expression. (□ True □ False)

122. Choose True or False for the following questions about the difference between expressions and equations.

a. We can evaluate $-7x + 4 = 4x - 7$ when $x = 1$ (□ True □ False)

b. $4x - 7$ is an equation. (□ True □ False)

c. $-7x + 4$ is an expression. (□ True □ False)

d. $-7x + 4 = 4x - 7$ is an expression. (□ True □ False)

e. We can check whether $x = 1$ is a solution of $-7x + 4$. (□ True □ False)

f. $-7x + 4 = 4x - 7$ is an equation. (□ True □ False)

g. We can evaluate $-7x + 4$ when $x = 1$ (□ True □ False)

h. We can check whether $x = 1$ is a solution of $-7x + 4 = 4x - 7$. (□ True □ False)

Solving Linear Inequalities

For the following exercises: Solve this inequality.

123. $4x + 4 > 12$

In set-builder notation, the solution set is [ ].

In interval notation, the solution set is [ ].

124. $5x + 10 > 35$

In set-builder notation, the solution set is [ ].

In interval notation, the solution set is [ ].

125. $50 \geq 6x - 4$

In set-builder notation, the solution set is ☐.

In interval notation, the solution set is ☐.

126. $25 \geq 7x - 3$

In set-builder notation, the solution set is ☐.

In interval notation, the solution set is ☐.

127. $74 \leq 10 - 8x$

In set-builder notation, the solution set is ☐.

In interval notation, the solution set is ☐.

128. $24 \leq 6 - 9x$

In set-builder notation, the solution set is ☐.

In interval notation, the solution set is ☐.

129. $-10x - 3 < -63$

In set-builder notation, the solution set is ☐.

In interval notation, the solution set is ☐.

130. $-2x - 9 < -29$

In set-builder notation, the solution set is ☐.

In interval notation, the solution set is ☐.

131. $4 \geq -4x + 4$

In set-builder notation, the solution set is ☐.

In interval notation, the solution set is ☐.

132. $2 \geq -5x + 2$

In set-builder notation, the solution set is ☐.

In interval notation, the solution set is ☐.

133. $-8 > 2 - x$

In set-builder notation, the solution set is ☐.

In interval notation, the solution set is ☐.

134. $-5 > 3 - x$

In set-builder notation, the solution set is ☐.

In interval notation, the solution set is ☐.

135. $7(x + 1) \geq 28$

In set-builder notation, the solution set is ☐.

In interval notation, the solution set is ☐.

136. $8(x + 5) \geq 112$

In set-builder notation, the solution set is ☐.

In interval notation, the solution set is ☐.

137. $10t + 6 < 5t + 31$

In set-builder notation, the solution set is ☐.

In interval notation, the solution set is ☐.

138. $10t + 2 < 2t + 18$

In set-builder notation, the solution set is ☐.

In interval notation, the solution set is ☐.

139. $-4z + 10 \le -z - 8$

In set-builder notation, the solution set is $\boxed{}$.

In interval notation, the solution set is $\boxed{}$.

140. $-4z + 5 \le -z - 16$

In set-builder notation, the solution set is $\boxed{}$.

In interval notation, the solution set is $\boxed{}$.

141. $a - 4 - 2a > -6 - 4a + 29$

In set-builder notation, the solution set is $\boxed{}$.

In interval notation, the solution set is $\boxed{}$.

142. $a - 5 - 8a > -10 - 10a + 26$

In set-builder notation, the solution set is $\boxed{}$.

In interval notation, the solution set is $\boxed{}$.

143. $-5p + 5 - 8p \ge 6p + 5$

In set-builder notation, the solution set is $\boxed{}$.

In interval notation, the solution set is $\boxed{}$.

144. $-8p + 2 - 5p \ge 7p + 2$

In set-builder notation, the solution set is $\boxed{}$.

In interval notation, the solution set is $\boxed{}$.

145. $112 < -8(p - 7)$

In set-builder notation, the solution set is $\boxed{}$.

In interval notation, the solution set is $\boxed{}$.

146. $18 < -9(p - 4)$

In set-builder notation, the solution set is $\boxed{}$.

In interval notation, the solution set is $\boxed{}$.

147. $-(x - 10) \ge 12$

In set-builder notation, the solution set is $\boxed{}$.

In interval notation, the solution set is $\boxed{}$.

148. $-(x - 7) \ge 17$

In set-builder notation, the solution set is $\boxed{}$.

In interval notation, the solution set is $\boxed{}$.

149. $-22 \le 2 - 4(z - 1)$

In set-builder notation, the solution set is $\boxed{}$.

In interval notation, the solution set is $\boxed{}$.

150. $133 \le 3 - 10(z - 8)$

In set-builder notation, the solution set is $\boxed{}$.

In interval notation, the solution set is $\boxed{}$.

151. $2 - (y + 8) < -10$

In set-builder notation, the solution set is $\boxed{}$.

In interval notation, the solution set is $\boxed{}$.

152. $3 - (y + 7) < 5$

In set-builder notation, the solution set is $\boxed{}$.

In interval notation, the solution set is $\boxed{}$.

153. $3 + 10(x - 7) < -26 - (1 - 5x)$

In set-builder notation, the solution set is [ ].

In interval notation, the solution set is [ ].

154. $4 + 9(x - 10) < -67 - (1 - 3x)$

In set-builder notation, the solution set is [ ].

In interval notation, the solution set is [ ].

Application Problems for Linear Inequalities

155. You are riding in a taxi and can only pay with cash. You have to pay a flat fee of \$40, and then pay \$2.70 per mile. You have a total of \$148 in your pocket.

Let x be the number of miles the taxi will drive you. You want to know how many miles you can afford. Write an inequality to represent this situation in terms of how many miles you can afford:

[ ] [ ] [ ]

Solve this inequality. At most how many miles can you afford?

You can afford at most [ ] miles.

Use interval notation to express the number of miles you can afford.

156. You are riding in a taxi and can only pay with cash. You have to pay a flat fee of \$40, and then pay \$3.40 per mile. You have a total of \$244 in your pocket.

Let x be the number of miles the taxi will drive you. You want to know how many miles you can afford. Write an inequality to represent this situation in terms of how many miles you can afford:

[ ] [ ] [ ]

Solve this inequality. At most how many miles can you afford?

You can afford at most [ ] miles.

Use interval notation to express the number of miles you can afford.

3.2 Linear Equations and Inequalities with Fractions

In this section, we will learn how to solve linear equations and inequalities with fractions.

3.2.1 Introduction

So far, in our last step of solving for a variable we have divided each side of the equation by a constant, as in:

$$\begin{aligned} 2x &= 10 \\ \frac{2x}{2} &= \frac{10}{2} \\ x &= 5 \end{aligned}$$

If we have a coefficient that is a fraction, we *could* proceed in exactly the same manner:

$$\begin{aligned} \frac{1}{2}x &= 10 \\ \frac{\frac{1}{2}x}{\frac{1}{2}} &= \frac{10}{\frac{1}{2}} \\ x &= 10 \cdot \frac{2}{1} = 20 \end{aligned}$$

What if our equation or inequality was more complicated though, for example $\frac{1}{4}x + \frac{2}{3} = \frac{1}{6}$? We would have to first do a lot of fraction arithmetic in order to then divide each side by the coefficient of x. An alternate approach is to instead *multiply* each side of the equation by a chosen constant that eliminates the denominator. In the equation $\frac{1}{2}x = 10$, we could simply multiply each side of the equation by 2, which would eliminate the denominator of 2:

$$\begin{aligned} \frac{1}{2}x &= 10 \\ 2 \cdot \left(\frac{1}{2}x\right) &= 2 \cdot 10 \\ x &= 20 \end{aligned}$$

For more complicated equations, we will multiply each side of the equation by the least common denominator (LCD) of all fractions contained in the equation.

3.2.2 Eliminating Denominators

Example 3.2.2.

A 4-foot tree was planted. The tree grows $\frac{2}{3}$ of a foot every year. How many years later will the tree reach 10 feet?

Since the tree grows $\frac{2}{3}$ of a foot every year, we can use a table to help write a formula modeling the tree's growth:

Years Passed	Tree's Height (ft)
0	4
1	$4 + \frac{2}{3}$
2	$4 + \frac{2}{3} \cdot 2$
$\vdots$	$\vdots$
y	$4 + \frac{2}{3}y$

From this, we've determined that y years since the tree was planted, the tree's height will be $4 + \frac{2}{3}y$ feet.

To find when the tree will be 10 feet tall, we write and solve this equation:

$$\begin{aligned} 4 + \frac{2}{3}y &= 10 \\ 3 \cdot \left(4 + \frac{2}{3}y\right) &= 3 \cdot 10 \\ 3 \cdot 4 + 3 \cdot \frac{2}{3}y &= 30 \\ 12 + 2y &= 30 \\ 2y &= 18 \\ y &= 9 \end{aligned}$$

Now we will check the solution 9 in the equation $4 + \frac{2}{3}y = 10$:

$$\begin{aligned} 4 + \frac{2}{3}y &= 10 \\ 4 + \frac{2}{3}(9) &\stackrel{?}{=} 10 \\ 4 + 6 &\stackrel{\checkmark}{=} 10 \\ 10 &\stackrel{\checkmark}{=} 10 \end{aligned}$$

In summary, the tree will be 10 feet tall 9 years later.

Let's look at a few more examples.

Example 3.2.3. Solve for x in $\frac{1}{4}x + \frac{2}{3} = \frac{1}{6}$.

Solution. To solve this equation, we first need to identify the LCD of all fractions in the equation. On the left side we have $\frac{1}{4}$ and $\frac{2}{3}$. On the right side we have $\frac{1}{6}$. The LCD of $3, 4$, and 6 is 12, so we will multiply each side of the equation by 12 in order to eliminate *all* of the denominators:

$$\begin{aligned}
\frac{1}{4}x + \frac{2}{3} &= \frac{1}{6}\\
12 \cdot \left(\frac{1}{4}x + \frac{2}{3}\right) &= 12 \cdot \frac{1}{6}\\
12 \cdot \left(\frac{1}{4}x\right) + 12 \cdot \left(\frac{2}{3}\right) &= 12 \cdot \frac{1}{6}\\
3x + 8 &= 2\\
3x &= -6\\
\frac{3x}{3} &= \frac{-6}{3}\\
x &= -2
\end{aligned}$$

Checking the solution -2 in the equation $\frac{1}{4}x + \frac{2}{3} = \frac{1}{6}$:

$$\begin{aligned}
\frac{1}{4}x + \frac{2}{3} &= \frac{1}{6}\\
\frac{1}{4}(-2) + \frac{2}{3} &\stackrel{?}{=} \frac{1}{6}\\
-\frac{2}{4} + \frac{2}{3} &\stackrel{?}{=} \frac{1}{6}\\
-\frac{6}{12} + \frac{8}{12} &\stackrel{?}{=} \frac{1}{6}\\
\frac{2}{12} &\stackrel{\checkmark}{=} \frac{1}{6}
\end{aligned}$$

The solution is therefore -2 and the solution set is $\{-2\}$.

Example 3.2.4. Solve for z in $-\frac{2}{5}z - \frac{3}{2} = -\frac{1}{2}z + \frac{4}{5}$.

Solution.

The first thing we need to do is identify the LCD of all denominators in this equation. Since the denominators are 2 and 5, the LCD is 10. So as our first step, we will multiply each side of the equation by 10 in order to eliminate all denominators:

$$\begin{aligned}
-\frac{2}{5}z - \frac{3}{2} &= -\frac{1}{2}z + \frac{4}{5}\\
10 \cdot \left(-\frac{2}{5}z - \frac{3}{2}\right) &= 10 \cdot \left(-\frac{1}{2}z + \frac{4}{5}\right)\\
10\left(-\frac{2}{5}z\right) - 10\left(\frac{3}{2}\right) &= 10\left(-\frac{1}{2}z\right) + 10\left(\frac{4}{5}\right)\\
-4z - 15 &= -5z + 8\\
z - 15 &= 8\\
z &= 23
\end{aligned}$$

Checking the solution 23 in the equation $-\frac{2}{5}z - \frac{3}{2} = -\frac{1}{2}z + \frac{4}{5}$, we replace every instance of z with 23 and find:

$$\begin{aligned}
-\frac{2}{5}z - \frac{3}{2} &= -\frac{1}{2}z + \frac{4}{5}\\
-\frac{2}{5}(23) - \frac{3}{2} &\stackrel{?}{=} -\frac{1}{2}(23) + \frac{4}{5}\\
-\frac{46}{5} - \frac{3}{2} &\stackrel{?}{=} -\frac{23}{2} + \frac{4}{5}\\
-\frac{46}{5} \cdot \frac{2}{2} - \frac{3}{2} \cdot \frac{5}{5} &\stackrel{?}{=} -\frac{23}{2} \cdot \frac{5}{5} + \frac{4}{5} \cdot \frac{2}{2}\\
-\frac{92}{10} - \frac{15}{10} &\stackrel{?}{=} -\frac{115}{10} + \frac{8}{10}\\
-\frac{107}{10} &\stackrel{\checkmark}{=} -\frac{107}{10}
\end{aligned}$$

Thus the solution is 23 and so the solution set is $\{23\}$.

Example 3.2.5. Solve for a in the equation $\frac{2}{3}(a + 1) + 5 = \frac{1}{3}$.

Solution.

$$\begin{aligned}
\frac{2}{3}(a+1)+5 &= \frac{1}{3}\\
3\cdot\left(\frac{2}{3}(a+1)+5\right) &= 3\cdot\frac{1}{3}\\
3\cdot\frac{2}{3}(a+1)+3\cdot 5 &= 1\\
2(a+1)+15 &= 1\\
2a+2+15 &= 1\\
2a+17 &= 1\\
2a &= -16\\
a &= -8
\end{aligned}$$

Check the solution -8 in the equation $\frac{2}{3}(a+1)+5=\frac{1}{3}$, we find that:

$$\begin{aligned}
\frac{2}{3}(a+1)+5 &= \frac{1}{3}\\
\frac{2}{3}(-8+1)+5 &\stackrel{?}{=} \frac{1}{3}\\
\frac{2}{3}(-7)+5 &\stackrel{?}{=} \frac{1}{3}\\
-\frac{14}{3}+\frac{15}{3} &\stackrel{?}{=} \frac{1}{3}\\
\frac{1}{3} &\stackrel{\checkmark}{=} \frac{1}{3}
\end{aligned}$$

The solution is therefore -8 and the solution set is $\{-8\}$.

Example 3.2.6. Solve for b in the equation $\frac{2b+1}{3}=\frac{2}{5}$.

Solution.

$$\begin{aligned}
\frac{2b+1}{3} &= \frac{2}{5}\\
15\cdot\frac{2b+1}{3} &= 15\cdot\frac{2}{5}\\
5(2b+1) &= 6\\
10b+5 &= 6\\
10b &= 1\\
b &= \frac{1}{10}
\end{aligned}$$

Checking the solution $\frac{1}{10}$ in the equation $\frac{2b+1}{3}=\frac{2}{5}$, we find that:

$$\begin{aligned}
\frac{2b+1}{3} &= \frac{2}{5}\\
\frac{2\left(\frac{1}{10}\right)+1}{3} &\stackrel{?}{=} \frac{2}{5}\\
\frac{\frac{1}{5}+1}{3} &\stackrel{?}{=} \frac{2}{5}\\
\frac{\frac{1}{5}+\frac{5}{5}}{3} &\stackrel{?}{=} \frac{2}{5}\\
\frac{\frac{6}{5}}{3} &\stackrel{?}{=} \frac{2}{5}\\
\frac{6}{5}\cdot\frac{1}{3} &\stackrel{?}{=} \frac{2}{5}\\
\frac{2}{5} &\stackrel{\checkmark}{=} \frac{2}{5}
\end{aligned}$$

The solution is $\frac{1}{10}$ and the solution set is $\left\{\frac{1}{10}\right\}$.

Remark 3.2.7. Some of you might solve Example 3.2.6 with a property called **cross-multiplication**.

Cross-multiplication is a specialized application of the process of clearing the denominators from an equation. This process will be covered in Section 3.4, as there are some restrictions on the process we need to be aware of before making significant use of cross-multiplication.

Let's look at a one more application problem involving equations with fractions.

Example 3.2.8. In a science lab, a container had 21 ounces of water at 9:00AM. Water has been evaporating at the rate of 3 ounces every 5 minutes. When will there be 8 ounces of water left?

Solution. Since the container has been losing 3 oz of water every 5 minutes, it loses $\frac{3}{5}$ oz every minute. In m minutes since 9:00AM, the container would lose $\frac{3}{5}m$ oz of water. Since the container had 21 oz of water at the beginning, the amount of water in the container can be modeled by $21 - \frac{3}{5}m$ (in oz).

To find when there would be 8 oz of water left, we write and solve this equation:

$$
\begin{aligned}
21 - \frac{3}{5}m &= 8 \\
5 \cdot (21 - \frac{3}{5}x) &= 5 \cdot 8 \\
5 \cdot 21 - 5 \cdot \frac{3}{5}x &= 40 \\
105 - 3m &= 40 \\
105 - 3m - 105 &= 40 - 105 \\
-3m &= -55 \\
\frac{-3m}{-3} &= \frac{-55}{-3} \\
m &= \frac{55}{3} \\
m &= 18\frac{1}{3}
\end{aligned}
$$

Checking the solution $\frac{55}{3}$ in the equation $21 - \frac{3}{5}m = 8$:

$$
\begin{aligned}
21 - \frac{3}{5}m &= 8 \\
21 - \frac{3}{5}\left(\frac{55}{3}\right) &\stackrel{?}{=} 8 \\
21 - 11 &\stackrel{?}{=} 8 \\
8 &\stackrel{\checkmark}{=} 8
\end{aligned}
$$

Therefore the solution is $\frac{55}{3}$ or $18\frac{1}{3}$. In context, this means that shortly after 9:18AM (or at 9:18:20AM), the container will have 8 ounces of water left.

Exercise 3.2.9. Solve the equation.

$$27 = \frac{x}{7} + \frac{x}{2}$$

Solution. To clear fractions in an equation, we multiply each term by a common denominator.

For this problem, a common denominator is 14.

$$\begin{aligned}
27 &= \frac{x}{7} + \frac{x}{2} \\
14 \cdot 27 &= 14 \cdot \frac{x}{7} + 14 \cdot \frac{x}{2} \\
378 &= 2x + 7x \\
378 &= 9x \\
\frac{378}{9} &= \frac{9x}{9} \\
42 &= x \\
x &= 42
\end{aligned}$$

The solution to this equation is 42. To stress that this is a value assigned to x, some report $x = 42$. We can also say that the solution set is $\{42\}$, or that $x \in \{42\}$. If we substitute 42 in for x in the original equation $27 = \frac{x}{7} + \frac{x}{2}$, the equation will be true. Please check this on your own; it is an important habit.

3.2.3 Solving Inequalities with Fractions

We can also solve linear inequalities involving fractions by multiplying each side of the inequality by the LCD of all fractions within the inequality. Remember that with inequalities, everything works exactly the same except that the inequality sign reverses direction whenever we multiply each side of the inequality by a negative number.

Example 3.2.10. Solve for x in the inequality $\frac{3}{4}x - 2 > \frac{4}{5}x$. Write the solution set in both set-builder notation and interval notation.

Solution.

$$\begin{aligned}
\frac{3}{4}x - 2 &> \frac{4}{5}x \\
20 \cdot \left(\frac{3}{4}x - 2\right) &> 20 \cdot \frac{4}{5}x \\
20 \cdot \frac{3}{4}x - 20 \cdot 2 &> 16x \\
15x - 40 &> 16x \\
15x - 40 - 15x &> 16x - 15x \\
-40 &> x \\
x &< -40
\end{aligned}$$

The solution set in set-builder notation is $\{x \mid x < -40\}$. Note that it's equivalent to write

$\{x \mid -40 > x\}$, but it's easier to understand if we write x first in an inequality.

The solution set in interval notation is $(-\infty, -40)$.

Example 3.2.11. Solve for y in the inequality $\frac{4}{7} - \frac{4}{3}y \leq \frac{2}{3}$. Write the solution set in both set-builder notation and interval notation.

Solution.

$$\begin{aligned}
\frac{4}{7} - \frac{4}{3}y &\leq \frac{2}{3} \\
21 \cdot \left(\frac{4}{7} - \frac{4}{3}y\right) &\leq 21 \cdot \left(\frac{2}{3}\right) \\
21\left(\frac{4}{7}\right) - 21\left(\frac{4}{3}y\right) &\leq 21\left(\frac{2}{3}\right) \\
12 - 28y &\leq 14 \\
-28y &\leq 2 \\
\frac{-28y}{-28} &\geq \frac{2}{-28} \\
y &\geq -\frac{1}{14}
\end{aligned}$$

Note that when we divided each side of the inequality by -28, the inequality symbol reversed direction.

The solution set in set-builder notation is $\left\{y \mid y \geq -\frac{1}{14}\right\}$.

The solution set in interval notation is $\left[-\frac{1}{14}, \infty\right)$.

Example 3.2.12. In a certain class, a student's grade is calculated by the average of their scores on 3 tests. Ronda scored 78% and 54% on the first two tests. If Ronda wants to earn at least a grade of C (70%), what's the lowest score she needs to earn on the third exam?

Solution. Assume Ronda will score x% on the third test. To make her average test score greater than or equal to 70%, we write and solve this inequality:

$$\begin{aligned}
\frac{78 + 54 + x}{3} &\geq 70 \\
\frac{132 + x}{3} &\geq 70 \\
3 \cdot \frac{132 + x}{3} &\geq 3 \cdot 70 \\
132 + x &\geq 210 \\
x &\geq 78
\end{aligned}$$

To earn at least a C grade, Ronda needs to score at least 78% on the third test.

3.2.4 Exercises

Solving Linear Equations with Fractions Exercises

For the following exercises: Solve the equation.

1. $\frac{r}{3} + 44 = 4r$

2. $\frac{t}{9} + 34 = 2t$

3. $\frac{b}{6} + 10 = 14$

4. $\frac{c}{3} + 7 = 13$

5.

a. $-\frac{B}{10} + 7 = 4$

b. $\frac{-m}{10} + 7 = 4$

c. $\frac{t}{-10} + 7 = 4$

d. $\frac{-b}{-10} + 7 = 4$

6.

a. $-\frac{C}{9} + 10 = 5$

b. $\frac{-b}{9} + 10 = 5$

c. $\frac{y}{-9} + 10 = 5$

d. $\frac{-p}{-9} + 10 = 5$

7. $1 - \frac{n}{2} = -5$

8. $8 - \frac{p}{6} = 4$

9. $-20 = 4 - \frac{4x}{9}$

10. $-11 = 1 - \frac{2y}{5}$

11. $5t = \frac{9t}{8} + 155$

12. $5b = \frac{3b}{4} + 68$

13. $220 = \frac{2}{7}c + 6c$

14. $125 = \frac{10}{3}B + 5B$

15. $126 - \frac{5}{4}C = 4C$

16. $195 - \frac{9}{10}n = 3n$

17. $3p = \frac{2}{5}p + 6$

18. $6x = \frac{10}{7}x + 3$

19. $\frac{3}{10} - 6y = 8$

20. $\frac{7}{6} - 4t = 8$

21. $\frac{9}{4} - \frac{1}{4}b = 10$

22. $\frac{7}{10} - \frac{1}{10}c = 8$

23. $\frac{8B}{7} - 6 = -\frac{10}{7}$

24. $\frac{2C}{3} - 4 = -\frac{4}{3}$

25. $\frac{10}{9} + \frac{8}{9}n = 5n$

26. $\frac{2}{7} + \frac{6}{7}p = 7p$

27. $\frac{2x}{7} - \frac{15}{7} = -\frac{3}{7}x$

28. $\frac{5y}{7} - \frac{24}{7} = -\frac{1}{7}y$

29. $\frac{2t}{5} + \frac{7}{8} = t$

30. $\frac{2a}{3} + \frac{7}{6} = a$

31. $\frac{2c}{7} - 25 = -\frac{3}{2}c$

32. $\frac{4B}{5} - 49 = -\frac{5}{6}B$

33. $-\frac{9}{2}C + 50 = \frac{7C}{4}$

34. $-\frac{7}{8}n + 34 = \frac{3n}{16}$

35. $\frac{9p}{4} - 8p = \frac{3}{8}$

36. $\frac{3x}{10} - 8x = \frac{9}{20}$

37. $\frac{7y}{4} + \frac{6}{7} = \frac{5}{6}y$

38. $\frac{3t}{2} + \frac{6}{5} = \frac{5}{8}t$

39. $\frac{4}{7}a = \frac{1}{4} + \frac{5a}{3}$

40. $\frac{3}{5}c = \frac{4}{7} + \frac{5c}{6}$

41. $\frac{5}{4} = \frac{B}{16}$

42. $\frac{7}{10} = \frac{C}{20}$

43. $-\frac{n}{42} = \frac{2}{7}$

44. $-\frac{p}{6} = \frac{8}{3}$

45. $-\frac{x}{30} = -\frac{9}{10}$

46. $-\frac{y}{12} = -\frac{5}{6}$

47. $-\frac{8}{3} = \frac{5t}{8}$

48. $-\frac{4}{9} = \frac{5a}{6}$

49. $\frac{7}{6} = \frac{c+3}{18}$

50. $\frac{3}{2} = \frac{A+10}{6}$

51. $\frac{5}{8} = \frac{C-4}{7}$

52. $\frac{5}{6} = \frac{n-8}{5}$

53. $\frac{p-3}{2} = \frac{p+2}{4}$

54. $\frac{x-7}{6} = \frac{x+9}{8}$

55. $\frac{y+2}{4} - \frac{y-6}{8} = \frac{3}{4}$

56. $\frac{t+6}{2} - \frac{t-3}{4} = \frac{5}{2}$

57. $\frac{10a+7}{4} - \frac{2-a}{8} = \frac{7}{9}$

58. $\frac{8c+7}{2} - \frac{4-c}{4} = \frac{2}{5}$

59. $34 = \frac{A}{9} + \frac{A}{8}$

60. $9 = \frac{C}{7} + \frac{C}{2}$

61. $\frac{n}{3} - 15 = \frac{n}{8}$

62. $\frac{p}{7} - 4 = \frac{p}{9}$

63. $\frac{x}{4} - 2 = \frac{x}{7} + 4$

64. $\frac{y}{2} - 2 = \frac{y}{9} + 5$

65. $\frac{1}{2}t + \frac{5}{3} = \frac{5}{3}t + \frac{2}{3}$

66. $4a + 1 = \frac{5}{2}a + 3$

67. $-\frac{1}{2}c - \frac{2}{5} = -\frac{3}{8}c + 4$

68. $\frac{9}{10}A - \frac{5}{3} = -A - \frac{3}{10}$

69. $\frac{6}{5}C - \frac{8}{3} = -\frac{3}{8}C - \frac{7}{9}$

70. $-\frac{1}{3}m - \frac{5}{9} = -m + \frac{5}{4}$

Application Problems

71. Izabelle is jogging in a straight line. She got a head start of 2 meters from the starting line, and she ran 5 meters every 6 seconds. After how many seconds will Izabelle be 12 meters away from the starting line?

Izabelle will be 12 meters away from the starting line [ ] seconds since she started running.

72. Rita is jogging in a straight line. She started at a place 31 meters from the starting line, and ran toward the starting line at the speed of 2 meters every 9 seconds. After how many seconds will Rita be 23 meters away from the starting line?

Rita will be 23 meters away from the starting line [ ] seconds since she started running.

73. Wendy had only $6.00 in her piggy bank, and she decided to start saving more. She saves $3.00 every 7 days. After how many days will she have $15.00 in the piggy bank?

Wendy will save $15.00 in her piggy bank after [ ] days.

74. Marc has saved $40.00 in his piggy bank, and he decided to start spending them. He spends $2.00 every 9 days. After how many days will he have $34.00 left in the piggy bank?

Marc will have $34.00 left in his piggy bank after [ ] days.

Solving Inequalities with Fractions Exercises

For the following exercises: Solve this inequality.

75. $\frac{x}{5} + 56 \geq 3x$

In set-builder notation, the solution set is [ ].

In interval notation, the solution set is [ ].

76. $\frac{x}{6} + 58 \geq 5x$

In set-builder notation, the solution set is [ ].

In interval notation, the solution set is [ ].

77. $\frac{5}{4} - 3y < 4$

In set-builder notation, the solution set is [ ].

In interval notation, the solution set is [ ].

78. $\frac{5}{6} - 2y < 6$

In set-builder notation, the solution set is [ ].

In interval notation, the solution set is [ ].

79. $-\frac{5}{2}t > \frac{2}{7}t - 117$

In set-builder notation, the solution set is [ ].

In interval notation, the solution set is [ ].

80. $-\frac{3}{4}t > \frac{6}{7}t - 135$

In set-builder notation, the solution set is [ ].

In interval notation, the solution set is [ ].

81. $\frac{5}{2} \geq \frac{x}{12}$

In set-builder notation, the solution set is [ ].

In interval notation, the solution set is [ ].

82. $\frac{9}{2} \geq \frac{x}{6}$

In set-builder notation, the solution set is [ ].

In interval notation, the solution set is [ ].

83. $-\frac{z}{20} < -\frac{7}{4}$

In set-builder notation, the solution set is [ ].

In interval notation, the solution set is [ ].

84. $-\frac{z}{8} < -\frac{5}{4}$

In set-builder notation, the solution set is [ ].

In interval notation, the solution set is [ ].

85. $\frac{x}{10} - 2 \leq \frac{x}{5}$

In set-builder notation, the solution set is [ ].

In interval notation, the solution set is [ ].

86. $\frac{x}{9} - 4 \leq \frac{x}{5}$

In set-builder notation, the solution set is [ ].

In interval notation, the solution set is [ ].

87. $\frac{y-3}{6} \geq \frac{y+5}{4}$

In set-builder notation, the solution set is ______.

In interval notation, the solution set is ______.

88. $\frac{y-10}{8} \geq \frac{y+9}{6}$

In set-builder notation, the solution set is ______.

In interval notation, the solution set is ______.

89. $\frac{7}{12} < \frac{x+6}{6} - \frac{x-3}{12}$

In set-builder notation, the solution set is ______.

In interval notation, the solution set is ______.

90. $\frac{5}{3} < \frac{x+3}{6} - \frac{x-7}{12}$

In set-builder notation, the solution set is ______.

In interval notation, the solution set is ______.

Application Problems

91. Your grade in a class is determined by the average of three test scores. You scored 70 and 90 on the first two tests. To earn at least 77 for this course, how much do you have to score on the third test?

Let x be the score you will earn on the third test. Write an inequality to represent this situation.

Solve this inequality. What is the minimum that you have to earn on the third test in order to earn a 77 for the course?

You cannot score over 100 on the third test. Use interval notation to represent the range of scores you can earn on the third test in order to earn at least 77 for this course.

92. Your grade in a class is determined by the average of three test scores. You scored 71 and 88 on the first two tests. To earn at least 80 for this course, how much do you have to score on the third test?

Let x be the score you will earn on the third test. Write an inequality to represent this situation.

Solve this inequality. What is the minimum that you have to earn on the third test in order to earn a 80 for the course?

You cannot score over 100 on the third test. Use interval notation to represent the range of scores you can earn on the third test in order to earn at least 80 for this course.

3.3 Isolating a Linear Variable

In this section, we will learn how to solve linear equations and inequalities with more than one variables.

3.3.1 Solving for a Variable

The formula of calculating a rectangle's area is $A = lw$, where l stands for the rectangle's length, and w stands for width. When a rectangle's length and width are given, we can easily calculate its area.

What if a rectangle's area and length are given, and we need to calculate its width?

If a rectangle's area is given as $12\,\text{m}^2$, and it's length is given as 4 m, we could find its width this way:

$$\begin{aligned} A &= lw \\ 12 &= 4w \\ \frac{12}{4} &= \frac{4w}{4} \\ 3 &= w \\ w &= 3 \end{aligned}$$

If we need to do this many times, we would love to have an easier way, without solving an equation each time. We will solve for w in the formula $A = lw$:

$$\begin{aligned} A &= lw \\ \frac{A}{l} &= \frac{lw}{l} \\ \frac{A}{l} &= w \\ w &= \frac{A}{l} \end{aligned}$$

Now if we want to find the width when $l = 4$ is given, we can simply replace l with 4 and simplify.

We solved for w in the formula $A = lw$ once, and we could use the new formula $w = \frac{A}{l}$ again and again saving us a lot of time down the road. Let's look at a few examples.

Remark 3.3.2. Note that in solving for A, we divided each side of the equation by l. The operations that we apply, and the order in which we do them, are determined by the operations in the original equation. In the original equation $A = lw$, we saw that w was *multiplied* by l, and so we knew that in order to "undo" that operation, we would need to *divide* each side by l. We will see this process of "un-doing" the operations throughout this section.

Example 3.3.3. Solve for R in $P = R - C$. (This is the relationship between profit, revenue, and cost.)

To solve for R, we first want to note that C is *subtracted* from R. To "undo" this, we will need to *add* C to each side of the equation:

$$P = \overset{\downarrow}{R} - C$$

$$P + C = \overset{\downarrow}{R} - C + C$$

$$P + C = \overset{\downarrow}{R}$$

$$R = P + C$$

Example 3.3.4. Solve for x in $y = mx + b$. (This is a line's equation in slope-intercept form.)

In the equation $y = mx + b$, we see that x is multiplied by m and then b is added to that. Our first step will be to isolate mx, which we'll do by subtracting b from each side of the equation:

$$y = m\overset{\downarrow}{x} + b$$

$$y - b = m\overset{\downarrow}{x} + b - b$$

$$y - b = m\overset{\downarrow}{x}$$

Now that we have mx on it's own, we'll note that x is multiplied by m. To "undo" this, we'll need to divide each side of the equation by m:

$$\frac{y - b}{m} = \frac{m\overset{\downarrow}{x}}{m}$$

$$\frac{y - b}{m} = \overset{\downarrow}{x}$$

$$x = \frac{y - b}{m}$$

Warning 3.3.5. It's important to note in Example 3.3.4 that each *side* was divided by m. We can't simply divide y by m, as the equation would no longer be equivalent.

Example 3.3.6. Solve for b in $A = \frac{1}{2}bh$. (This is the area formula for a triangle.)

To solve for b, we need to determine what operations need to be "undone." The expression $\frac{1}{2}bh$ has multiplication between $\frac{1}{2}$ and b and h. As a first step, we will multiply each side of the equation by 2 in order to eliminate the denominator of 2:

$$A = \frac{1}{2}\overset{\downarrow}{b}h$$

$$2 \cdot A = 2 \cdot \frac{1}{2}\overset{\downarrow}{b}h$$

$$2A = \overset{\downarrow}{b}h$$

As a last step, we will "undo" the multiplication between b and h by dividing each side by h:

$$\begin{aligned}\frac{2A}{h} &= \frac{\overset{\downarrow}{b}h}{h}\\ \frac{2A}{h} &= \overset{\downarrow}{b}\\ b &= \frac{2A}{h}\end{aligned}$$

Example 3.3.7. Solve for y in $2x + 5y = 10$. (This is a linear equation in standard form.)

To solve for y in the equation $2x + 5y = 10$, we will first have to solve for $5y$. We'll do so by subtracting $2x$ from each side of the equation. After that, we'll be able to divide each side by 5 to finish solving for y:

$$\begin{aligned}2x + 5\overset{\downarrow}{y} &= 10\\ 2x + 5\overset{\downarrow}{y} - 2x &= 10 - 2x\\ 5\overset{\downarrow}{y} &= 10 - 2x\\ \frac{5\overset{\downarrow}{y}}{5} &= \frac{10 - 2x}{5}\\ y &= \frac{10 - 2x}{5}\end{aligned}$$

Remark 3.3.8. As we will learn in later sections, the result in Example 3.3.7 can also be written as $y = \frac{10}{5} - \frac{2x}{5}$ which can then be written as $y = 2 - \frac{2}{5}x$.

Example 3.3.9. Solve for F in $C = \frac{5}{9}(F - 32)$. (This represents the relationship between temperature in degrees Celsius and degrees Fahrenheit.)

To solve for F, we first need to see that it is contained inside a set of parentheses. To get the expression $F - 32$ on its own, we'll need to eliminate the $\frac{5}{9}$ outside those parentheses. One way we can "undo" this multiplication is by dividing each side by $\frac{5}{9}$. As we learned in Section 3.2 though, a better approach is to instead multiply each side by the reciprocal of $\frac{9}{5}$:

$$\begin{aligned}C &= \frac{5}{9}(\overset{\downarrow}{F} - 32)\\ \frac{9}{5}\cdot C &= \frac{9}{5}\cdot\frac{5}{9}(\overset{\downarrow}{F} - 32)\\ \frac{9}{5}C &= \overset{\downarrow}{F} - 32\end{aligned}$$

Now that we have $F - 32$, we simply need to add 32 to each side to finish solving for F:

$$\frac{9}{5}C + 32 = \overset{\downarrow}{F} - 32 + 32$$

$$\frac{9}{5}C + 32 = \overset{\downarrow}{F}$$

$$F = \frac{9}{5}C + 32$$

3.3.2 Exercises

Solving for a Variable

1.

a. Solve this linear equation for y.

$y + 1 = 7$

b. Solve this linear equation for x.

$x + q = n$

2.

a. Solve this linear equation for y.

$y + 7 = 15$

b. Solve this linear equation for t.

$t + b = m$

3.

a. Solve this linear equation for r.

$r - 5 = 3$

b. Solve this linear equation for y.

$y - m = 3$

4.

a. Solve this linear equation for r.

$r - 7 = 1$

b. Solve this linear equation for x.

$x - m = 1$

5.

a. Solve this linear equation for t.

$-t + 1 = -9$

b. Solve this linear equation for r.

$-r + p = m$

6.

a. Solve this linear equation for t.

$-t + 7 = 1$

b. Solve this linear equation for y.

$-y + x = B$

7.

a. Solve this linear equation for t.

$7t = 14$

b. Solve this linear equation for x.

$cx = b$

8.

a. Solve this linear equation for x.

$7x = 42$

b. Solve this linear equation for r.

$pr = y$

9.

a. Solve this linear equation for x.

$\frac{x}{7} = 2$

b. Solve this linear equation for y.

$\frac{y}{t} = n$

10.

a. Solve this linear equation for y.

$\frac{y}{7} = 6$

b. Solve this linear equation for x.

$\frac{x}{C} = A$

11.

a. Solve this linear equation for y.

$5y + 8 = 48$

b. Solve this linear equation for r.

$tr + C = a$

12.

a. Solve this linear equation for r.

$7r + 9 = 44$

b. Solve this linear equation for y.

$Ay + q = C$

13.

a. Solve this linear equation for r.

$rt = q$

b. Solve this linear equation for t.

$rt = q$

14.

a. Solve this linear equation for t.

$tr = b$

b. Solve this linear equation for r.

$tr = b$

15.

a. Solve this linear equation for t.

$t + y = m$

b. Solve this linear equation for y.

$t + y = m$

16.

a. Solve this linear equation for t.

$t + x = p$

b. Solve this linear equation for x.

$t + x = p$

17.

a. Solve this linear equation for y.

$Br + y = x$

b. Solve this linear equation for B.

$Br + y = x$

18.

a. Solve this linear equation for r.

$tx + r = a$

b. Solve this linear equation for t.

$tx + r = a$

19.

a. Solve this linear equation for a.

$y = pa + x$

b. Solve this linear equation for p.

$y = pa + x$

20.

a. Solve this linear equation for c.

$t = Ac + m$

b. Solve this linear equation for A.

$t = Ac + m$

21. Solve this linear equation for x:

$$y = mx - b$$

22. Solve this linear equation for x:

$$y = -mx + b$$

23.

a. Solve this equation for b:

$28 = \frac{1}{2}b \cdot 8$

b. Solve this equation for b:

$A = \frac{1}{2}b \cdot h$

24.

a. Solve this equation for b:

$20 = \frac{1}{2}b \cdot 10$

b. Solve this equation for b:

$A = \frac{1}{2}b \cdot h$

25. Solve this linear equation for r:

$$C = 2\pi r$$

26. Solve this linear equation for h:

$$V = \pi r^2 h$$

27. Solve these linear equations for x.

a. $\frac{x}{2} + 5 = 9$

b. $\frac{x}{t} + 5 = n$

28. Solve these linear equations for y.

a. $\frac{y}{3} + 6 = 9$

b. $\frac{y}{t} + 6 = r$

29. Solve this linear equation for y:

$$\frac{y}{r} + t = b$$

30. Solve this linear equation for r:

$$\frac{r}{x} + q = b$$

31. Solve this linear equation for r:

$$\frac{r}{5} + y = C$$

32. Solve this linear equation for t:

$$\frac{t}{5} + y = p$$

33. Solve this linear equation for p:

$$r = a - \frac{5p}{x}$$

34. Solve this linear equation for q:

$$b = p - \frac{6q}{C}$$

35. Solve this linear equation for x:

$$Ax + By = C$$

Note that the variables are upper case A, B, and C and lower case x and y.

36. Solve this linear equation for y:

$$Ax + By = C$$

Note that the variables are upper case A, B, and C and lower case x and y.

3.4 Ratios and Proportions

3.4.1 Introduction

A **ratio** is a means of comparing two things using division. One common example is a unit price. For example, if a box of cereal costs \$3.99 and weighs 21 ounces then we can write this ratio as:

$$\frac{\$3.99}{21\,\text{oz}}$$

If we want to know the unit price (that is, how much each individual ounce costs), then we can divide \$3.99 by 21 ounces and obtain \$0.19 per ounce. These two ratios are equivalent, and the equation showing that they are equal is a **proportion**. In this case, we could write the following proportion:

$$\frac{\$3.99}{21\,\text{oz}} = \frac{\$0.19}{1\,\text{oz}}$$

In this section, we will extend this concept and write proportions where one quantity is unknown and solve for that unknown.

Remark 3.4.2. Sometimes ratios are stated using a colon instead of a fraction. For example, the ratio $\frac{2}{1}$ can be written as $2 : 1$.

Example 3.4.3. Let's now assume that we wanted to know what a proportional cost would be for a box of cereal that weighs 18 ounces. Letting C be this unknown cost (in dollars), we could set up the following proportion:

$$\frac{\text{cost in dollars}}{\text{weight in oz}} = \frac{\text{cost in dollars}}{\text{weight in oz}}$$
$$\frac{\$3.99}{21\,\text{oz}} = \frac{\$C}{18\,\text{oz}}$$

To solve this proportion, we will first note that it will be easier to solve without units:

$$\frac{3.99}{21} = \frac{C}{18}$$

Next we want to recognize that each side contains a fraction. Our standard approach for solving this type of equation is to multiply each side by the least common denominator (LCD). In this case, the LCD of 21 and 18 is 126. As with many other proportions we solve, it is often easier to just multiply each side by the common denominator of $18 \cdot 21$, which we know will make each denominator cancel:

$$\frac{3.99}{21} = \frac{C}{18}$$
$$18 \cdot 21 \cdot \frac{3.99}{21} = \frac{C}{18} \cdot 18 \cdot 21$$

$$\begin{aligned}18 \cdot \cancel{21}\frac{3.99}{\cancel{21}} &= \frac{C}{\cancel{18}} \cdot \cancel{18} \cdot 21 \\ 71.82 &= 21C \\ \frac{71.82}{21} &= \frac{21C}{21} \\ C &= 3.42\end{aligned}$$

Thus a proportional price for an 18 ounce box of cereal would be \$3.42.

3.4.2 Solving Proportions

Solving proportions uses the process of clearing denominators that we covered in Section 3.2. Because a proportion is exactly one fraction equal to another, we can simplify the process of clearing the denominators simply by multiplying both sides of the equation by both denominators. In other words, we don't specifically need the LCD to clear the denominators.

Example 3.4.4. Solve $\frac{x}{8} = \frac{15}{12}$ for x.

Instead of finding the LCD of the two fractions, we'll simply multiply both sides of the equation by 8 and by 12. This will still have the effect of canceling the denominators on both sides of the equation.

$$\begin{aligned}\frac{x}{8} &= \frac{15}{12} \\ 12 \cdot 8 \cdot \frac{x}{8} &= \frac{15}{12} \cdot 12 \cdot 8 \\ 12 \cdot \cancel{8} \cdot \frac{x}{\cancel{8}} &= \frac{15}{\cancel{12}} \cdot \cancel{12} \cdot 8 \\ 12 \cdot x &= 15 \cdot 8 \\ 12x &= 120 \\ \frac{12x}{12} &= \frac{120}{12} \\ x &= 10\end{aligned}$$

Our work indicates 10 is the solution. We can check this as we would for any equation, by substituting 10 for x and verifying we obtain a true statement:

$$\begin{aligned}\frac{10}{8} &\stackrel{?}{=} \frac{15}{12} \\ \frac{5}{4} &\stackrel{\checkmark}{=} \frac{5}{4}\end{aligned}$$

Since both fractions reduce to $\frac{5}{4}$, we know the solution to the equation $\frac{x}{8} = \frac{15}{12}$ is 10 and the solution set is $\{10\}$.

When solving proportions, we can use the name **cross-multiplication** to describe the process of what just occurred. Say we have a proportion

$$\frac{a}{b} = \frac{c}{d}$$

To remove fractions, we multiply both sides with the common denominator, bd, and we have:

$$\begin{aligned}\frac{a}{b} &= \frac{c}{d}\\ bd \cdot \frac{a}{b} &= \frac{c}{d} \cdot bd\\ \cancel{b}d \cdot \frac{a}{\cancel{b}} &= \frac{c}{\cancel{d}} \cdot b\cancel{d}\\ ad &= bc\end{aligned}$$

Since a and d are diagonally across the equals sign from each other in $\frac{a}{b} = \frac{c}{d}$, as are b and c, we call this approach **cross-multiplication.**

$$\text{If } \frac{a}{b} = \frac{c}{d}, \text{ then } ad = bc.$$

If we understand cross-multiplication, we are able to rewrite a proportion $\frac{a}{b} = \frac{c}{d}$ in an equivalent form that does not have any fractions, $ad = bc$, as our first step of work. If we had used this skill in Example 3.4.4, we would have had:

$$\begin{aligned}\frac{x}{8} &= \frac{15}{12}\\ 12 \cdot x &= 15 \cdot 8\\ 12x &= 120\end{aligned}$$

Notice this is the same equation we had in the fifth line of our work in solving Example 3.4.4, but we obtained it without having to contemplate what we need to multiply by to clear the fractions.

We are able to use cross-multiplication when solving proportions, but it is extremely important to note that cross-multiplication only works when we are solving a proportion, an equation that has one ratio or fraction equal to another ratio or fraction. If an equation has anything more than one ratio or fraction on a single side of an equation, we cannot use cross-multiplication. For example, we cannot use cross-multiplication to solve $\frac{3}{4}x - \frac{2}{5} = \frac{9}{4}$, unless we first manipulate the equation to have exactly one fraction and nothing else on each side of the equation.

It is also important to be aware of the fact that cross-multiplication is a special version of our general process of clearing fractions: multiplying both sides of an equation by a common denominator of all the fractions in an equation.

Example 3.4.5. Solve $\frac{t}{5} = \frac{t+2}{3}$ for t.

Solution. Again this equation is a proportion, so we are able to multiply both sides of the equation by both denominators to clear the fractions:

$$\begin{aligned}\frac{t}{5} &= \frac{t+2}{3}\\ 5 \cdot 3 \cdot \frac{t}{5} &= \frac{t+2}{3} \cdot 5 \cdot 3\end{aligned}$$

$$\cancel{5}\cdot 3\cdot\frac{t}{\cancel{5}}=\frac{t+2}{\cancel{3}}\cdot 5\cdot\cancel{3}$$
$$3\cdot t=5\cdot(t+2)$$

It is critical that we include the parentheses around $t+2$, so that we are multiplying 5 against the entire numerator.

$$\begin{aligned}3t&=5(t+2)\\3t&=5t+10\\3t-5t&=5x+10-5t\\-2t&=10\\\frac{-2t}{-2}&=\frac{10}{-2}\\t&=-5\end{aligned}$$

We should check that this value -5 is actually the solution of the equation:

$$\begin{aligned}\frac{-5}{5}&\stackrel{?}{=}\frac{-5+2}{3}\\-1&\stackrel{?}{=}\frac{-3}{3}\\-1&\stackrel{\checkmark}{=}-1\end{aligned}$$

Since we have verified that -5 is the solution for $\frac{t}{5}=\frac{t+2}{3}$, we know that the solution set is $\{-5\}$.

Example 3.4.6. Solve $\frac{r+7}{8}=-\frac{9}{4}$ for r.

Solution. This proportion is a bit different in the fact that one fraction is negative. The key to working with a negative fraction is to attach the negative sign to either the numerator or denominator, but not both:

$$\frac{-9}{4}=-\frac{9}{4}\quad\text{and}\quad\frac{9}{-4}=-\frac{9}{4},\quad\text{but}\quad\frac{-9}{-4}=+\frac{9}{4}$$

Since we're trying to eliminate the fractions, it will likely make the work a bit easier to attach the negative to the numerator.

We'll work with the equation in the form $\frac{r+7}{8}=\frac{-9}{4}$

$$\begin{aligned}\frac{r+7}{8}&=\frac{-9}{4}\\8\cdot 4\cdot\frac{r+7}{8}&=\frac{-9}{4}\cdot 8\cdot 4\\\cancel{8}\cdot 4\cdot\frac{r+7}{\cancel{8}}&=\frac{-9}{\cancel{4}}\cdot 8\cdot\cancel{4}\\4\cdot(r+7)&=8\cdot(-9)\\4r+28&=-72\\4r+28-28&=-72-28\\4r&=-100\end{aligned}$$

$$\frac{4r}{4} = \frac{-100}{4}$$
$$r = -25$$

We should check that this value -25 is actually the solution of the equation:

$$\frac{-25+7}{8} \stackrel{?}{=} -\frac{9}{4}$$
$$\frac{-18}{8} \stackrel{?}{=} -\frac{9}{4}$$
$$-\frac{9}{4} \stackrel{\checkmark}{=} -\frac{9}{4}$$

Since we have verified that -25 is the solution for $\frac{r+7}{8} = -\frac{9}{4}$, we know that the solution set is $\{-25\}$.

Example 3.4.7. Solve $\frac{x}{15} = \frac{40}{25}$ for x.

Solution. To solve this proportion, begin by multiplying both sides by both denominators.

$$\frac{x}{15} = \frac{40}{25}$$
$$15 \cdot 25 \cdot \frac{x}{15} = \frac{40}{25} \cdot 15 \cdot 25$$
$$\cancel{15} \cdot 25 \cdot \frac{x}{\cancel{15}} = \frac{40}{\cancel{25}} \cdot 15 \cdot \cancel{25}$$
$$25 \cdot x = 40 \cdot 15$$
$$25x = 600$$
$$\frac{25x}{25} = \frac{600}{25}$$
$$x = 24$$

You can easily verify that this value 24 is actually the solution of the equation:

$$\frac{24}{15} \stackrel{?}{=} \frac{40}{25}$$
$$\frac{8}{5} \stackrel{\checkmark}{=} \frac{8}{5}$$

Since we have verified that 24 is the solution for $\frac{x}{15} = \frac{40}{25}$, we know that the solution set is $\{24\}$.

Example 3.4.8. Solve $\frac{x-4}{6} = \frac{x+3}{4}$ for x.

Solution. To solve this proportion, begin by multiplying both sides by both denominators.

$$\begin{aligned}
\frac{x-4}{6} &= \frac{x+3}{4} \\
6 \cdot 4 \cdot \frac{x-4}{6} &= \frac{x+3}{4} \cdot 6 \cdot 4 \\
\cancel{6} \cdot 4 \cdot \frac{x-4}{\cancel{6}} &= \frac{x+3}{\cancel{4}} \cdot 6 \cdot \cancel{4} \\
4 \cdot (x-4) &= (x+3) \cdot 6 \\
4x - 16 &= 6x + 18 \\
4x - 16 + 16 &= 6x + 18 + 16 \\
4x &= 6x + 34 \\
4x - 6x &= 6x + 34 - 6x \\
-2x &= 34 \\
\frac{-2x}{-2} &= \frac{34}{-2} \\
x &= -17
\end{aligned}$$

We can check that this value is correct by substituting it back into the original equation:

$$\begin{aligned}
\frac{x-4}{6} &= \frac{x+3}{4} \\
\frac{-17-4}{6} &\stackrel{?}{=} \frac{-17+3}{4} \\
\frac{-21}{6} &\stackrel{?}{=} \frac{-14}{4} \\
\frac{-7}{2} &\stackrel{\checkmark}{=} \frac{-7}{2}
\end{aligned}$$

Since we have verified that -17 is the solution for $\frac{x-4}{6} = \frac{x+3}{4}$, we know that the solution set is $\{-17\}$.

3.4.3 Proportionality in Similar Triangles

One really useful example of ratios and proportions involves similar triangles. Two triangles are considered **similar** if they have the same angles and their side lengths are proportional, as shown in Figure 3.4.9:

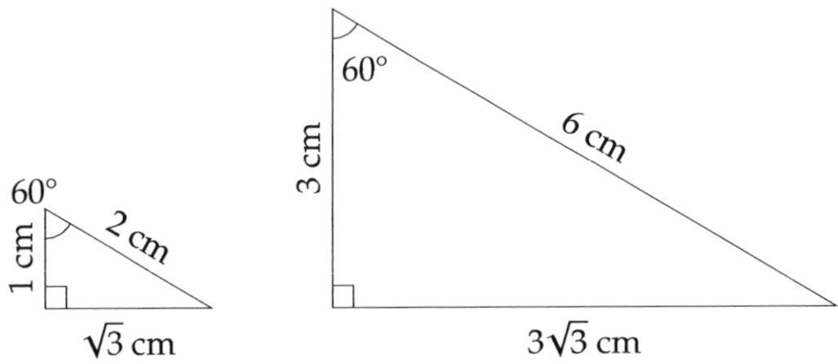

Figure 3.4.9: Similar Triangles

In the first triangle in Figure 3.4.9, the ratio of the left side length to the hypotenuse length is $\frac{1\,\text{cm}}{2\,\text{cm}}$; in the second triangle, the ratio of the left side length to the hypotenuse length is $\frac{3\,\text{cm}}{6\,\text{cm}}$. Since both reduce to $\frac{1}{2}$, we can write the following proportion:

$$\frac{1\,\text{cm}}{2\,\text{cm}} = \frac{3\,\text{cm}}{6\,\text{cm}}$$

If we extend this concept, we can use it to solve for an unknown side length. Consider the two similar triangles in the next example.

Example 3.4.10.

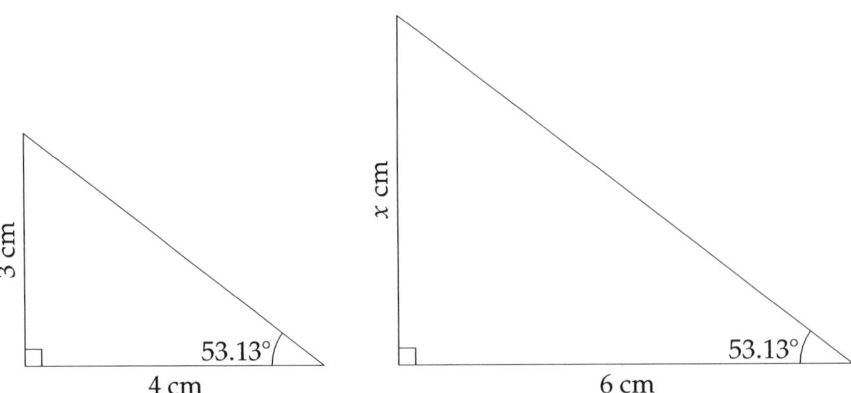

Figure 3.4.11: Similar Triangles

Since the two triangles are similar, we know that their side length should be proportional. To determine the unknown length, we can set up a proportion and solve for x:

$$\frac{\text{bigger triangle's left side length in cm}}{\text{bigger triangle's bottom side length in cm}} = \frac{\text{smaller triangle's left side length in cm}}{\text{smaller triangle's bottom side length in cm}}$$

$$\begin{aligned}
\frac{x\,\text{cm}}{6\,\text{cm}} &= \frac{3\,\text{cm}}{4\,\text{cm}} \\
\frac{x}{6} &= \frac{3}{4} \\
12 \cdot \frac{x}{6} &= 12 \cdot \frac{3}{4} && \text{(12 is the least common denominator)} \\
2x &= 9 \\
\frac{2x}{2} &= \frac{9}{2} \\
x &= \frac{9}{2} \text{ or } 4.5
\end{aligned}$$

The unknown side length is then 4.5 cm.

Remark 3.4.12. Looking at the triangles in Figure 3.4.9, you may notice that there are many different proportions you could set up, such as:

$$\begin{aligned}
\frac{2\,\text{cm}}{1\,\text{cm}} &= \frac{6\,\text{cm}}{3\,\text{cm}} \\
\frac{2\,\text{cm}}{6\,\text{cm}} &= \frac{1\,\text{cm}}{3\,\text{cm}} \\
\frac{6\,\text{cm}}{2\,\text{cm}} &= \frac{3\,\text{cm}}{1\,\text{cm}} \\
\frac{3\sqrt{3}\,\text{cm}}{\sqrt{3}\,\text{cm}} &= \frac{3\,\text{cm}}{1\,\text{cm}}
\end{aligned}$$

This is often the case when we set up ratios and proportions.

If we take a second look at Figure 3.4.11, there are also several other proportions we could have used to find the value of x.

$$\frac{\text{bigger triangle's left side length}}{\text{smaller triangle's left side length}} = \frac{\text{bigger triangle's bottom side length}}{\text{smaller triangle's bottom side length}}$$

$$\frac{\text{smaller triangle's bottom side length}}{\text{bigger triangle's bottom side length}} = \frac{\text{smaller triangle's left side length}}{\text{bigger triangle's left side length}}$$

$$\frac{\text{bigger triangle's bottom side length}}{\text{smaller triangle's bottom side length}} = \frac{\text{bigger triangle's left side length}}{\text{smaller triangle's left side length}}$$

Written as algebraic proportions, these three equations would, respectively, be

$$\frac{x\,\text{cm}}{3\,\text{cm}} = \frac{6\,\text{cm}}{4\,\text{cm}}, \qquad \frac{4\,\text{cm}}{6\,\text{cm}} = \frac{3\,\text{cm}}{x\,\text{cm}}, \qquad \frac{6\,\text{cm}}{4\,\text{cm}} = \frac{x\,\text{cm}}{3\,\text{cm}}$$

While these are only a few of the possibilities, if we clear the denominators from any properly designed proportion, every one is equivalent to $x = 4.5$.

3.4.4 Creating and Solving Proportions

Proportions can be used to solve many real-life applications. The key to using proportions to solve such applications is to first set up a ratio where all values are known. We then set up a second ratio that will be proportional to the first, but has one value in the ratio unknown. Let's look at a few examples.

Example 3.4.13. Property taxes for a residential property are proportional to the assessed value of the property. Assume that a certain property in a given neighborhood is assessed at \$234,100 and its annual property taxes are \$2,518.92. What are the annual property taxes for a house that is assessed at \$287,500?

Solution. Let T be the annual property taxes (in dollars) for a property assessed at \$287,500. We can write and solve this proportion:

$$\begin{aligned}\frac{\text{tax}}{\text{property value}} &= \frac{\text{tax}}{\text{property value}}\\ \frac{2518.92}{234100} &= \frac{T}{287500}\end{aligned}$$

The least common denominator of this proportion is rather large, so we will instead multiply each side by 234100 and 287500 and simplify from there:

$$\begin{aligned}\frac{2518.92}{234100} &= \frac{T}{287500}\\ 234100 \cdot 287500 \cdot \frac{2518.92}{234100} &= \frac{T}{287500} \cdot 234100 \cdot 287500\\ 287500 \cdot 2518.92 &= T \cdot 234100\\ \frac{287500 \cdot 2518.92}{234100} &= \frac{234100T}{234100}\\ T &\approx 3093.50\end{aligned}$$

The property taxes for a property assessed at $\$287,500$ are $\$3,093.50$.

Example 3.4.14. Tagging fish is a means of estimating the size of the population of fish in a body of water (such as a lake). A sample of fish is taken, tagged, and then redistributed into the lake. When another sample is taken, the proportion of fish that are tagged out of that sample are assumed to be proportional to the total number of fish tagged out of the entire population of fish in the lake.

$$\frac{\text{number of tagged fish in sample}}{\text{number of fish in sample}} = \frac{\text{number of tagged fish total}}{\text{number of fish total}}$$

Assume that 90 fish are caught and tagged. Once they are redistributed, a sample of 200 fish is taken. Of these, 7 are tagged. Estimate how many fish total are in the lake.

Solution. Let n be the number of fish in the lake. We can set up this proportion to represent the scenario:

$$\frac{7}{200} = \frac{90}{n}$$

To solve for n, which is in a denominator, we'll need to multiply each side by both 200 and n:

$$\begin{aligned}
\frac{7}{200} &= \frac{90}{n} \\
200 \cdot n \cdot \frac{7}{200} &= \frac{90}{n} \cdot 200 \cdot n \\
\cancel{200} \cdot n \cdot \frac{7}{\cancel{200}} &= \frac{90}{\cancel{n}} \cdot 200 \cdot \cancel{n} \\
7n &= 1800 \\
\frac{7n}{7} &= \frac{1800}{7} \\
n &\approx 2471.4286
\end{aligned}$$

According to this sample, we can estimate that there are about 2,471 fish in the lake.

Example 3.4.15. Infant Tylenol contains 160 mg of acetaminophen in each 5 mL of liquid medicine. If an infant is prescribed 60 mg of acetaminophen, how many milliliters of liquid medicine should he/she receive?

Solution. Assume the infant should receive q milliliters of liquid medicine, and we can set up the following proportion:

$$\begin{aligned}
\frac{\text{amount of liquid medicine in mL}}{\text{amount of acetaminophen in mg}} &= \frac{\text{amount of liquid medicine in mL}}{\text{amount of acetaminophen in mg}} \\
\frac{5\,\text{mL}}{160\,\text{mg}} &= \frac{q\,\text{mL}}{60\,\text{mg}} \\
\frac{5}{160} &= \frac{q}{60} \\
160 \cdot 60 \cdot \frac{5}{160} &= \frac{q}{60} \cdot 160 \cdot 60 \\
60 \cdot 5 &= q \cdot 160 \\
300 &= 160q \\
\frac{300}{160} &= \frac{160q}{160} \\
q &= 1.875
\end{aligned}$$

So to receive 60 mg of acetaminophen, an infant should receive 1.875 mL of liquid medicine.

Example 3.4.16. An architect is making a scale model of a building. The actual building will be 30 ft tall. In the model, the height of the building will be 2 in. How tall should a person who is 5 ft 6 in be made in the model so that the model is to scale?

Solution. Let h be the height of the person in the model, which we'll measure in inches. We'll create a proportion that compares the building and person's heights in the model to their heights in real life:

$$\frac{\text{height of model building in inches}}{\text{height of actual building in feet}} = \frac{\text{height of model person in inches}}{\text{height of actual person in feet}}$$
$$\frac{2\,\text{in}}{30\,\text{ft}} = \frac{h\,\text{in}}{5\,\text{ft}\,6\text{in}}$$

Before we can just eliminate the units, we'll need to convert 5 ft 6 in to feet:

$$\frac{2\,\text{in}}{30\,\text{ft}} = \frac{h\,\text{in}}{5.5\,\text{ft}}$$

Now we can remove the units and continue solving:

$$\begin{aligned}
\frac{2}{30} &= \frac{h}{5.5}\\
30 \cdot 5.5 \cdot \frac{2}{30} &= \frac{h}{5.5} \cdot 30 \cdot 5.5\\
5.5 \cdot 2 &= h \cdot 30\\
11 &= 30h\\
\frac{11}{30} &= \frac{30h}{30}\\
\frac{11}{30} &= h\\
h &\approx 0.3667
\end{aligned}$$

A person who is 5 ft 6 in tall should be $\frac{11}{30}$ inches (about 0.3667 inches) tall in the model.

3.4.5 Exercises

Solving Proportions

1. Solve $\frac{x}{28} = \frac{18}{63}$ for x.

The solution is ______.

2. Solve $\frac{x}{14} = \frac{24}{42}$ for x.

The solution is ______.

3. Solve $\frac{18}{x} = \frac{63}{28}$ for x.

The solution is ______.

4. Solve $\frac{24}{x} = \frac{20}{10}$ for x.

The solution is ______.

5. Solve $\frac{x}{4} = \frac{x-20}{9}$ for x.

The solution is ______.

6. Solve $\frac{x}{11} = \frac{x+2}{12}$ for x.

The solution is ______.

7. Solve $\frac{x}{7} = \frac{x+8}{9}$ for x.

The solution is ______.

8. Solve $\frac{x}{5} = \frac{x-12}{7}$ for x.

The solution is ______.

9. Solve $\frac{x+2}{2} = \frac{x-4}{6}$ for x.

The solution is ______.

10. Solve $\frac{x-9}{3} = \frac{x+11}{9}$ for x.

The solution is ______.

11. Solve $\frac{x-3}{5} = \frac{x-6}{8}$ for x.

The solution is ______.

12. Solve $\frac{x-10}{9} = \frac{x-14}{15}$ for x.

The solution is ______.

13. Solve $\frac{x}{21} = -\frac{72}{56}$ for x.

The solution is ______.

14. Solve $\frac{x}{18} = -\frac{10}{6}$ for x.

The solution is ______.

15. Solve $\frac{x+2}{48} = -\frac{14}{12}$ for x.

The solution is ______.

16. Solve $\frac{x-2}{27} = -\frac{40}{24}$ for x.

The solution is ______.

Applications

17. The following two triangles are similar to each other. Find the length of the missing side.

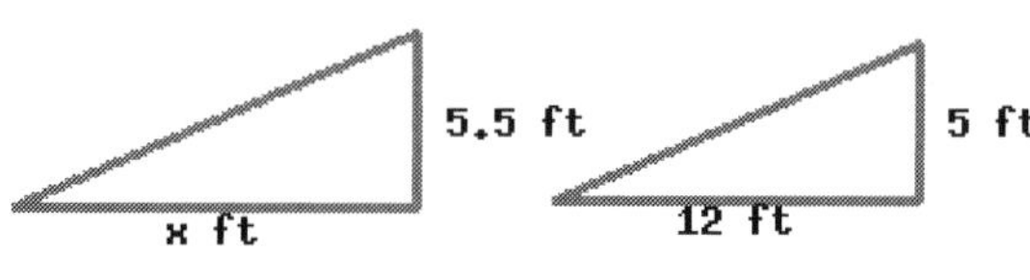

The missing side's length is ______

18. The following two triangles are similar to each other. Find the length of the missing side.

The missing side's length is [ ]

19. The following two triangles are similar to each other. Find the length of the missing side.

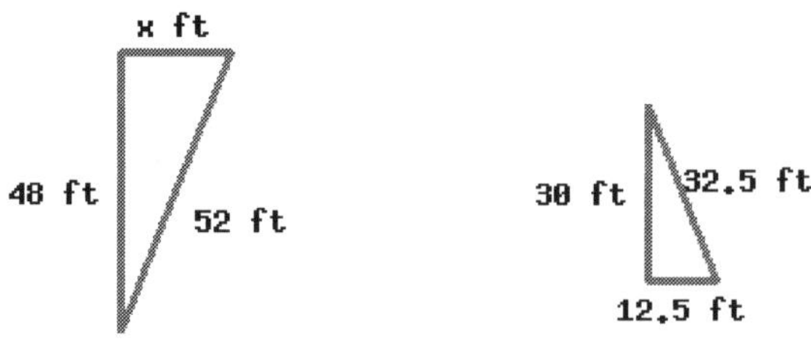

The missing side's length is [ ]

20. The following two triangles are similar to each other. Find the length of the missing side.

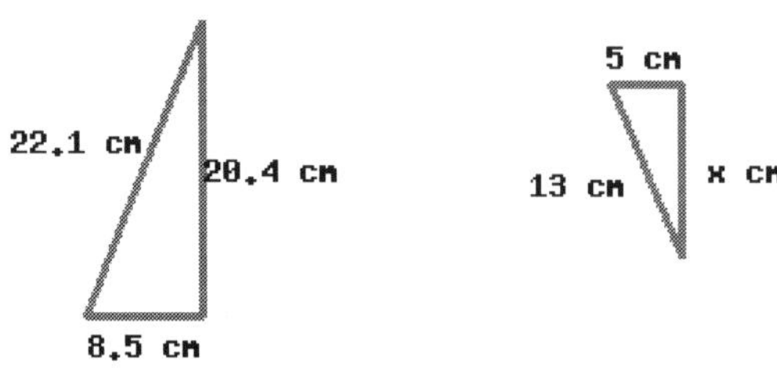

The missing side's length is [ ]

21. According to a salad recipe, each serving requires 4 teaspoons of vegetable oil and 20 teaspoons of vinegar. If 14 teaspoons of vegetable oil were used, how many teaspoons of vinegar should be used?

If 14 teaspoons of vegetable oil were used, [ ] teaspoons of vinegar should be used.

22. According to a salad recipe, each serving requires 3 teaspoons of vegetable oil and 18 teaspoons of vinegar. If 108 teaspoons of vinegar were used, how many teaspoons of vegetable oil should be used?

If 108 teaspoons of vinegar were used, ______ teaspoons of vegetable oil should be used.

23. Adrian makes $92 every eight hours she works. How much will she make if she works twenty-two hours this week?

If Adrian works twenty-two hours this week, she will make ______.

24. Michele makes $174 every twelve hours she works. How much will she make if she works twenty-six hours this week?

If Michele works twenty-six hours this week, she will make ______.

25. A mutual fund consists of 87% stock and 13% bond. In other words, for each 87 dollars of stock, there are 13 dollars of bond. For a mutual fund with $2,310.00 of stock, how many dollars of bond are there?

Fill in the blank with decimal rounded to the hundredth place.

For a mutual fund with $2,310.00 of stock, there are approximately ______ of bond.

26. A mutual fund consists of 17% stock and 83% bond. In other words, for each 17 dollars of stock, there are 83 dollars of bond. For a mutual fund with $2,960.00 of bond, how many dollars of stock are there?

Fill in the blank with decimal rounded to the hundredth place.

For a mutual fund with $2,960.00 of bond, there are approximately ______ of stock.

27. Tammy jogs every day. Last month, she jogged 17 hours for a total of 10.2 miles. At this speed, if Tammy runs 47 hours, how far can she run?

At this speed, Tammy can run ______ in 47 hours.

28. Julie jogs every day. Last month, she jogged 8 hours for a total of 10.4 miles. At this speed, how long would it take Julie to run 53.95 miles?

At this speed, Julie can run 53.95 mi in ______.

29. Eric purchased 4.3 pounds of apples at the total cost of $16.77. If he purchases 9.9 pounds of apples at this store, how much would it cost?

It would cost ______ to purchase 9.9 pounds of apples.

30. Neil purchased 2.1 pounds of apples at the total cost of \$7.35. If the price doesn't change, how many pounds of apples can Neil purchase with \$32.90?

With \$32.90, Neil can purchase [ ] of apples.

31. Tammy collected a total of 1482 stamps over the past 13 years. At this rate, how many stamps would she collect in 25 years?

At this rate, Tammy would collect [ ] stamps in 25 years.

32. Stephen collected a total of 2250 stamps over the past 18 years. At this rate, how many years would it take he to collect 3500 stamps?

At this rate, Stephen can collect 3500 stamps in [ ] years.

33. In a city, the owner of a house valued at 250 thousand dollars needs to pay \$1,382.50 in property tax. At this tax rate, how much property tax should the owner pay if a house is valued at 880 thousand dollars?

The owner of a 880-thousand-dollar house should pay [ ] in property tax.

34. In a city, the owner of a house valued at 470 thousand dollars needs to pay \$2,476.90 in property tax. At this tax rate, if the owner of a house paid \$4,584.90 of property tax, how much is the house worth?

If the owner of a house paid \$4,584.90 of property tax, the house is worth [ ] thousand dollars.

35. To try to determine the health of the Rocky Mountain elk population in the Wenaha Wildlife Area, the Oregon Department of Fish and Wildlife caught, tagged, and released 39 Rocky Mountain elk. A week later, they returned and observed 42 Rocky Mountain elk, 9 of which had tags. Approximately how many Rocky Mountain elk are in the Wenaha Wildlife Area?

There are approximately [ ] elk in the wildlife area.

36. To try to determine the health of the black-tailed deer population in the Jewell Meadow Wildlife Area, the Oregon Department of Fish and Wildlife caught, tagged, and released 28 black-tailed deer. A week later, they returned and observed 63 black-tailed deer, 18 of which had tags. Approximately how many black-tailed deer are in the Jewell Meadow Wildlife Area?

There are approximately [ ] deer in the wildlife area.

37. A restaurant used 1048 lb of vegetable oil in 40 days. At this rate, how many pounds of vegetable oil will be used in 42 days?

The restaurant will use [ ] of vegetable oil in 42 days.

Use *lb* for pounds.

38. A restaurant used 806.4 lb of vegetable oil in 28 days. At this rate, 1612.8 lb of oil will last how many days?

The restaurant will use 1612.8 lb of vegetable oil in ______ days.

3.5 Special Solution Sets

Most of the time, a linear equation's final equivalent equation looks like $x = 3$, and the solution set is written to show that there is only one solution: $\{3\}$. Similarly, a linear inequality's final equivalent equation looks like $x < 5$, and the solution set is represented with either $(-\infty, 5)$ in interval notation or $\{x|x < 5\}$ in set-builder notation. It's possible that both linear equations and inequalities have all real numbers as possible solutions, and it's possible that no real numbers are solutions to each. In this section, we will explore these special solution sets.

3.5.1 Special Solution Sets

Recall that for the equation $x + 2 = 5$, there is only one number which will make the equation true: 3. This means that our solution is 3, and we write the **solution set** as $\{3\}$. We say the equation's solution set has one **element**, 3.

We'll now explore equations that have all real numbers as possible solutions or no real numbers as possible solutions.

Example 3.5.2. Solve for x in $3x = 3x + 4$.

To solve this equation, we need to move all terms containing x to one side of the equals sign:

$$\begin{aligned} 3x &= 3x + 4 \\ 3x - 3x &= 3x + 4 - 3x \\ 0 &= 4 \end{aligned}$$

Notice that x is no longer present in the equation. What value can we substitute into x to make $0 = 4$ true? Nothing! We say this equation has no solution. Or, the equation has an empty solution set. We can write this as $\emptyset$, which is the symbol for the empty set.

The equation $0 = 4$ is known a **false statement**, as it never holds true.

Example 3.5.3. Solve for x in $2x + 1 = 2x + 1$.

We will move all terms containing x to one side of the equals sign:

$$\begin{aligned} 2x + 1 &= 2x + 1 \\ 2x + 1 - 2x &= 2x + 1 - 2x \\ 1 &= 1 \end{aligned}$$

At this point, x is no longer contained in the equation. What value can we substitute into x to make $1 = 1$ true? Any number! This means that all real numbers are possible solutions to the equation $2x + 1 = 2x + 1$. We say this equation's solution set contains *all real numbers*. We can write this set using set-builder notation as $\{x \mid x \text{ is a real number}\}$ or using interval notation as $(-\infty, \infty)$.

The equation $1 = 1$ is known as an **identity** as it is always true.

Remark 3.5.4. What would have happened if we had continued solving after we obtained $1 = 1$ in Example 3.5.3?

$$\begin{aligned} 1 &= 1 \\ 1 - 1 &= 1 - 1 \\ 0 &= 0 \end{aligned}$$

As we can see, all we found was another identity — a different equation that is true for all values of x.

Warning 3.5.5. Note that there is a very important difference between ending with an equivalent equation of $0 = 0$ and $x = 0$. The first holds true for all real numbers, and the solution set is $\{x \mid x \text{ is a real number}\}$. The second has only one solution: 0. We write that solution set to show that only the number zero is the solution: $\{0\}$.

Example 3.5.6. Solve for t in the inequality $4t + 5 > 4t + 2$.

To solve for t, we will first subtract $4t$ from each side to get all terms containing t on one side:

$$\begin{aligned} 4t + 5 &> 4t + 2 \\ 4t + 5 - 4t &> 4t + 2 - 4t \\ 5 &> 2 \end{aligned}$$

Notice that again, the variable t is no longer contained in the inequality. We then need to consider which values of t make the inequality true. The answer is *all values*, so our solution set is all real numbers, which we can write as $\{t \mid t \text{ is a real number}\}$.

Example 3.5.7. Solve for x in the inequality $-5x + 1 \leq -5x$.

To solve for x, we will first add $5x$ to each side to get all terms containing x on one side:

$$\begin{aligned} -5x + 1 &\leq -5x \\ -5x + 1 + 5x &\leq -5x + 5x \\ 1 &\leq 0 \end{aligned}$$

Once more, the variable x is absent. So we can ask ourselves, "For which values of x is $1 \leq 0$ true?" The answer is *none*, and so there is no solution to this inequality. We can write the solution set using $\emptyset$.

Remark 3.5.8. Again consider what would have happened if we had continued solving after we obtained $1 \leq 0$ in Example 3.5.7.

$$1 \leq 0$$

$$1 - 1 \leq 0 - 1$$
$$0 \leq -1$$

As we can see, all we found was another false statement—a different equation that is not true for any real number.

Let's summarize the two special cases when solving linear equations and inequalities:

All Real Numbers When the equivalent equation or inequality is an *identity* such as $2 = 2$ or $0 < 2$, all real numbers are solutions. We write this solution set as either $\{x \mid x \text{ is a real number}\}$ or $(-\infty, \infty)$.

No Solution When the equivalent equation or inequality is a *false statement* such as $0 = 2$ or $0 > 2$, no real number is a solution. We write this solution set as either $\{\ \}$ or $\emptyset$ or write the words "no solution exists."

List 3.5.9: Special Solution Sets for Equations and Inequalities

3.5.2 Solving Equations and Inequalities with Special Solution Sets

Let's look at a few more complicated examples.

Example 3.5.10. Solve for a in $\frac{2}{3}(a+1) - \frac{5}{6} = \frac{2}{3}a$.

To solve this equation for a, we'll want to recall the technique of multiplying each side of the equation by the LCD of all fractions. Here, this means that we will multiply each side by 6 as our first step. After that, we'll be able to simplify each side of the equation and continue solving for a:

$$\begin{aligned}
\frac{2}{3}(a+1) - \frac{5}{6} &= \frac{2}{3}a \\
6 \cdot \left(\frac{2}{3}(a+1) - \frac{5}{6}\right) &= 6 \cdot \frac{2}{3}a \\
6 \cdot \frac{2}{3}(a+1) - 6 \cdot \frac{5}{6} &= \frac{2}{3}a \\
4(a+1) - 5 &= 4a \\
4a + 4 - 5 &= 4a \\
4a - 1 &= 4a \\
4a - 1 - 4a &= 4a - 4a \\
-1 &= 0
\end{aligned}$$

The statement $-1 = 0$ is false, so the equation has no solution. We state the empty set as the solution set: $\emptyset$.

Example 3.5.11. Solve for x in the equation $3(x + 2) - 8 = (5x + 4) - 2(x + 1)$.

To solve for x, we will first need to simplify the left side and right side of the equation as much as possible by distributing and combining like terms:

$$\begin{aligned} 3(x + 2) - 8 &= (5x + 4) - 2(x + 1) \\ 3x + 6 - 8 &= 5x + 4 - 2x - 2 \\ 3x - 2 &= 3x + 2 \end{aligned}$$

From here, we'll want to subtract $3x$ from each side:

$$\begin{aligned} 3x - 2 - 3x &= 3x + 2 - 3x \\ -2 &= 2 \end{aligned}$$

As the equation $-2 = 2$ is not true for any value of x, there is no solution to this equation. We write the solution set as: $\emptyset$.

Example 3.5.12. Solve for z in the inequality $\frac{3z}{5} + \frac{1}{2} \leq \left(\frac{z}{10} + \frac{3}{4}\right) + \left(\frac{z}{2} - \frac{1}{4}\right)$.

To solve for z, we will first need to multiply each side of the inequality by the LCD, which is 40. After that, we'll finish solving by putting all terms containing a variable on one side of the inequality:

$$\begin{aligned} \frac{3z}{5} + \frac{1}{2} &\leq \left(\frac{z}{10} + \frac{3}{4}\right) + \left(\frac{z}{2} - \frac{1}{4}\right) \\ 40 \cdot \left(\frac{3z}{5} + \frac{1}{2}\right) &\leq 40 \cdot \left(\left(\frac{z}{10} + \frac{3}{4}\right) + \left(\frac{z}{2} - \frac{1}{4}\right)\right) \\ 40 \cdot \left(\frac{3z}{5}\right) + 40 \cdot \left(\frac{1}{2}\right) &\leq 40 \cdot \left(\frac{z}{10} + \frac{3}{4}\right) + 40 \cdot \left(\frac{z}{2} - \frac{1}{4}\right) \\ 40 \cdot \left(\frac{3z}{5}\right) + 40 \cdot \left(\frac{1}{2}\right) &\leq 40 \cdot \left(\frac{z}{10}\right) + 40 \cdot \left(\frac{3}{4}\right) + 40 \cdot \left(\frac{z}{2}\right) - 40 \cdot \left(\frac{1}{4}\right) \\ 24z + 20 &\leq 4z + 30 + 20z - 10 \\ 24z + 20 &\leq 24z + 20 \\ 24z + 20 - 24z &\leq 24z + 20 - 24z \\ 20 &\leq 20 \end{aligned}$$

As the equation $20 \leq 20$ is true for all values of z, all real numbers are solutions to this inequality. Thus the solution set is $\{z \mid z \text{ is a real number}\}$.

3.5.3 Exercises

Solving Equations with Special Solution Sets

For the following exercises: Solve the equation.

1. $6A = 6A + 8$

2. $2B = 2B + 2$

3. $8m + 6 = 8m + 6$

4. $6n + 1 = 6n + 1$

5. $2q - 6 - 3q = -9 - q + 3$

6. $8y - 9 - 9y = -10 - y + 1$

7. $-7 - 6r + 6 = -r + 15 - 5r$

8. $-5 - 9a + 3 = -a + 15 - 8a$

9.

a. $8b + 3 = 3b + 3$

b. $8b + 3 = 8b + 3$

c. $8b + 3 = 8b + 7$

10.

a. $6A + 6 = 2A + 6$

b. $6A + 6 = 6A + 6$

c. $6A + 6 = 6A + 10$

11. $5(9 - 8B) - (8B - 9) = 23 - 2(10 + 24B)$

12. $4(4 - 6m) - (8m - 4) = 15 - 2(1 + 16m)$

13. $30 - 4(9 + 3n) = -13n - (6 - n)$

14. $12 - 6(3 + 6q) = -37q - (6 - q)$

15. $6(y - 8) = 6(y - 3)$

16. $4(r - 5) = 4(r - 2)$

Solving Inequalities with Special Solution Sets

For the following exercises: Solve this inequality. Answer using interval notation.

17. $4x > 4x + 8$

18. $6x > 6x + 5$

19. $-6x \le -6x - 9$

20. $-8x \le -8x - 2$

21. $-2 + 8x + 7 \ge 8x + 5$

22. $-6 + 10x + 8 \ge 10x + 2$

23. $-10 + 10x + 18 < 10x + 8$

24. $-4 + 2x + 9 < 2x + 5$

25. $-6 - 5z + 5 > -z + 18 - 4z$

26. $-6 - 9z + 2 > -z + 11 - 8z$

27. $3(4 - 8m) - (4m - 5) > 8 - 2(9 + 14m)$

28. $4(1 - 2m) - (2m - 5) > 9 - 2(4 + 5m)$

29. $8(k - 9) \le 8(k - 2)$

30. $8(k - 7) \le 8(k - 4)$

31. $10x \le 10x + 1$

32. $10x \le 10x + 7$

3.6 Chapter Review

3.6.1 Solving Multistep Linear Equations and Inequalities Review

Steps to Solving Linear Equations

Simplify Simplify the expressions on each side of the equation by distributing and combining like terms.

Isolate Use addition or subtraction to separate the variable terms and constant terms (numbers) so that they are on different sides of the equation.

Eliminate Use multiplication or division to eliminate the variable term's coefficient.

Check Check the solution.

Summarize State the solution set or (in the case of an application problem) summarize the result in a complete sentence using appropriate units.

Example 3.6.1. Solve for a in $4 - (3 - a) = -2 - 2(2a + 1)$.

To solve this equation, we will simplify each side of the equation, manipulate it so that all variable terms are on one side and all constant terms are on the other, and then solve for a:

$$\begin{aligned} 4 - (3 - a) &= -2 - 2(2a + 1) \\ 4 - 3 + a &= -2 - 4a - 2 \\ 1 + a &= -4 - 4a \\ 1 + a + 4a &= -4 - 4a + 4a \\ 1 + 5a &= -4 \\ 1 + 5a - 1 &= -4 - 1 \\ 5a &= -5 \\ \frac{5a}{5} &= \frac{-5}{5} \\ a &= -1 \end{aligned}$$

Checking the solution -1 in the original equation, we get:

$$\begin{aligned} 4 - (3 - a) &= -2 - 2(2a + 1) \\ 4 - (3 - (-1)) &\stackrel{?}{=} -2 - 2(2(-1) + 1) \\ 4 - (4) &\stackrel{?}{=} -2 - 2(-1) \\ 0 &\stackrel{\checkmark}{=} 0 \end{aligned}$$

Therefore the solution to the equation is -1 and the solution set is $\{-1\}$.

Differentiating among Simplifying Expressions, Evaluating Expressions and Solving Equations Let's summarize the differences among simplifying expressions, evaluating expressions and solving equations:

- An expression like $10 - 3(x + 2)$ can be simplified to $-3x + 4$ (as in Example 3.1.15), but we cannot solve for x in an expression.
- As x takes different values, an expression has different values. In Example 3.1.16, when $x = 2$, $10 - 3(x + 2) = -2$; but when $x = 3$, $10 - 3(x + 2) = -5$.
- An equation connects two expressions with an equals sign. In Example 3.1.17, $10 - 3(x + 2) = x - 16$ has the expression $10 - 3(x + 2)$ on the left side of equals sign, and the expression $x - 16$ on the right side.
- When we solve the equation $10 - 3(x + 2) = x - 16$, we are looking for a number which makes those two expressions have the same value. In Example 3.1.17, we found the solution to be $x = 5$, which makes both $10 - 3(x + 2) = -11$ and $x - 16 = -11$, as shown in the checking part.

Solving Multistep Inequalities When solving a linear inequality, we follow the same steps in List 3.1.4. The only difference in our steps to solving is that when we multiply or divide by a negative number on both sides of an inequality, the direction of the inequality symbol must switch.

Example Solve for x in $-2 - 2(2x + 1) > 4 - (3 - x)$. Write the solution set in both set-builder notation and interval notation.

$$\begin{aligned}
-2 - 2(2x + 1) &> 4 - (3 - x) \\
-2 - 4x - 2 &> 4 - 3 + x \\
-4x - 4 &> x + 1 \\
-4x - 4 - x &> x + 1 - x \\
-5x - 4 &> 1 \\
-5x - 4 + 4 &> 1 + 4 \\
-5x &> 5 \\
\frac{-5x}{-5} &< \frac{5}{-5} \\
x &< -1
\end{aligned}$$

Note that when we divided both sides of the inequality by -5, we had to switch the direction of the inequality symbol.

The solution set in set-builder notation is $\{x \mid x < -1\}$.

The solution set in interval notation is $(-\infty, -1)$.

3.6.2 Solving Linear Equations and Inequalities with Fractions Review

Clearing Fractions When there are fractions in a linear equation or inequality, we will multiply each side of the equation by the least common denominator (LCD) of all fractions.

Once the LCD is distributed and all fractions are reduced, we will have an equation that is equivalent to the original, yet without any fractions

Example Solve for x in $\frac{1}{4}x + \frac{2}{3} = \frac{1}{6}$.

We'll solve by multiplying each side of the equation by 12:

$$\begin{aligned} \frac{1}{4}x + \frac{2}{3} &= \frac{1}{6} \\ 12 \cdot \left(\frac{1}{4}x + \frac{2}{3}\right) &= 12 \cdot \frac{1}{6} \\ 12 \cdot \left(\frac{1}{4}x\right) + 12 \cdot \left(\frac{2}{3}\right) &= 12 \cdot \frac{1}{6} \\ 3x + 8 &= 2 \\ 3x &= -6 \\ \frac{3x}{3} &= \frac{-6}{3} \\ x &= -2 \end{aligned}$$

Checking the solution:

$$\begin{aligned} \frac{1}{4}x + \frac{2}{3} &= \frac{1}{6} \\ \frac{1}{4}(-2) + \frac{2}{3} &\stackrel{?}{=} \frac{1}{6} \\ -\frac{2}{4} + \frac{2}{3} &\stackrel{?}{=} \frac{1}{6} \\ -\frac{6}{12} + \frac{8}{12} &\stackrel{?}{=} \frac{1}{6} \\ \frac{2}{12} &\stackrel{?}{=} \frac{1}{6} \\ \frac{1}{6} &\stackrel{\checkmark}{=} \frac{1}{6} \end{aligned}$$

The solution is therefore -2. We write the solution set s $\{-2\}$.

3.6.3 Isolating a Linear Variable Review

Isolating a Linear Variable When we isolate a linear variable, we follow the same steps of solving linear equations, except we treat all the other variables as numbers. For example, when we solve for x in $y = mx + b$, we follow the same steps as solving for x in $1 = 2x + 3$.

Example Solve for x in $y = mx + b$.

$$\begin{aligned} y &= mx + b \\ y - b &= mx + b - b \\ y - b &= mx \\ \frac{y - b}{m} &= \frac{mx}{m} \\ \frac{y - b}{m} &= x \end{aligned}$$

3.6.4 Ratios and Proportions Review

Ratios and Proportions A ratio is a comparison of two values, usually written as a fraction.

A proportion is a statement that two ratios are equal. Many applications require solving for a variable in a proportion.

Example Property taxes for a residential property are proportional to the assessed value of the property. Assume that a certain property in a given neighborhood is assessed at $234,100 and its annual property taxes are $2,518.92. What are the annual property taxes for a house that is assessed at $287,500?

Let T be the annual property taxes (in dollars) for a property assessed at $287,500. We can write and solve this proportion:

$$\begin{aligned}
\frac{\text{tax}}{\text{property value}} &= \frac{\text{tax}}{\text{property value}} \\
\frac{2518.92}{234100} &= \frac{T}{287500} \\
234100 \cdot 287500 \cdot \frac{2518.92}{234100} &= \frac{T}{287500} \cdot 234100 \cdot 287500 \\
287500 \cdot 2518.92 &= T \cdot 234100 \\
\frac{287500 \cdot 2518.92}{234100} &= \frac{234100T}{234100} \\
T &\approx 3093.50
\end{aligned}$$

The property taxes for a property assessed at $287,500 are $3,093.50.

3.6.5 Special Solution Sets Review

Special Solution Sets Most of the time, a linear equation has only one solution. It's possible that the equation has no solution and it's also possible that every real number is a solution.

When solving linear inequalities, it's also possible that no solution exists or that all real numbers are solutions.

Examples

Solve for x in $3x = 3x + 4$.

To solve this equation, we need to move all terms containing x to one side of the equals sign:

$$\begin{aligned} 3x &= 3x + 4 \\ 3x - 3x &= 3x + 4 - 3x \\ 0 &= 4 \end{aligned}$$

This equation has no solution. We write the solution set as $\emptyset$, which is the symbol for the empty set.

Solve for t in the inequality $4t + 5 > 4t + 2$.

To solve for t, we will first subtract $4t$ from each side to get all terms containing t on one side:

$$\begin{aligned} 4t + 5 &> 4t + 2 \\ 4t + 5 - 4t &> 4t + 2 - 4t \\ 5 &> 2 \end{aligned}$$

All values of the variable t make the inequality true. The solution set is all real numbers, which we can write as $\{t \mid t \text{ is a real number}\}$ in set notation, or $(-\infty, \infty)$ in interval notation.

3.6.6 Exercises

1.

a. Solve the following linear equation:

$$3(r + 7) + 6 = 54$$

b. Evaluate the following expression when $r = 9$:

$3(r + 7) + 6 =$ ______

c. Simplify the following expression:

$3(r + 7) + 6 =$ ______

2. Solve the equation.
$26 = -10r - 7 - r$

3. Solve the equation.

$$3 + 9(t - 8) = -20 - (9 - 4t)$$

4. Solve the equation.

$$-10 - 8q + 7 = -q + 16 - 7q$$

5. Solve the equation.

$$14 = \frac{C}{3} + \frac{C}{4}$$

6. Solve the equation.

$$\frac{m - 7}{6} = \frac{m + 3}{8}$$

7. Solve this inequality.

$$2-(y+6)<6$$

In set-builder notation, the solution set is [ ].

In interval notation, the solution set is [ ].

8. Solve this inequality. Answer using interval notation.

$$2(k-8)\le 2(k-4)$$

9. Solve this inequality.

$$3+9(x-6)<-78-(9-5x)$$

In set-builder notation, the solution set is [ ].

In interval notation, the solution set is [ ].

10. Solve this inequality.

$$-\frac{5}{2}t>\frac{2}{7}t-117$$

In set-builder notation, the solution set is [ ].

In interval notation, the solution set is [ ].

11. Solve this linear equation for y:

$$Ax+By=C$$

Note that the variables are upper case A, B, and C and lower case x and y.

12. Solve this linear equation for p:

$$c=n-\frac{8p}{r}$$

13. Holli has $74 in her piggy bank. She plans to purchase some Pokemon cards, which costs $1.35 each. She plans to save $64.55 to purchase another toy. At most how many Pokemon cards can he purchase?
Write an equation to solve this problem.

Holli can purchase at most [ ] Pokemon cards.

14. Use a linear equation to solve the word problem.
Thanh has $95.00 in his piggy bank, and he spends $5.00 every day.
Sharell has $4.00 in her piggy bank, and she saves $1.50 every day.
If they continue to spend and save money this way, how many days later would they have the same amount of money in their piggy banks?

[ ] days later, Thanh and Sharell will have the same amount of money in their piggy banks.

15. A hockey team played a total of 112 games last season. The number of games they won was 16 more than three times of the number of games they lost.
Write and solve an equation to answer the following questions.

The team lost ______ games.

The team won ______ games.

16. A rectangle's perimeter is 162 ft. Its length is 3 ft longer than two times of its width. Use an equation to find the rectangle's length and width.

Its width is ______.

Its length is ______.

17. Lily has saved \$47.00 in her piggy bank, and she decided to start spending them. She spends \$2.00 every 3 days. After how many days will she have \$41.00 left in the piggy bank?

Lily will have \$41.00 left in her piggy bank after ______ days.

18. The following two triangles are similar to each other. Find the length of the missing side.

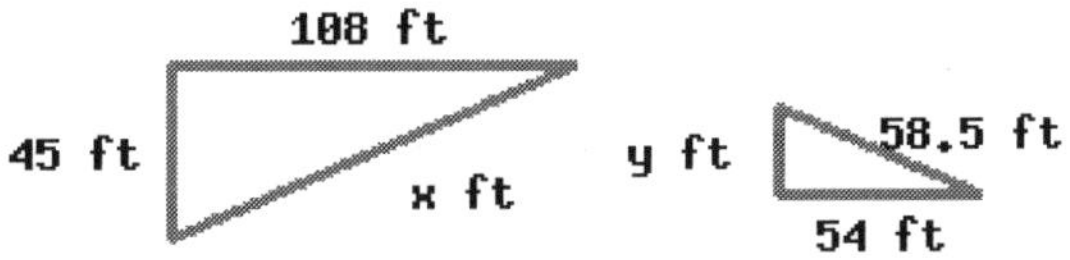

108 ft
45 ft
x ft
y ft
58.5 ft
54 ft

The first missing side's length is $x =$ ______.

The second missing side's length is $y =$ ______.

19. A restaurant used 765 lb of vegetable oil in 34 days. At this rate, 1327.5 lb of oil will last how many days?

The restaurant will use 1327.5 lb of vegetable oil in ______ days.

CHAPTER 4

Graphing Lines

4.1 Cartesian Coordinates

When we model relationships with graphs, we use the **Cartesian coordinate system**. This section covers the basic vocabulary and ideas that come with the Cartesian coordinate system.

René Descartes Several conventions used in mathematics are attributed to (or at least named after) René Descartes. The Cartesian coordinate system is one of these. You can read about René and these conventions at en.wikipedia.org/wiki/René_Descartes.

The Cartesian coordinate system identifies the location of every point in a plane. Basically, the system gives every point in a plane its own "address" in relation to a starting point. We'll use a street grid as an analogy. Here is a map with Carl's home at the center. The map also shows some nearby businesses. Assume each unit in the grid represents one city block.

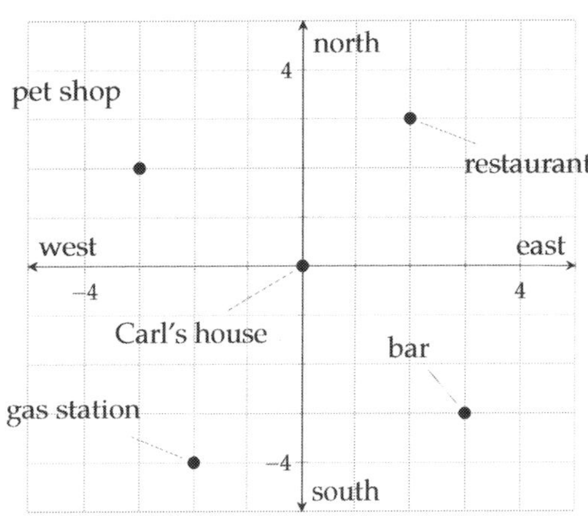

Figure 4.1.2: Carl's neighborhood

If Carl has an out-of-town guest who asks him how to get to the restaurant, Carl could say:

"First go 2 blocks east, then go 3 blocks north."

cc BY

Carl uses two numbers to locate the restaurant. In the Cartesian coordinate system, these numbers are called **coordinates** and they are written as the **ordered pair** $(2, 3)$. The first coordinate, 2, represents distance traveled from Carl's house to the east (or to the right horizontally on the graph). The second coordinate, 3, represents distance to the north (up vertically on the graph).

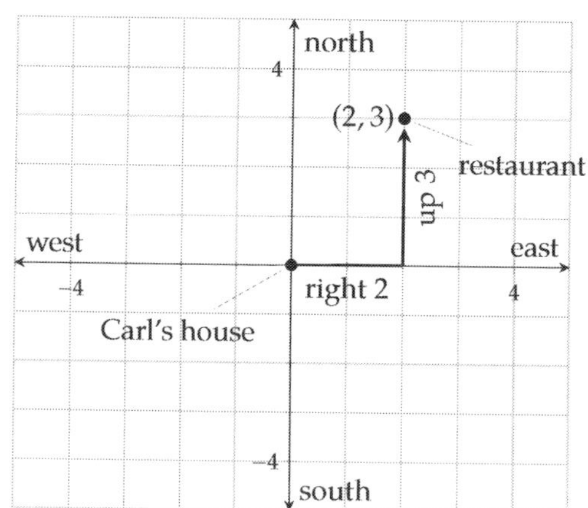

Figure 4.1.3: Carl's neighborhood

Alternatively, to travel from Carl's home to the pet shop, he would go 3 blocks west, and then 2 blocks north.

In the Cartesian coordinate system, the *positive* directions are to the *right* horizontally and *up* vertically. The *negative* directions are to the *left* horizontally and *down* vertically. So the pet shop's Cartesian coordinates are $(-3, 2)$.

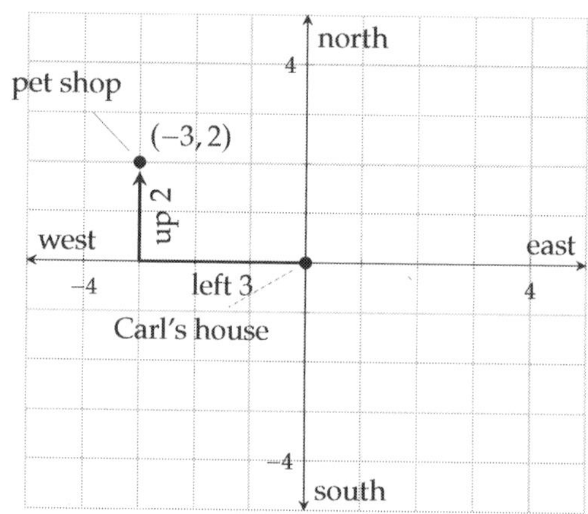

Figure 4.1.4: Carl's neighborhood

Remark 4.1.5. It's important to know that the order of Cartesian coordinates is (horizontal, vertical). This idea of communicating horizontal information *before* vertical information is consistent throughout most of mathematics.

Notation Issue: Coordinates or Interval? Unfortunately, the notation for an ordered pair looks exactly like interval notation for an open interval. *Context* will help you understand if $(2, 3)$ indicates the point 2 units right of the origin and 3 units up, or if $(2, 3)$ indicates the interval of all real numbers between 2 and 3.

Exercise 4.1.6. Use Figure 4.1.2 to answer the following questions.

1. What are the coordinates of the bar?

2. What are the coordinates of the gas station?

3. What are the coordinates of Carl's house?

Traditionally, the variable x represents numbers on the horizontal axis, so it is called the x-**axis**. The variable y represents numbers on the vertical axis, so it is called the y-**axis**. The axes meet at the point $(0,0)$, which is called the **origin**. Every point in the plane is represented by an **ordered pair**, (x,y).

In a Cartesian coordinate system, the map of Carl's neighborhood would look like this:

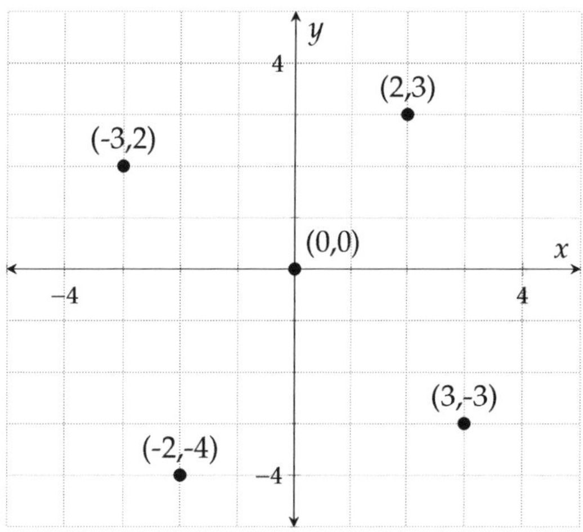

Figure 4.1.7: Carl's Neighborhood in a Cartesian Coordinate System

Definition 4.1.8 (Cartesian Coordinate System)**.** A Cartesian coordinate system is a coordinate system that specifies each point uniquely in a plane by a pair of numerical coordinates, which are the signed (positive/negative) distances to the point from two fixed perpendicular directed lines, measured in the same unit of length. Those two reference lines are called the **horizontal axis** and **vertical axis**, and the point where they meet is the **origin**. The horizontal and vertical axes are often called the x-**axis** and y-**axis**. (Visit en.wikipedia.org/wiki/Cartesian_coordinate_system for more.)

The plane based on the x-axis and y-axis is called a **coordinate plane**. The ordered pair used to locate a point is called the point's **coordinates**, which consists of an x-**coordinate** and a y-**coordinate**. For example, for the point $(1,2)$, its x-coordinate is 1, and its y-coordinate is 2. The origin has coordinates $(0,0)$.

A Cartesian coordinate system is divided into four **quadrants**, as shown in Figure 4.1.9. The quadrants are traditionally labeled with Roman numerals.

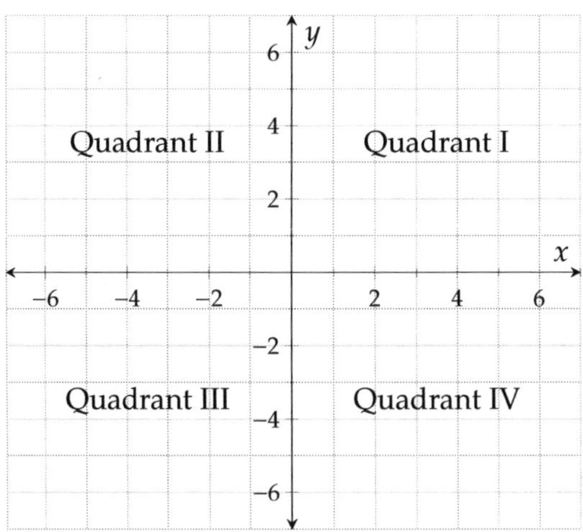

Figure 4.1.9: A Cartesian grid with four quadrants marked

Example 4.1.10. On paper, sketch a Cartesian coordinate system with units, and then plot the following points: $(3,2), (-5,-1), (0,-3), (4,0)$.

Solution.

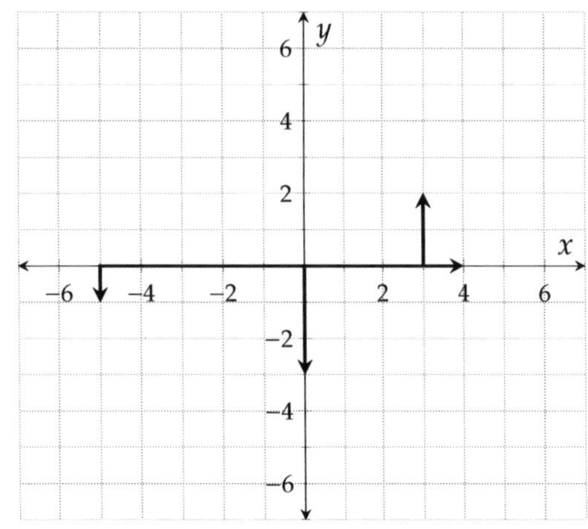

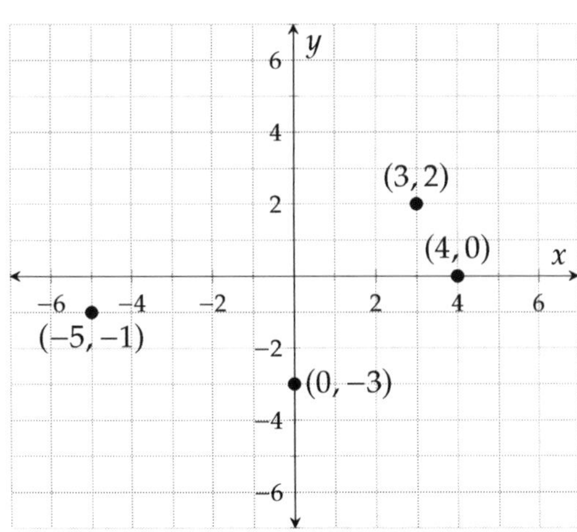

4.1.1 Exercises

Identify coordinates.

For the following exercises: Locate each point in the graph:

1.

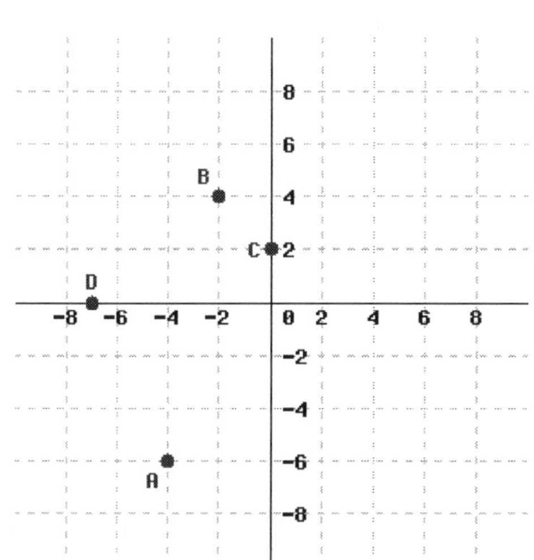

Write each point's position as an ordered pair, like $(1, 2)$.

$A =$ ______ $B =$ ______
$C =$ ______ $D =$ ______

2.

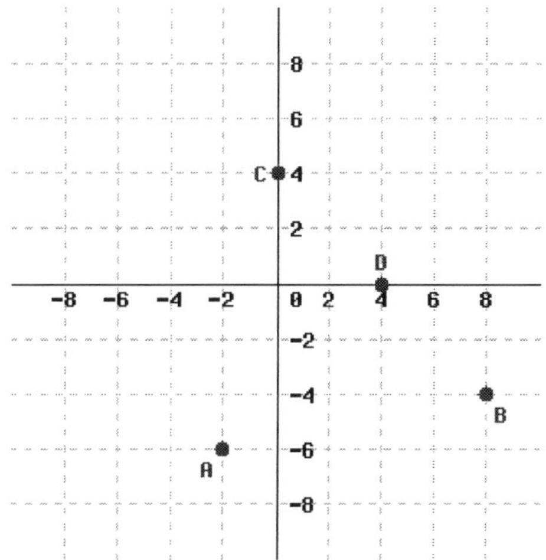

Write each point's position as an ordered pair, like $(1, 2)$.

$A =$ ______ $B =$ ______
$C =$ ______ $D =$ ______

Make some sketches.

3. Sketch the points $(8, 2)$, $(5, 5)$, $(-3, 0)$, and $(2, -6)$ on a Cartesian plane.

4. Sketch the points $(1, -4)$, $(-3, 5)$, $(0, 4)$, and $(-2, -6)$ on a Cartesian plane.

5. Sketch the points $(208, -50)$, $(97, 112)$, $(-29, 103)$, and $(-80, -172)$ on a Cartesian plane.

6. Sketch the points $(110, 38)$, $(-205, 52)$, $(-52, 125)$, and $(-172, -80)$ on a Cartesian plane.

7. Sketch the points $(5.5, 2.7)$, $(-7.3, 2.75)$, $(-\frac{10}{3}, \frac{1}{2})$, and $(-\frac{28}{5}, -\frac{29}{4})$ on a Cartesian plane.

8. Sketch the points $(1.9, -3.3)$, $(-5.2, -8.11)$, $(\frac{7}{11}, \frac{15}{2})$, and $(-\frac{16}{3}, \frac{19}{5})$ on a Cartesian plane.

9. Sketch a Cartesian plane and shade the quadrants where the x-coordinate is negative.

10. Sketch a Cartesian plane and shade the quadrants where the y-coordinate is positive.

11. Sketch a Cartesian plane and shade the quadrants where the x-coordinate has the same sign as the y-coordinate.

12. Sketch a Cartesian plane and shade the quadrants where the x-coordinate and the y-coordinate have opposite signs.

These exercises have Cartesian plots with some context.

13. This graph gives the minimum estimates of the wolf population in Washington from 2008 through 2015. (Source: http://wdfw.wa.gov/publications/01793/wdfw01793.pdf)

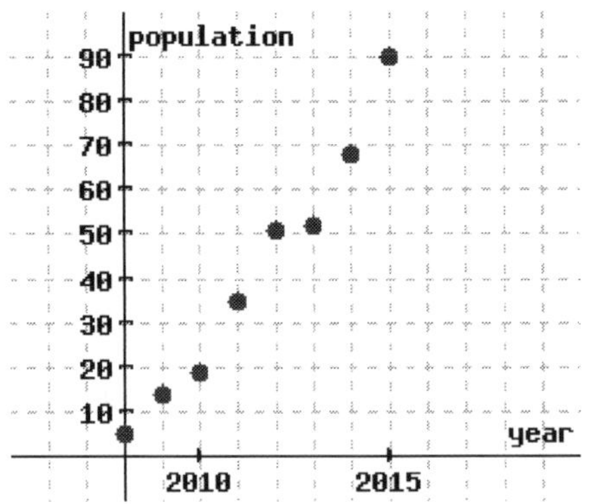

What are the Cartesian coordinates for the point representing the year 2011? [ ]

Between 2011 and 2012, the wolf population grew by [ ] wolves.

List at least three ordered pairs in the graph.

14. Here is a graph of the foreign-born US population (in millions) during Census years 1960 to 2010. (Source: http://www.pewhispanic.org/2015/09/28/chapter-5-u-s-foreign-born-population-trends/.)

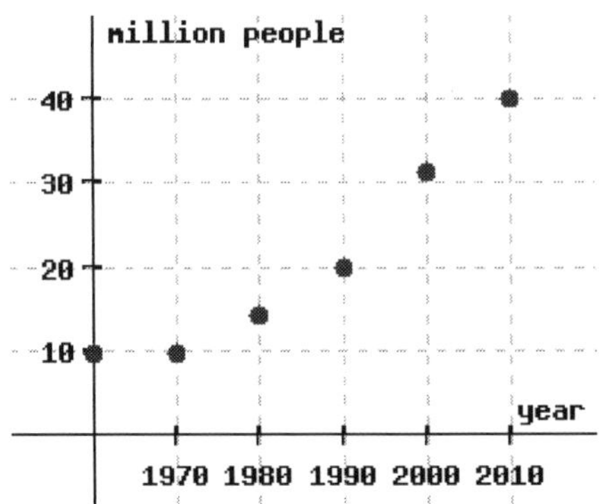

What are the Cartesian coordinates for the point representing the year 1980? [ ]

Between 1980 and 2000, the US population that is foreign-born increased by [ ] million people.

List at least three ordered pairs in the graph.

Regions in the Cartesian plane.

15. The point $(5, 10)$ is in Quadrant (□ I □ II □ III □ IV) .

The point $(-7, 10)$ is in Quadrant (□ I □ II □ III □ IV) .

The point $(-10, -8)$ is in Quadrant (□ I □ II □ III □ IV) .

The point $(6, -7)$ is in Quadrant (□ I □ II □ III □ IV) .

16. The point $(7, 3)$ is in Quadrant (□ I □ II □ III □ IV) .

The point $(-9, -5)$ is in Quadrant (□ I □ II □ III □ IV) .

The point $(-8, 1)$ is in Quadrant (□ I □ II □ III □ IV) .

The point $(1, -9)$ is in Quadrant (□ I □ II □ III □ IV) .

17. Answer the following questions on the coordinate system:

For the point (x, y), if $y = 0$, then the point is in/on (□ Quadrant I □ Quadrant II □ Quadrant III □ Quadrant IV □ the x-axis □ the y-axis) .

For the point (x, y), if $x < 0$ and $y > 0$, then the point is in/on (□ Quadrant I □ Quadrant II □ Quadrant III □ Quadrant IV □ the x-axis □ the y-axis) .

For the point (x, y), if $x < 0$ and $y < 0$, then the point is in/on (□ Quadrant I □ Quadrant II □ Quadrant III □ Quadrant IV □ the x-axis □ the y-axis) .

For the point (x, y), if $x > 0$ and $y > 0$, then the point is in/on (□ Quadrant I □ Quadrant II □ Quadrant III □ Quadrant IV □ the x-axis □ the y-axis) .

For the point (x, y), if $x = 0$, then the point is in/on (□ Quadrant I □ Quadrant II □ Quadrant III □ Quadrant IV □ the x-axis □ the y-axis) .

For the point (x, y), if $x > 0$ and $y < 0$, then the point is in/on (□ Quadrant I □ Quadrant II □ Quadrant III □ Quadrant IV □ the x-axis □ the y-axis) .

18. Answer the following questions on the coordinate system:

For the point (x, y), if $x > 0$ and $y > 0$, then the point is in/on (□ Quadrant I □ Quadrant II □ Quadrant III □ Quadrant IV □ the x-axis □ the y-axis) .

For the point (x, y), if $y = 0$, then the point is in/on (□ Quadrant I □ Quadrant II □ Quadrant III □ Quadrant IV □ the x-axis □ the y-axis) .

For the point (x, y), if $x < 0$ and $y < 0$, then the point is in/on (□ Quadrant I □ Quadrant II □ Quadrant III □ Quadrant IV □ the x-axis □ the y-axis) .

For the point (x, y), if $x < 0$ and $y > 0$, then the point is in/on (□ Quadrant I □ Quadrant II □ Quadrant III □ Quadrant IV □ the x-axis □ the y-axis) .

For the point (x, y), if $x > 0$ and $y < 0$, then the point is in/on (□ Quadrant I □ Quadrant II □ Quadrant III □ Quadrant IV □ the x-axis □ the y-axis) .

For the point (x, y), if $x = 0$, then the point is in/on (□ Quadrant I □ Quadrant II □ Quadrant III □ Quadrant IV □ the x-axis □ the y-axis) .

19. Assume the point (x, y) if in Quadrant I, locate the following points:

The point $(-x, y)$ is in Quadrant (□ I □ II □ III □ IV) .

The point $(x, -y)$ is in Quadrant (□ I □ II □ III □ IV) .

The point $(-x, -y)$ is in Quadrant (□ I □ II □ III □ IV) .

20. Assume the point (x, y) if in Quadrant II, locate the following points:

The point $(-x, y)$ is in Quadrant (□ I □ II □ III □ IV) .

The point $(x, -y)$ is in Quadrant (□ I □ II □ III □ IV) .

The point $(-x, -y)$ is in Quadrant (□ I □ II □ III □ IV) .

Writing questions.

21. What would be the difficulty with trying to plot $(12, 4)$, $(13, 5)$, and $(310, 208)$ all on the same graph?

22. The points $(3, 5)$, $(5, 6)$, $(7, 7)$, and $(9, 8)$ all lie on a straight line. What can go wrong if you make a plot of a Cartesion plane with these points marked, and you don't have tick marks that are evely spaced apart?

4.2 Graphing Equations

We have graphed *points* in a coordinate system, and now we will graph *lines* and *curves*.

A **graph** of an equation is a picture of that equation's solution set. For example, the graph of $y = -2x+3$ is shown in Figure 4.2.3(c). The graph plots the ordered pairs whose coordinates make $y = -2x + 3$ true. Table 4.2.2 shows a few points that make the equation true.

$y = -2x + 3$	(x, y)
$5 \stackrel{\checkmark}{=} -2(-1) + 3$	$(-1, 5)$
$3 \stackrel{\checkmark}{=} -2(0) + 3$	$(0, 3)$
$1 \stackrel{\checkmark}{=} -2(1) + 3$	$(1, 1)$
$-1 \stackrel{\checkmark}{=} -2(2) + 3$	$(2, -1)$
$-3 \stackrel{\checkmark}{=} -2(3) + 3$	$(3, -3)$
$-5 \stackrel{\checkmark}{=} -2(4) + 3$	$(4, -5)$

Table 4.2.2

Table 4.2.2 tells us that the points $(-1,5)$, $(0,3)$, $(1,1)$, $(2,-1)$, $(3,-3)$, and $(4,-5)$ are all solutions to the equation $y = -2x + 3$, and so they should all be shaded as part of that equation's graph. You can see them in Figure 4.2.3(a). But there are many more points that make the equation true. More points are plotted in Figure 4.2.3(b). Even more points are plotted in Figure 4.2.3(c) — so many, that together the points look like a straight line.

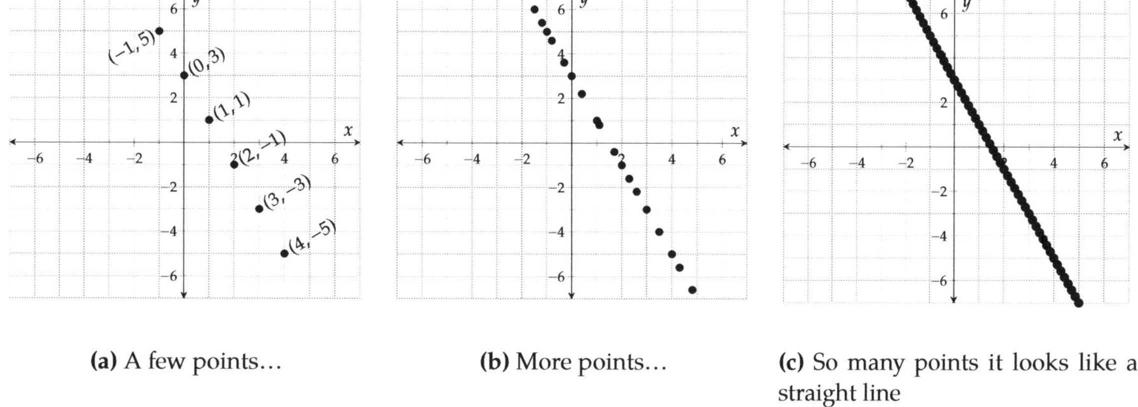

(a) A few points...

(b) More points...

(c) So many points it looks like a straight line

Figure 4.2.3: Graphs of the Equation $y = -2x + 3$

Generalizing...

Remark 4.2.4. The graph of an equation shades all the points (x, y) that make the equation true once the x- and y-values are substituted in. Typically, there are *so many* points shaded, that the final graph appears to be a continuous line or curve that you could draw with one stroke of a pen.

Exercise 4.2.5. The point $(4, -5)$ is on the graph in Figure 4.2.3.(c). What happens when you substitute these values into the equation $y = -2x + 3$?

$$\begin{aligned} y &= -2x+3 \\ \underline{\quad} &= \underline{\qquad\qquad\qquad} \end{aligned}$$

This equation is (□ true □ false) .

Exercise 4.2.6. Decide whether the points $(5, -2)$ and $(-10, -7)$ will be on the graph of the equation $y = -\frac{3}{5}x + 1$.

At $(5, -2)$:

$$\begin{aligned} y &= -\tfrac{3}{5}x+1 \\ \underline{\quad} &= \underline{\qquad\qquad\qquad} \end{aligned}$$

This equation is (□ true □ false) , and $(5, -2)$ is (□ part of □ not part of) the graph of $y = -\frac{3}{5}x + 1$.

At $(-10, -7)$:

$$\begin{aligned} y &= -\tfrac{3}{5}x+1 \\ \underline{\quad} &= \underline{\qquad\qquad\qquad} \end{aligned}$$

This equation is (□ true □ false) , and $(5, -2)$ is (□ part of □ not part of) the graph of $y = -\frac{3}{5}x + 1$..

Solution. If the point $(5, -2)$ is on $y = -\frac{3}{5}x+1$, once we substitute $x = 5$ and $y = -2$ into the line's equation, the equation should be true. Let's try:

$$\begin{aligned} y &= -\frac{3}{5}x+1 \\ -2 &\stackrel{?}{=} -\frac{3}{5}(5)+1 \\ -2 &\stackrel{?}{=} -3+1 \\ -2 &\stackrel{\checkmark}{=} -2 \end{aligned}$$

Because this last equation is true, we can definitively say that $(5, -2)$ is on the graph of $y = -\frac{3}{5}x+1$.

However if we substitute $x = -10$ and $y = -7$ into the equation, it leads to $-7 = 7$, which is false. This definitively tells us that $(-10, -7)$ is not on the graph.

So to make our own graph of an equation with two variables x and y, we can choose some reasonable x-values, then calculate the corresponding y-values, and then plot the (x, y)-pairs as points. For many (not-so-complicated) algebraic equations, connecting those points with a smooth curve will produce an excellent graph.

Example 4.2.7. Let's plot a graph for the equation $y = -2x + 5$. We use a table to organize our work:

x	$y = -2x + 5$	Point
-2		
-1		
0		
1		
2		

(a) Set up the table

x	$y = -2x + 5$	Point
-2	$-2(-2) + 5 = 9$	$(-2, 9)$
-1	$-2(-1) + 5 = 7$	$(-1, 7)$
0	$-2(0) + 5 = 5$	$(0, 5)$
1	$-2(1) + 5 = 3$	$(1, 3)$
2	$-2(2) + 5 = 1$	$(2, 1)$

(b) Complete the table

Figure 4.2.8: Making a table for $y = -2x + 5$

We use points from the table to graph the equation. First, plot each point carefully. Then, connect the points with a smooth curve. Here, the curve is a straight line. Lastly, we can communicate that the graph extends further by sketching arrows on both ends of the line.

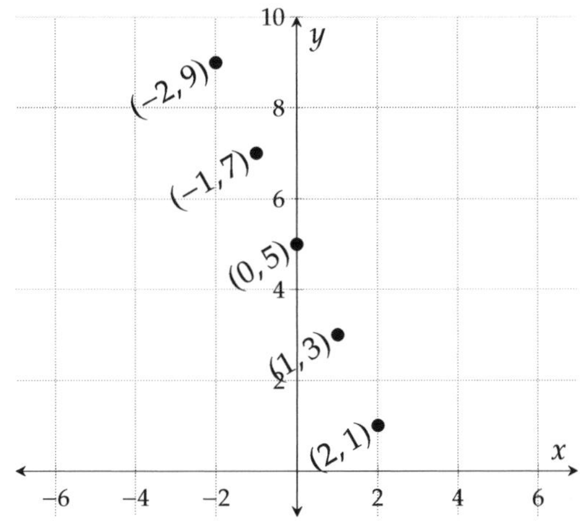

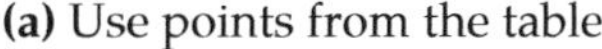
(a) Use points from the table

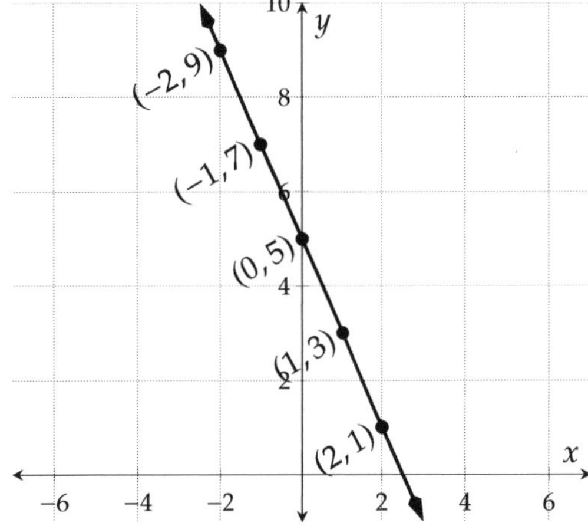

(b) Connect the points in whatever pattern is apparent

Figure 4.2.9: Graphing the Equation $y = -2x + 5$

Remark 4.2.10. Note that our choice of x-values is arbitrary. As long as we determine the coordinates of enough points to indicate the behavior of the graph, we may choose whichever x-values we like. For simpler caclulations, people often start with the integers from -2 to 2. However sometimes the equation has context that suggests using other x-values, as in the next example.

Example 4.2.11. One car's gas tank holds 14 gal of fuel. Over the course of a long road trip, that car uses its fuel at an average rate of $0.032\,\frac{\text{gal}}{\text{mi}}$. If a driver fills the tank at the beginning of a long trip, then the amount of fuel remaining in the tank, y, after driving x miles is given by the equation $y = 14 - 0.032x$. Make a suitable table of values and graph this equation.

Solution. Choosing x-values from -2 to 2, as in our previous example, wouldn't make sense here. Driving a negative number of miles is not possible, and any long road trip is longer than 2 miles. So in this context, choose x-values that reflect the number of miles one might drive in a day.

x	$y = 14 - 0.032x$	Point
20	13.36	$(20, 13.36)$
50	12.4	$(50, 12.4)$
80	11.44	$(80, 11.44)$
100	10.8	$(100, 10.8)$
200	7.6	$(200, 7.6)$

Table 4.2.12: Make the table

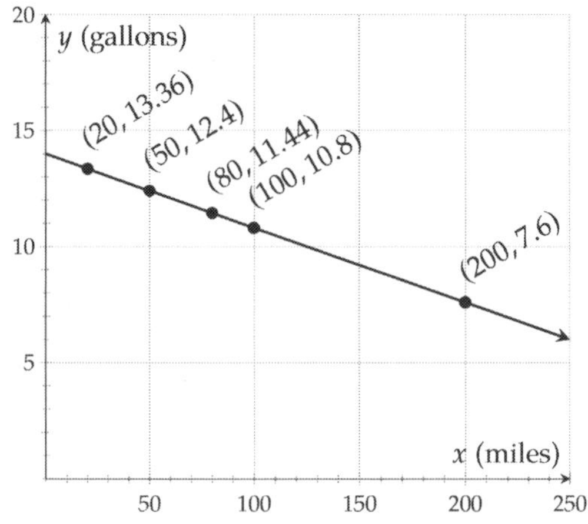

Figure 4.2.13: Make the graph

Units on axes labels In the graph from Example 4.2.11, notice how both axes indicate units that help describe the meaning of each variable. Whenever a graph has real-world context, be sure to label both axes clearly with both variable name (like x) and units.

Example 4.2.14. Build a table and graph the equation $y = x^2$. Use x-values from -3 to 3.

Solution.

x	$y = x^2$	Point
-3	$(-3)^2 = 9$	$(-3, 9)$
-2	$(-2)^2 = 4$	$(-2, 4)$
-1	$(-1)^2 = 1$	$(-1, 1)$
0	$(0)^2 = 0$	$(0, 0)$
1	$(1)^2 = 1$	$(0, 1)$
2	$(2)^2 = 4$	$(2, 4)$
3	$(3)^2 = 9$	$(3, 9)$

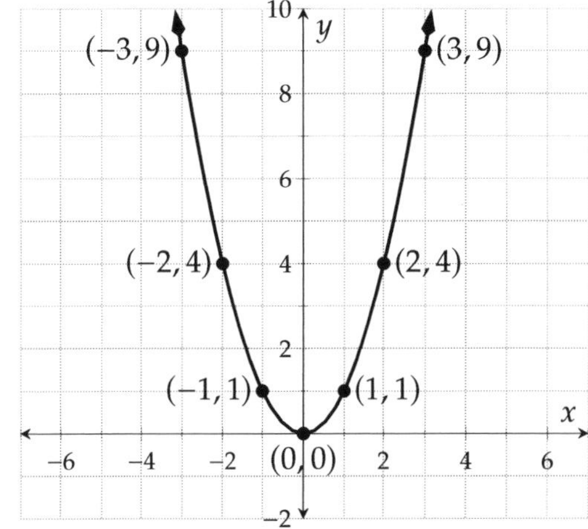

In this example, the points do not fall on a straight line. Many algebraic equations have graphs that are non-linear, where the points do not fall on a straight line. Since each x-value corresponds to a single y-value (the square of x) we connected the points with a smooth curve, sketching from left to right.

4.2.1 Exercises

Determine if the given points are to be included in the graph of the equation.

For the following exercises: Consider the equation

1. $y = 9x + 6$

Which of the following ordered pairs are solutions to the given equation? There may be more than one correct answer.

□ $(-9, -75)$ □ $(-2, -12)$ □ $(0, 10)$ □ $(4, 44)$

2. $y = 10x + 3$

Which of the following ordered pairs are solutions to the given equation? There may be more than one correct answer.

□ $(0, 5)$ □ $(-5, -47)$ □ $(-2, -17)$ □ $(8, 88)$

3. $y = -10x - 2$

Which of the following ordered pairs are solutions to the given equation? There may be more than one correct answer.

□ $(-2, 22)$ □ $(0, -2)$ □ $(3, -32)$ □ $(-4, 38)$

4. $y = -9x - 6$

Which of the following ordered pairs are solutions to the given equation? There may be more than one correct answer.

□ $(0, -6)$ □ $(-2, 14)$ □ $(7, -69)$ □ $(-7, 57)$

5. $y = \frac{2}{5}x - 5$

Which of the following ordered pairs are solutions to the given equation? There may be more than one correct answer.

□ $(-5, -3)$ □ $(-20, -13)$ □ $(5, -3)$
□ $(0, 0)$

6. $y = \frac{4}{5}x - 2$

Which of the following ordered pairs are solutions to the given equation? There may be more than one correct answer.

□ $(-5, -4)$ □ $(0, 0)$ □ $(-20, -18)$
□ $(25, 18)$

7. $y = -\frac{3}{4}x - 5$

Which of the following ordered pairs are solutions to the given equation? There may be more than one correct answer.

□ $(12, -9)$ □ $(-12, 8)$ □ $(-8, 1)$
□ $(0, -5)$

8. $y = -\frac{5}{6}x - 1$

Which of the following ordered pairs are solutions to the given equation? There may be more than one correct answer.

□ $(0, -1)$ □ $(18, -15)$ □ $(-6, 9)$
□ $(-24, 19)$

Make a table for the equation with x-values suggested.

9. Fill out this table for the equation $y = -x + 8$. The first row is an example.

x	y	Points
−3	11	$(-3, 11)$
−2	____	________
−1	____	________
0	____	________
1	____	________
2	____	________

10. Fill out this table for the equation $y = -x + 9$. The first row is an example.

x	y	Points
−3	12	$(-3, 12)$
−2	____	________
−1	____	________
0	____	________
1	____	________
2	____	________

11. Fill out this table for the equation $y = 6x + 1$. The first row is an example.

x	y	Points
−3	−17	$(-3, -17)$
−2	____	________
−1	____	________
0	____	________
1	____	________
2	____	________

12. Fill out this table for the equation $y = 2x + 7$. The first row is an example.

x	y	Points
−3	1	$(-3, 1)$
−2	____	________
−1	____	________
0	____	________
1	____	________
2	____	________

13. Fill out this table for the equation $y = -5x + 4$. The first row is an example.

x	y	Points
-3	19	$(-3, 19)$
-2	______	____________
-1	______	____________
0	______	____________
1	______	____________
2	______	____________

14. Fill out this table for the equation $y = -4x + 10$. The first row is an example.

x	y	Points
-3	22	$(-3, 22)$
-2	______	____________
-1	______	____________
0	______	____________
1	______	____________
2	______	____________

15. Fill out this table for the equation $y = \frac{3}{8}x - 7$. The first row is an example.

x	y	Points
-24	-16	$(-24, -16)$
-16	______	____________
-8	______	____________
0	______	____________
8	______	____________
16	______	____________

16. Fill out this table for the equation $y = \frac{5}{4}x + 1$. The first row is an example.

x	y	Points
-12	-14	$(-12, -14)$
-8	______	____________
-4	______	____________
0	______	____________
4	______	____________
8	______	____________

17. Fill out this table for the equation $y = -\frac{7}{10}x + 10$. The first row is an example.

x	y	Points
-30	31	$(-30, 31)$
-20	______	____________
-10	______	____________
0	______	____________
10	______	____________
20	______	____________

18. Fill out this table for the equation $y = -\frac{7}{6}x - 2$. The first row is an example.

x	y	Points
-18	19	$(-18, 19)$
-12	______	____________
-6	______	____________
0	______	____________
6	______	____________
12	______	____________

Make a table for the equation.

19. Make a table of solutions for the equation $y = 14x$.

x	y

20. Make a table of solutions for the equation $y = 18x$.

x	y

21. Make a table of solutions for the equation $y = -10x + 1$.

x	y

22. Make a table of solutions for the equation $y = -8x - 5$.

x	y

23. Make a table of solutions for the equation $y = \frac{3}{8}x - 5$.

x	y

24. Make a table of solutions for the equation $y = \frac{11}{10}x - 3$.

x	y

25. Make a table of solutions for the equation $y = -\frac{7}{12}x - 7$.

x	y

26. Make a table of solutions for the equation $y = \frac{13}{15}x + 2$.

x	y

These exercises have Cartesian plots with some context.

27. A certain water heater will cost you \$900 to buy and have installed. This water heater claims that its operating expense (money spent on electricity or gas) will be about \$31 per month. According to this information, the equation $y = 900 + 31x$ models the total cost of the water heater after x months, where y is in dollars. Make a table of at least five values and plot a graph of this equation.

28. You bought a new Toyota Corolla for \$18,600 with a zero interest loan over a five-year period. That means you'll have to pay \$310 each month for the next five years (sixty months) to pay it off. According to this information, the equation $y = 18600 - 310x$ models the loan balance after x months, where y is in dollars. Make a table of at least five values and plot a graph of this equation. Make sure to include a data point representing when you will have paid off the loan.

29. The pressure inside a full propane tank will rise and fall if the ambient temperature rises and falls. The equation $P = 0.1963(T + 459.67)$ models this relationship, where the temperature T is measured in °F and the pressure and the pressure P is measured in $\frac{\text{lb}}{\text{in}^2}$. Make a table of at least five values and plot a graph of this equation. Make sure to use T-values that make sense in context.

30. A beloved coworker is retiring and you want to give her a gift of week-long vacation rental at the coast that costs \$1400 for the week. You might end up paying for it yourself, but you ask around to see if the other 29 office coworkers want to split the cost evenly. The equation $y = \frac{1400}{x}$ models this situation, where x people contribute to the gift, and y is the dollar amount everyone contributes. Make a table of at least five values and plot a graph of this equation. Make sure to use x-values that make sense in context.

Plot some graphs.

31. Create a table of ordered pairs and then make a plot of the equation $y = 2x + 3$.

32. Create a table of ordered pairs and then make a plot of the equation $y = 3x + 5$.

33. Create a table of ordered pairs and then make a plot of the equation $y = -4x + 1$.

34. Create a table of ordered pairs and then make a plot of the equation $y = -x - 4$.

35. Create a table of ordered pairs and then make a plot of the equation $y = \frac{5}{2}x$.

36. Create a table of ordered pairs and then make a plot of the equation $y = \frac{4}{3}x$.

37. Create a table of ordered pairs and then make a plot of the equation $y = -\frac{2}{5}x - 3$.

38. Create a table of ordered pairs and then make a plot of the equation $y = -\frac{3}{4}x + 2$.

39. Create a table of ordered pairs and then make a plot of the equation $y = x^2 + 1$.

40. Create a table of ordered pairs and then make a plot of the equation $y = (x - 2)^2$. Use x-values from 0 to 4.

41. Create a table of ordered pairs and then make a plot of the equation $y = -3x^2$.

42. Create a table of ordered pairs and then make a plot of the equation $y = -x^2 - 2x - 3$.

4.3 Exploring Two-Variable Data and Rate of Change

This section is about examining data that has been plotted on a Cartesian coordinate system, and then making observations. In some cases, we'll be able to turn those observations into useful mathematical calculations.

4.3.1 Modeling data with two variables

Using mathematics, we can analyze real data from the world around us. We can use what we discover to better understand the world, and sometimes to make predictions. Here's an example of data about the economic situation in the US:

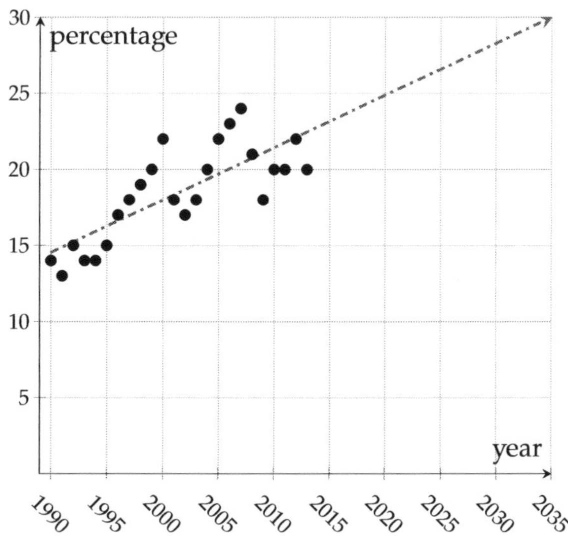

Figure 4.3.2: Share of all income held by the top 1 %, United States, 1990–2013 (www.epi.org)

If this trend continues, what percentage of all income will the top 1 % have in the year 2030? If we model data in the chart with the trend line, we can estimate the value to be 28.6 %. This is one way math is used in real life.

Does that trend line have an equation like those we looked at in Section 4.2? Is it even correct to look at this data set and decide that a straight line is a good model? These are some of the questions we want to consider as we begin this section. The answers will evolve through the next several sections.

4.3.2 Patterns in Tables

Example 4.3.3. Find a pattern in each table. What is the missing entry in each table? Can you describe each pattern in words and/or mathematics?

black	white
big	small
short	tall
few	

USA	Washington
UK	London
France	Paris
Mexico	

1	2
2	4
3	6
5	

Figure 4.3.4: Patterns in 3 tables

Solution.

black	white
big	small
short	tall
few	*many*

USA	Washington
UK	London
France	Paris
Mexico	*Mexico City*

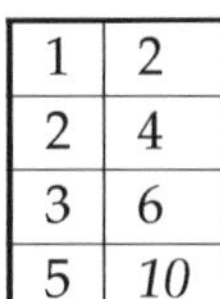

1	2
2	4
3	6
5	*10*

Figure 4.3.5: Patterns in 3 tables

First table Each word on the right has the opposite meaning of the word to its left.

Second table Each city on the right is the capital of the country to its left.

Third table Each number on the right is double the number to its left.

We can view each table as assigning each input in the left column a corresponding output in the right column. In the first table, for example, when the input "big" is on the left, the output "small" is on the right. The first table's function is to output a word with the opposite meaning of each input word. (This is not a numerical example.)

The third table *is* numerical. And its function is to take a number as input, and give twice that number as its output. Mathematically, we can describe the pattern as "$y = 2x$," where x represents the input, and y represents the output. Labeling the table mathematically, we have Table 4.3.6.

x (input)	y (output)
1	2
2	4
3	6
5	10
10	20
Pattern: $y = 2x$	

Table 4.3.6: Table with a mathematical pattern

The equation $y = 2x$ summarizes the pattern in the table. For each of the following tables, find an equation that describes the pattern you see. Numerical pattern recognition may or may not come naturally for you. Either way, pattern recognition is an important mathematical skill that anyone can develop. Solutions for these exercises provide some ideas for recognizing patterns.

Exercise 4.3.7.

x	y
0	10
1	11
2	12
3	13

Write an equation in the form $y = \ldots$ suggested by the pattern in the table.

Solution. Looking for a similar relationship in each row is one approach to pattern recognition. Here, the y-value in each row is 10 greater than its corresponding x-value. So the equation $y = x + 10$ describes the pattern. Of course, there are more complicated patterns to explore, as we'll see in the next exercise.

Exercise 4.3.8.

x	y
0	-1
1	2
2	5
3	8

Write an equation in the form $y = \ldots$ suggested by the pattern in the table.

Solution. The relationship between x and y in each row is not as clear here. Another popular approach for finding patterns: in each column, consider how the values change from one row to the next. From row to row, the x-value increases by 1. Also, the y-value increases by 3 from row to row.

	x	y	
	0	−1	
$+1 \rightarrow$	1	2	$\leftarrow +3$
$+1 \rightarrow$	2	5	$\leftarrow +3$
$+1 \rightarrow$	3	8	$\leftarrow +3$

Since row-to-row change is always 1 for x and is always 3 for y, the rate of change from one row to another row is always the same: 3 units of y for every 1 unit of x.

We know that the output for $x = 0$ is $y = -1$. And our observation about the constant rate of change tells us that if we increase the input by x units from 0, the ouput should increase by $\overbrace{3+3+\cdots+3}^{x\text{ times}}$, which is $3x$. So the output would be $-1 + 3x$. So the equation is $y = 3x - 1$.

Exercise 4.3.9.

x	y
0	0
1	1
2	4
3	9

Write an equation in the form $y = \ldots$ suggested by the pattern in the table.

Solution. Looking for a relationship in each row here, we see that each y-value is the square of the corresponding x-value. So the equation is $y = x^2$.

What if we had tried the approach we used in the previous exercise, comparing change from row to row in each column?

	x	y	
	0	0	
$+1 \rightarrow$	1	1	$\leftarrow +1$
$+1 \rightarrow$	2	4	$\leftarrow +3$
$+1 \rightarrow$	3	9	$\leftarrow +5$

Here, the rate of change is *not* constant from one row to the next. While the x-values are increasing by 1 from row to row, the y-values increase more and more from row to row. Notice that there is a pattern there as well? Mathematicians are fascinated by relationships that produce more complicated patterns. (We'll study more complicated patterns later.)

4.3.3 Rate of Change

For an hourly wage-earner, the amount of money they earn depends on how many hours they work. If a worker earns \$15 per hour, then 10 hours of work corresponds to \$150 of pay. Working *one* additional hour will change 10 hours to 11 hours; and this will cause the \$150 in pay to rise by *fifteen* dollars to \$165 in pay. Any time we compare how one amount changes (dollars earned) as a consequence of another amount changing (hours worked), we are talking about a **rate of change**.

Given a table of two-variable data, between any two rows we can compute a **rate of change**.

Example 4.3.10. The following data, given in both table and graphed form, gives the counts of invasive cancer diagnoses in Oregon over a period of time. (wonder.cdc.gov)

Year	Invasive Cancer Incidents
1999	17,599
2000	17,446
2001	17,847
2002	17,887
2003	17,559
2004	18,499
2005	18,682
2006	19,112
2007	19,376
2008	20,370
2009	19,909
2010	19,727
2011	20,636
2012	20,035
2013	20,458

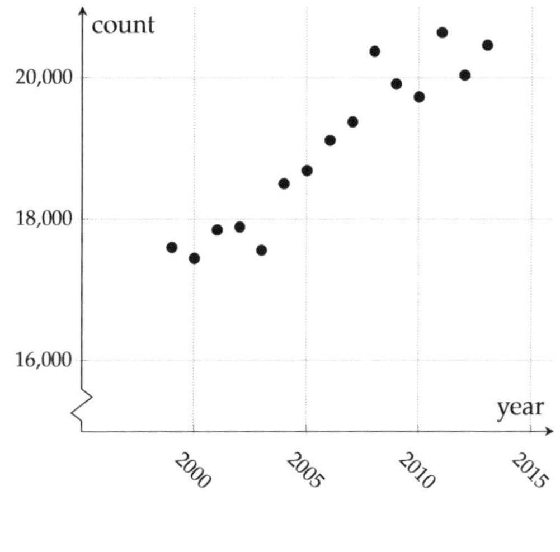

What is the **rate of change** in Oregon invasive cancer diagnoses between 2000 and 2010? The total (net) change in diagnoses over that timespan is

$$19727 - 17446 = 2281.$$

Since 10 years passed (which you can calculate as 2010 – 2000), the rate of change is 2281 diagnoses per 10 years, or

$$\frac{2281 \text{ diagnoses}}{10 \text{ year}} = 228.1 \frac{\text{diagnoses}}{\text{year}}.$$

We read that last quantity as "228.1 diagnoses per year". This rate of change means that between the years 2000 and 2010, there were 228.1 more diagnoses each year, on average. (Notice that there was no single year in that span when diagnoses increased by 228.1.)

Let's practice calculating rates of change over different timespans:

Exercise 4.3.11. Use the data in Example 4.3.10 to find the rate of change in Oregon invasive cancer diagnoses between 1999 and 2002. Just give the numerical value; the units are provided.

[ ] $\frac{\text{diagnoses}}{\text{year}}$

And what was the rate of change between 2003 and 2011?

[ ] $\frac{\text{diagnoses}}{\text{year}}$

Solution. To find the rate of change between 1999 and 2002, calculate

$$\frac{17887 - 17599}{2002 - 1999} = 96.$$

To find the rate of change between 2003 and 2011, calculate

$$\frac{20636 - 17559}{2011 - 2003} = 384.625.$$

We are ready to give a formal defintion for **rate of change**. Considering our work from Example 4.3.10 and Exercise 4.3.11, we settle on:

Definition 4.3.12 (Rate of Change). If (x_1, y_1) and (x_2, y_2) are two data points from a set of two-variable data, then the **rate of change** between them is

$$\frac{\text{change in } y}{\text{change in } x} = \frac{\Delta y}{\Delta x} = \frac{y_2 - y_1}{x_2 - x_1}.$$

(The Greek letter delta, Δ, is used to represent "change in" since it is the first letter of the Greek word for "difference".)

In Example 4.3.10 and Exercise 4.3.11 we found three rates of change. Figure 4.3.13 highlights the three pairs of points that were used to make these calculations.

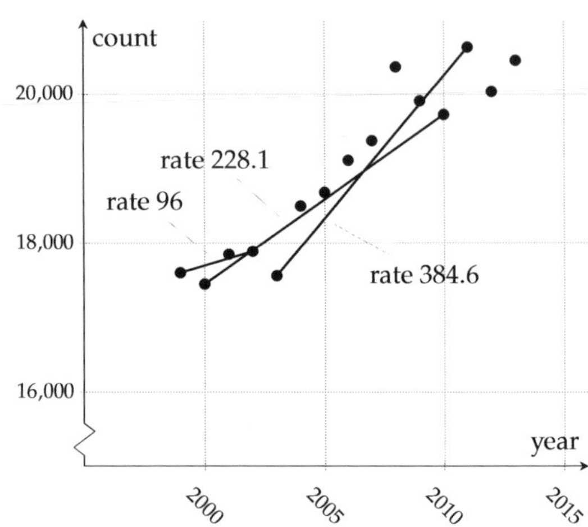

Figure 4.3.13

Note how the larger the numerical rate of change between two points, the steeper the line is that connects them. This is such an important observation, we'll put it in an official remark.

Remark 4.3.14. The rate of change between two data points is intimately related to the steepness of the line segment that connects those points.

1. The steeper the line, the larger the rate of change, and vice versa.
2. If one rate of change between two data points equals another rate of change between two different data points, then the corresponding line segments will have the same steepness.
3. When a line segment between two data points slants down from left to right, the rate of change between those points will be negative.

In the solution to Exercise 4.3.8, the key observation was that the **rate of change** from one row to the next was constant: 3 units of increase in y for every 1 unit of increase in x. Graphing this pattern in Figure 4.3.15, we see that every line segment here has the same steepness, so the entire graph is a line.

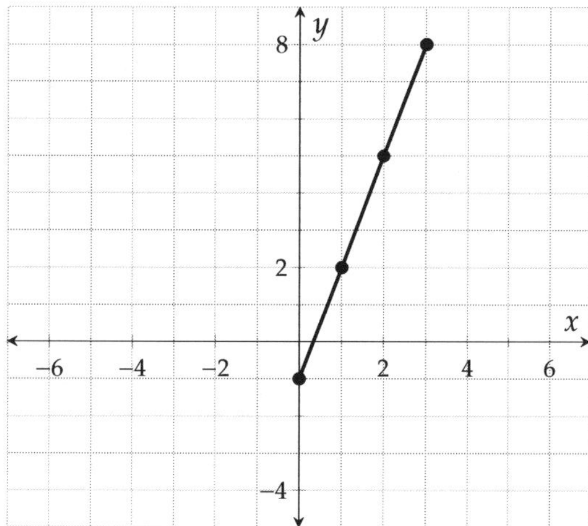

Figure 4.3.15

Whenever the rate of change is constant no matter which two (x, y)-pairs (or data pairs) are chosen from a data set, then you can conclude the graph will be a straight line *even without making the graph*. For obvious reasons, we call this kind of relationship a **linear** relationship. We'll study linear relationships in more detail throughout this chapter. Right now in this section, we feel it is important to simply identify if data has a linear relationship or not.

Identify if each pattern below describes a linear relationship or not.

Exercise 4.3.16. Is there a linear relationship in the table?

x	y
-8	3.1
-5	2.1
-2	1.1
1	0.1

(□ The relationship is linear □ The relationship is not linear)

Solution. From one x-value to the next, the change is always 3. From one y-value to the next, the change is alwasy -1. So the rate of change is always $\frac{-1}{3} = -\frac{1}{3}$. Since the rate of change is constant, the data have a linear relationship.

Exercise 4.3.17. Is there a linear relationship in the table?

x	y
11	208
13	210
15	214
17	220

(□ The relationship is linear □ The relationship is not linear)

Solution. The rate of change between the first two points is $\frac{210-208}{13-11} = 1$. The rate of change between the last two points is $\frac{220-214}{17-15} = 3$. This is one way to demonstrate that the rate of change differs for different pairs of points, so this pattern is not linear.

Exercise 4.3.18. Is there a linear relationship in the table?

x	y
3	-2
6	-8
8	-12
12	-20

(□ The relationship is linear □ The relationship is not linear)

Solution. The changes in x from one row to the next are +3,+2, and +8. That's not a consistent pattern, but we need to consider rates of change between points. The rate of change between the first two points is $\frac{-8-(-2)}{6-3} = -2$. The rate of change between the next two points is $\frac{-12-(-8)}{8-6} = -2$. And the rate of change between the last two points is $\frac{-20-(-12)}{12-8} = -2$. So the rate of change, -2, is constant regardless of which pairs we choose. That means these pairs describe a linear relationship.

Let's return to the data that we opened the section with, in Figure 4.3.2. Is that data linear? Well, yes and no. To be completely honest, it's not linear. It's easy to pick out pairs of points where the steepness changes from one pair to the next. In other words, the points do not all fall into a single line.

However if we stand back, there does seem to be an overall upward trend that is captured by the line someone has drawn over the data. Points *on this line* do have a linear pattern. Let's estimate the rate of change between some points on this line. We are free to use any points to do this, so let's make this calculation easier by choosing points we can clearly identify on the graph: (1991, 15) and (2020, 25).

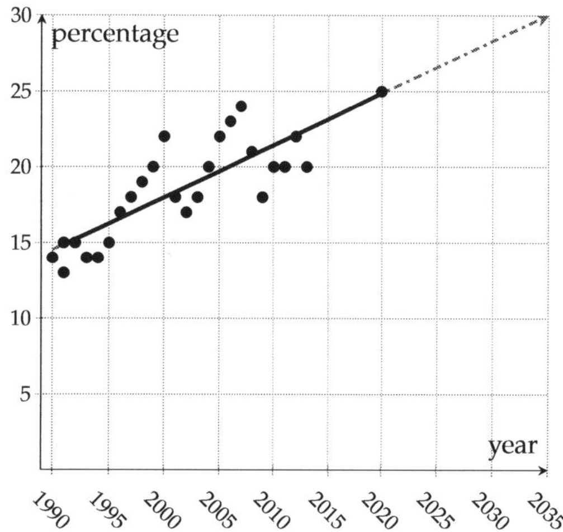

Figure 4.3.19: Share of all income held by the top 1%, United States, 1990–2013 (www.epi.org)

The rate of change between those two points is

$$\frac{25-15}{2020-1991} = \frac{10}{29} \approx 0.3448.$$

So we might say that *on average* the rate of change expressed by this data is $0.3448\,\frac{\%}{\text{yr}}$.

4.3.4 Exercises

Find a formula relating y to x using a table. The pattern should be relatively easy to spot. Later sections will address more specific ways to approach this task in general.

For the following exercises: Write an equation in the form $y = \ldots$ suggested by the pattern in the table.

1.

x	y
−2	−8
−1	−4
0	0
1	4
2	8

2.

x	y
3	12
4	16
5	20
6	24
7	28

3.

x	y
6	11
7	12
8	13
9	14
10	15

4.

x	y
7	16
8	17
9	18
10	19
11	20

5.

x	y
18	11
12	5
14	7
19	12
7	0

6.

x	y
0	2
6	8
3	5
14	16
10	12

7.

x	y
1	1
25	5
9	3
4	2
16	4

8.

x	y
−4	4
−2	2
−1	1
−4	4
−2	2

9.

x	y
0.1	0.01
0.3	0.09
0.5	0.25
0.7	0.49
0.9	0.81

10.

x	y
0.7	0.49
1	1
1.3	1.69
1.6	2.56
1.9	3.61

11.

x	y
59	$\frac{1}{59}$
54	$\frac{1}{54}$
24	$\frac{1}{24}$
31	$\frac{1}{31}$
41	$\frac{1}{41}$

12.

x	y
70	$\frac{1}{70}$
20	$\frac{1}{20}$
65	$\frac{1}{65}$
4	$\frac{1}{4}$
40	$\frac{1}{40}$

Calculate some rates of change.

13. This table gives population estimates for Portland, Oregon from 1990 through 2014 (www.google.com/publicdata as of Oct 11, 2016; Google may update with better estimates).

Year	Population	Year	Population
1990	487849	2003	539546
1991	491064	2004	533120
1992	493754	2005	534112
1993	497432	2006	538091
1994	497659	2007	546747
1995	498396	2008	556442
1996	501646	2009	566143
1997	503205	2010	585261
1998	502945	2011	593859
1999	503637	2012	602954
2000	529922	2013	609520
2001	535185	2014	619360
2002	538803		

Find the rate of change in Portand population between 1991 and 2007. Just give the numerical value; the units are provided.

And what was the rate of change between 2010 and 2011?

List all the years where there is a negative rate of change between that year and the next year.

14. This table and graph gives population estimates for Portland, Oregon from 1990 through 2014 (www.google.com/publicdata as of Oct 11, 2016; Google may update with better estimates).

Year	Population	Year	Population
1990	487849	2003	539546
1991	491064	2004	533120
1992	493754	2005	534112
1993	497432	2006	538091
1994	497659	2007	546747
1995	498396	2008	556442
1996	501646	2009	566143
1997	503205	2010	585261
1998	502945	2011	593859
1999	503637	2012	602954
2000	529922	2013	609520
2001	535185	2014	619360
2002	538803		

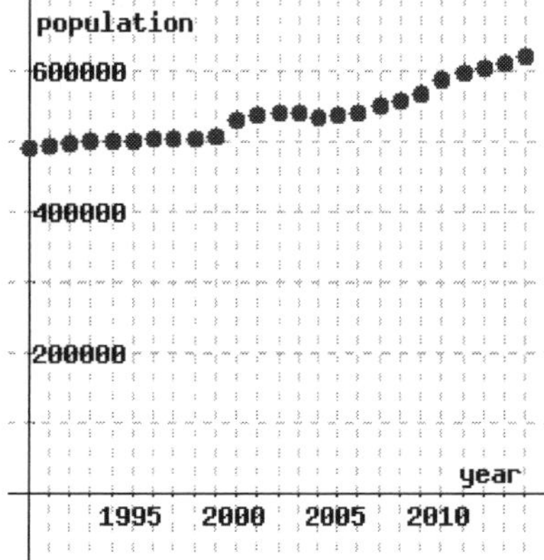

Between what two years that are two years apart was the rate of change highest?

What was that rate of change? Just give the numerical value; the units are provided.

$\square \ \frac{\text{people}}{\text{year}}$

4.4 Slope

In Section 4.3, we observed that a constant rate of change between points produces a linear relationship, whose graph is a straight line. Such a constant rate of change has a special name, **slope**, and we'll explore slope in more depth here.

4.4.1 What is slope?

When the **rate of change** from point to point never changes, those points must fall on a straight line, as in Figure 4.4.2, and there is a **linear relationship** between the variables x and y.

Rather than say "constant rate of change" in every such situation, mathematicians call that common rate of change **slope.**

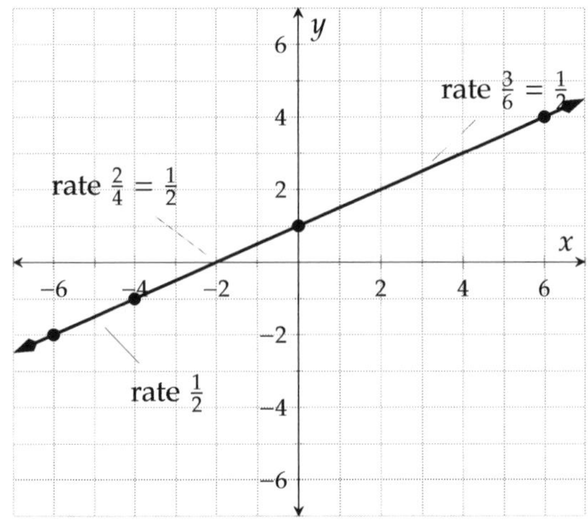

Figure 4.4.2: Between successive points, the rate of change is always 1/2.

Definition 4.4.3 (Slope). When x and y are two variables where the rate of change between any two points is always the same, we call this common rate of change the **slope**. Since having a constant rate of change means the graph will be a straight line, it's also called the **slope of the line**.

Considering the definition for Definition 4.3.12, this means that you can calculate slope, m, as

$$m = \frac{\text{change in } y}{\text{change in } x} = \frac{\Delta y}{\Delta x} \tag{4.4.1}$$

when x and y have a linear relationship.

A slope is a rate of change. So if there are units for the horizontal and vertical variables, then there will be units for the slope. The slope will be measured in $\frac{\text{vertical units}}{\text{horizontal units}}$.

If the slope is nonzero, we say that there is a **linear relationship** between x and y. When the slope is 0, we say that y is **constant** with respect to x.

Here are some scenarios with different slopes. Note that a slope is more meaningful with units.

- If a tree grows 2.5 feet every year, it's rate of change in height is the same from year to year. So the height and time have a linear relationship where the **slope** is $2.5\,\frac{\text{ft}}{\text{yr}}$.
- If a company loses 2 million dollars every year, it's rate of change in reserve funds is the same from year to year. So the company's reserve funds and time have a linear relationship where the **slope** is -2 million dollars per year.
- If Cara has stopped growing, her rate of change in height is the same from year to year—it's zero. So the **slope** is $0\,\frac{\text{in}}{\text{yr}}$. Cara's height is **constant** with respect to time. In a statistics course, you would say that height and time don't have a relationship at all, in the sense that information about Cara's height tells you nothing about her age.

Remark 4.4.4. A useful phrase for remembering the definition of **slope** is "rise over run". Here, "rise" refers to "change in y," Δy, and "run" refers to "change in x," Δx. Be careful though. As we have learned, the horizontal direction comes *first* in mathematics, followed by the vertical direction. The phrase "rise over run" reverses this. (It's a bit awkward to say, but the phrase "run under rise" puts the horizontal change first.)

Slope m Why is the letter m commonly used as the symbol for "slope"? Some believe that it comes from the French word "monter" which means "to climb." And French has a special place in this topic because Descartes was French. Interestingly, there are reports of French algebra classrooms using s to stand for slope. When asked why, some say that it's because of the *English* word "slope."

Example 4.4.5 (Matthew's Savings). On DEC. 31, Matthew had only \$50 in his savings account. For the the new year, he resolved to deposit \$20 into his savings account each week, without withdrawing any money from the account.

Matthew keeps his resolution, and his account balance increases steadily by \$20 each week. That's a constant rate of change, so his account balance and time have a linear relationship with slope 20 dollars/wk.

We can model the balance, y, in Matthew's savings account after x weeks with an equation. Since Matthew started with \$50 and adds \$20 each week, the account balance y after x weeks is

$$y = 50 + 20x \tag{4.4.2}$$

where y is a dollar amount. Notice that the slope, 20 dollars/wk, serves as the multiplier for x weeks.

We can also consider Matthew's savings using a table.

	x, weeks since DEC. 31	y, savings account balance (dollars)	
	0	50	
x increases by 1$\longrightarrow$	1	70	$\longleftarrow y$ increases by 20
x increases by 1$\longrightarrow$	2	90	$\longleftarrow y$ increases by 20
x increases by 2$\longrightarrow$	4	130	$\longleftarrow y$ increases by 40
x increases by 3$\longrightarrow$	7	190	$\longleftarrow y$ increases by 60
x increases by 5$\longrightarrow$	12	290	$\longleftarrow y$ increases by 100

Table 4.4.6: Matthew's savings

In first few rows of the table, we see that when the number of weeks x increases by 1, the balance y increases by 20. The row-to-row rate of change is $\frac{20}{1} = 20$, the slope. In any table for a linear relationship, whenever x increases by 1 unit, y will increase by the slope.

In further rows, notice that as row-to-row change in x increases, row-to-row change in y increases proportionally to preserve the constant rate of change. Looking at the change in the last two rows of the table, we see x increases by 5 and y increases by 20, which gives rate of change $\frac{100}{5} = 20$, the value of the slope again.

We can "see" the rates of change between consecutive rows of the table on a graph of Matthew's savings by including **slope triangles**.

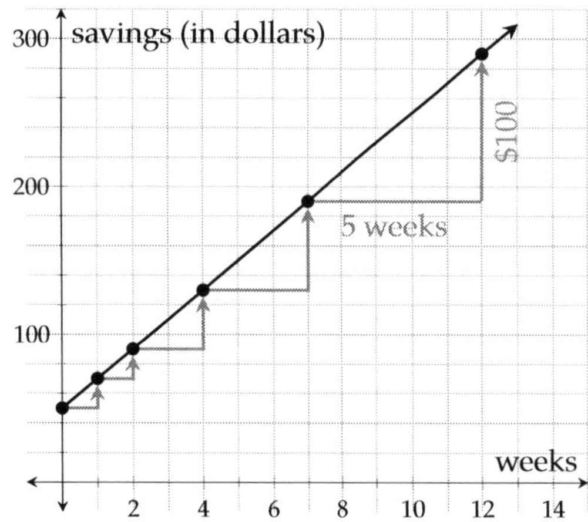

Figure 4.4.7: Matthew's savings

The large, labeled slope triangle indicates that when 5 weeks pass, Matthew saves \$100. This is the rate of change between the last two rows of the table, $\frac{100}{5} = 20\,{}^{\text{dollars}}/_{\text{wk}}$.

The smaller slope triangles indicate, from left to right, the rates of change $\frac{20}{1}$, $\frac{20}{1}$, $\frac{40}{2}$, and $\frac{60}{3}$ respectively. All of these rates simplify to the slope, $20\,{}^{\text{dollars}}/_{\text{wk}}$.

Every slope triangle on the graph of Matthew's savings has the same shape (geometrically, they are called similar triangles) since the ratio of vertical change to horizontal change is always $20\,{}^{\text{dollars}}/_{\text{wk}}$. On any graph of any line, we can draw a slope triangle and compute slope as "rise over run."

Of course, we could draw a slope triangle on the other side of the line:

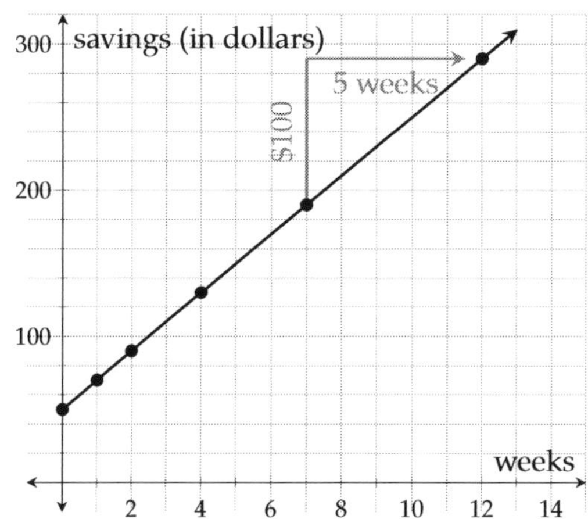

Figure 4.4.8: Matthew's savings

This slope triangle works just as well for identifying "rise" and "run," but it focuses on vertical change before horizontal change. For consistency with mathematical conventions, we will generally draw slope triangles showing horizontal change followed by vertical change, as in Figure 4.4.7.

Example 4.4.10. The following graph of a line models the amount of gas, in gallons, in a car's tank as it drives. Find the line's slope, and interpret its meaning in this context.

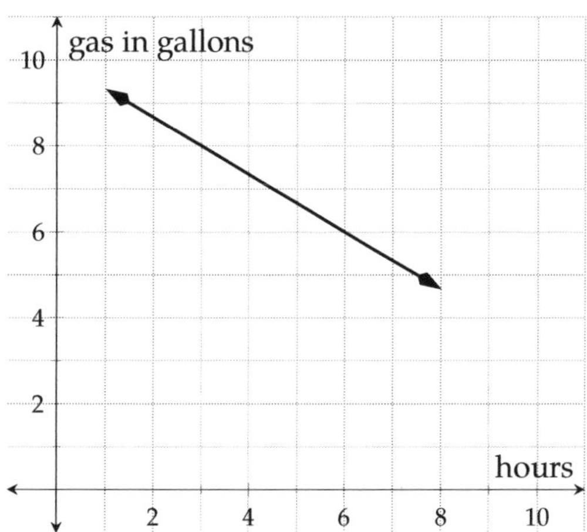

Figure 4.4.11: Amount of gas in a car's tank

Solution. To find a line's slope using its graph, we first identify two points on it, and then draw a slope triangle. Naturally, we would want to choose two points whose x- and y-coordinates are easy to identify exactly based on the graph. We will pick the two points where $x = 3$ and

$x = 6$, because they are right on the grid lines:

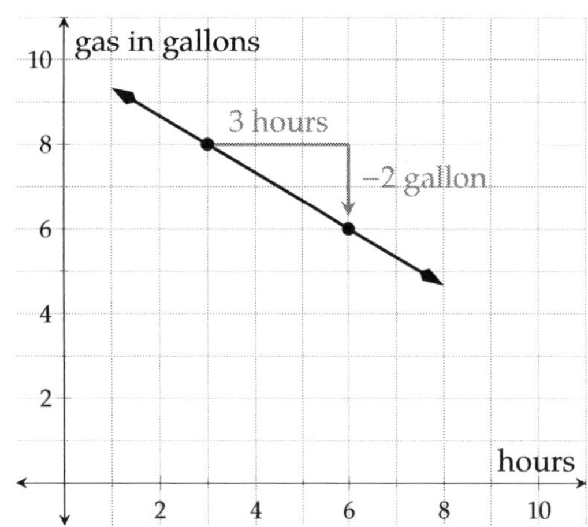

Notice that the *change* in y is negative, because the amount of gas is decreasing. Since we chose points with integer coordinates, we can easily calculate the slope:

$$\text{slope} = \frac{-2}{3} = -\frac{2}{3}.$$

Figure 4.4.12: A Good Slope Triangle

With units, the slope is $-\frac{2}{3}\,\frac{\text{gal}}{\text{h}}$. In the given context, this slope implies gas in the tank is *decreasing* at the rate of $\frac{2}{3}\,\frac{\text{gal}}{\text{h}}$. Since this slope is written as a fraction, there is another way to understand it: gas in the tank is decreasing by 2 gallons every 3 hours.

Exercise 4.4.13. The graph plots the number of invasive cancer diagnoses in Oregon over time, and a trend-line has been drawn.

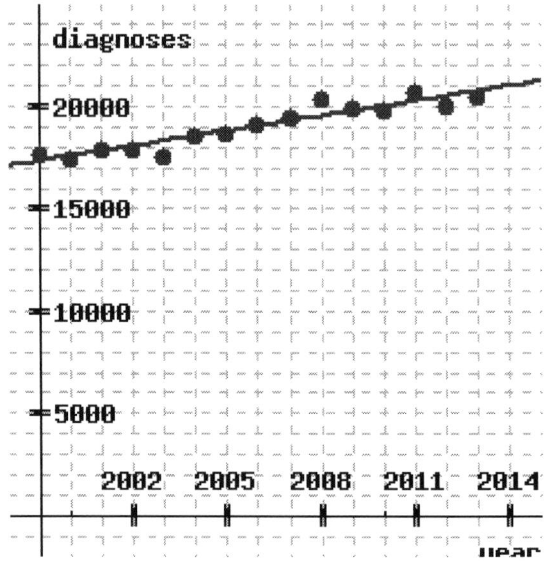

Estimate the slope of the trend-line. Just give the numerical value; the units are provided.

[ ] $\frac{\text{diagnoses}}{\text{year}}$

Solution. Do your best to identify two points on the line. With this line, it seems (2006, 19000) and (2014, 21000) are two such points.

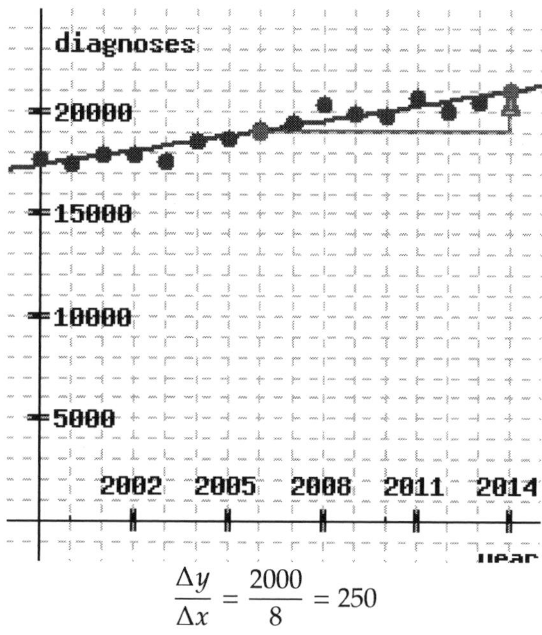

$$\frac{\Delta y}{\Delta x} = \frac{2000}{8} = 250$$

So the slope is about 250 diagnoses per year. This is only an estimate since we are not all certain the two points we chose are actually on the line.

4.4.2 Comparing Slopes

It's useful to understand what it means for different slopes to appear on the same coordinate system.

> **Example 4.4.14.** Effie, Ivan and Cleo are in a foot race. Figure 4.4.15 models the distance each has traveled in the first few seconds. Each runner takes a second to accelerate up to their running speed, but then runs at a constant speed. So they are then traveling with a constant rate of change, and the straight line portions of their graphs have a slope. Find each line's slope, and interpret its meaning in this context. What comparisons can you make with these runners?

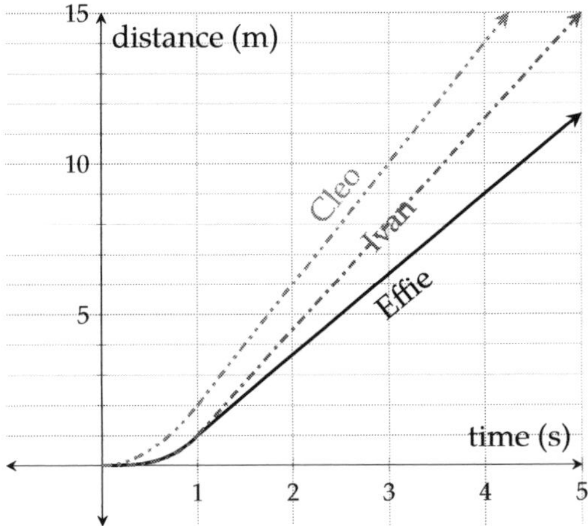

Figure 4.4.15: A three-way foot race

We will draw slope triangles to find each line's slope.

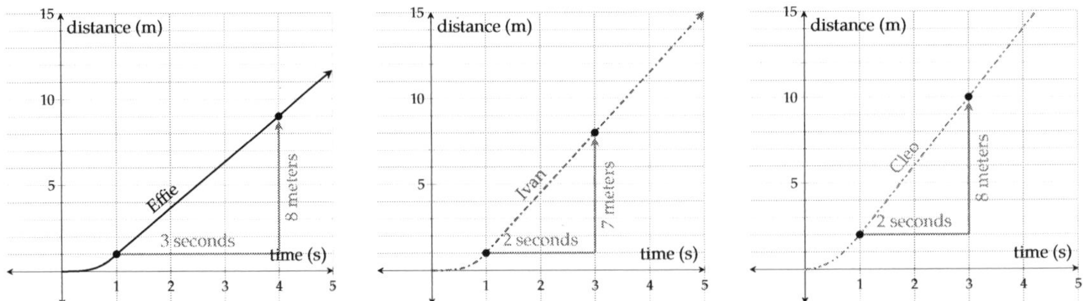

Figure 4.4.16: Find the Slope of Each Line

Using Formula (4.4.1), we have:

- Effie's slope is $\frac{8}{3} \approx 2.666$ meters per second
- Ivan's slope is $\frac{7}{2} = 3.5$ meters per second
- Cleo's slope is $\frac{8}{2} = 4$ meters per second

In a time-distance graph, the slope of a line represents speed. The slopes in these examples and the running speeds of these runners are both measured in $\frac{\text{m}}{\text{s}}$. Another important relationship we can see is that, the more sharply a line is slanted, the bigger the slope is. This should make sense because for each passing second, the faster person travels longer, making a slope triangle's height taller. This means that, numerically, we can tell that Cleo is the fastest runner (and Effie is the slowest) just by comparing the slopes $4 > 3.5 > 2.666$.

4.4.3 Finding Slope by Two Given Points

Several times in this section we computed a slope by drawing a slope triangle. That's not really necessary if you have coordinates for two points that a line passes through. In fact, sometimes it's impractical to draw a slope triangle.[1] Here we will stress how to find a line's slope without drawing a slope triangle.

Example 4.4.18. Your neighbor planted a sapling from Portland Nursery in his front yard. Ever since, for several years now, it has been growing at a constant rate. By the end of the third year, the tree was 15 ft tall; by the end of the sixth year, the tree was 27 ft tall. What's the tree's rate of growth (i.e. the slope)?

We *could* sketch a graph for this scenario, and include a slope triangle. If we did that, it would look like this:

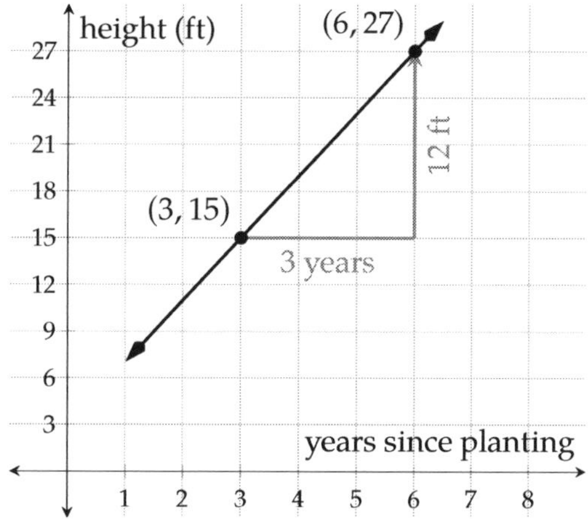

Figure 4.4.19: Height of a Tree

By the slope triangle and Equation (4.4.1) we have:

$$\begin{aligned} \text{slope} = m &= \frac{\Delta y}{\Delta x} \\ &= \frac{12}{3} \\ &= 4 \end{aligned}$$

So the tree is growing at a rate of $4\,\frac{\text{ft}}{\text{yr}}$.

But hold on. Did we really *need* this picture? The "rise" of 12 came from a subtraction of two y-values: $27 - 15$. And the "run" of 3 came from a subtraction of two x-values: $6 - 3$.

Here is a picture-free approach. We know that after 3 yr, the height is 15 ft. As an ordered pair, that information gives us the point $(3, 15)$ which we can label as $(\overset{x_1}{3}, \overset{y_1}{15})$. Similarly, the background information tells us to consider $(6, 27)$, which we label as $(\overset{x_2}{6}, \overset{y_2}{27})$. Here, x_1 and y_1 represent the first point's x-value and y-value, and x_2 and y_2 represent the second point's x-value and y-value.

It's important to use subscript instead of superscript here, because y^2 means to take the number y and square it. Whereas y_2 tells you that there are at least two y-values in the conversation,

[1]For instance if you only have specific information about two points that are too close together to draw a triangle, or if you cannot clearly see precise coordinates where you might start and stop your slope triangle.

and y_2 is the second of them.

Now we can write an alternative to Equation (4.4.1):

$$\text{slope} = m = \frac{\Delta y}{\Delta x} = \frac{y_2 - y_1}{x_2 - x_1} \tag{4.4.3}$$

This is known as the **slope formula**. The following graph will help you understand why this formula works. Basically, we are still using a slope triangle to calculate the slope.

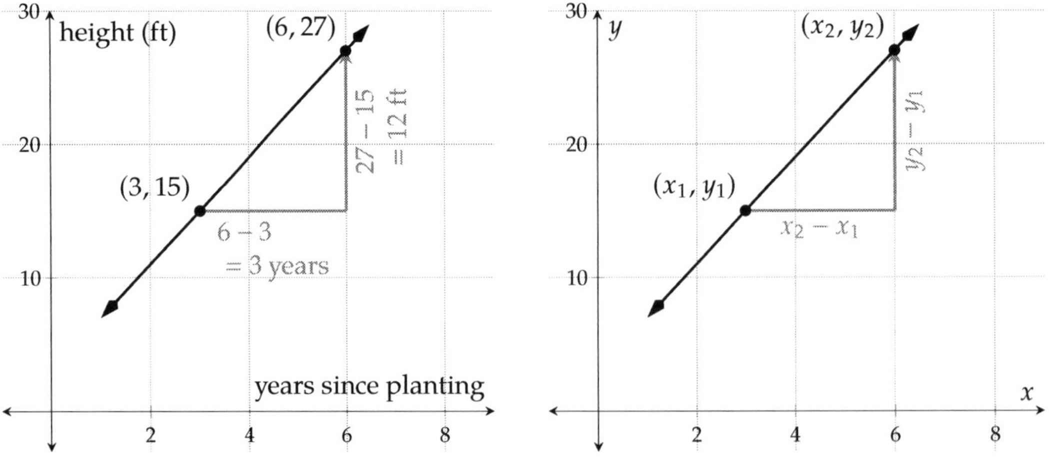

Figure 4.4.20: Understanding the slope formula

The beauty of the slope formula (4.4.3) is that to find a line's slope, we don't need to draw a slope triangle any more. Let's look at an example.

Example 4.4.21. A line passes the points $(-5, 25)$ and $(4, -2)$. Find this line's slope.

Solution. If you are new to this formula, it's important to label each number before using the formula. The two given points are:

$$(\overset{x_1}{-5}, \overset{y_1}{25}) \text{ and } (\overset{x_2}{4}, \overset{y_2}{-2})$$

Now apply the slope formula (4.4.3):

$$\begin{aligned} \text{slope} &= \frac{y_2 - y_1}{x_2 - x_1} \\ &= \frac{-2 - (25)}{4 - (-5)} \\ &= \frac{-27}{9} \end{aligned}$$

$$= -3$$

Note that we used parentheses when substituting in x_1 and y_1. This is a good habit to protect yourself from making errors with subtraction and double negatives.

Exercise 4.4.22 (Jogging on Mt. Hood). Tammy is training for a race up the slope of Mt. Hood, from Sandy to Government Camp, and then back. The graph below models her elevation from her starting point as time passes. Find the slopes of the three line segments, and interpret their meanings in this context.

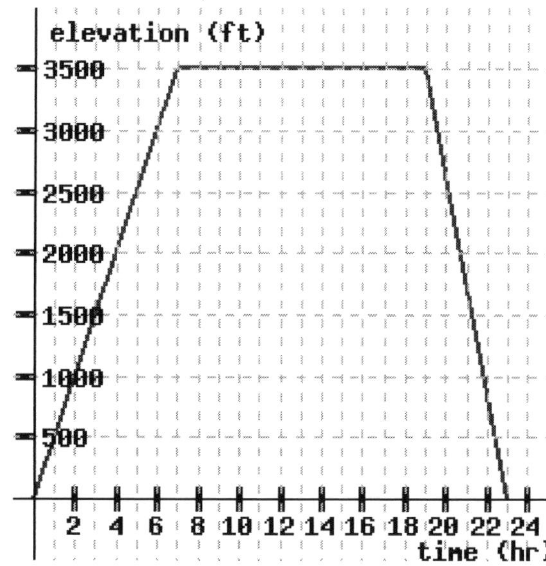

The first segment has slope ______.

The second segment has slope ______.

The third segment has slope ______.

Solution. The first segment started at $(0, 0)$ and stopped at $(7, 3500)$. This implies, Tammy started at the starting point, traveled 7 hours and reached a point 3500 feet higher in elevation from the starting point. The slope of the line is

$$\frac{\Delta y}{\Delta x} = \frac{3500}{7} = 500$$

and with units, that is 500 ft/hr. In context, Tammy was running, gaining 500 feet in elevation per hour.

The third segment started from $(19, 3500)$ and stopped at $(23, 0)$. This implies, Tammy started this part of her trip from a spot 3500 feet higher in elevation from the starting point, traveled for 4 hours and returned to the starting elevation. The slope of the line is

$$\frac{\Delta y}{\Delta x} = \frac{-3500}{4} = 875$$

and with units, that is −875 ft/hr. In context, Tammy was running, dropping in elevation by 875 feet per hour.

What happened in the second segment, which started at (7, 3500) and ended at (19, 3500)? This implies she started this portion 3500 feet higher in elevation from the starting point, and didn't change elevation for 19 hours. The slope of the line is

$$\frac{\Delta y}{\Delta x} = \frac{0}{19} = 0$$

and with units, that is 0 ft/hr. In context, Tammy was running but neither gaining nor losing elevation.This exercise demonstrates some facts. In a linear relationship, as the x-value increases

(in other words as you read its graph from left to right):

- if the y-values increase (in other words, the line goes upward), its slope is positive.
- if the y-values decrease (in other words, the line goes downward), its slope is negative.
- if the y-values don't change (in other words, the line is flat, or horizontal), its slope is 0.

4.4.4 Exercises

Find slope using a graph

1. A line's graph is shown below.

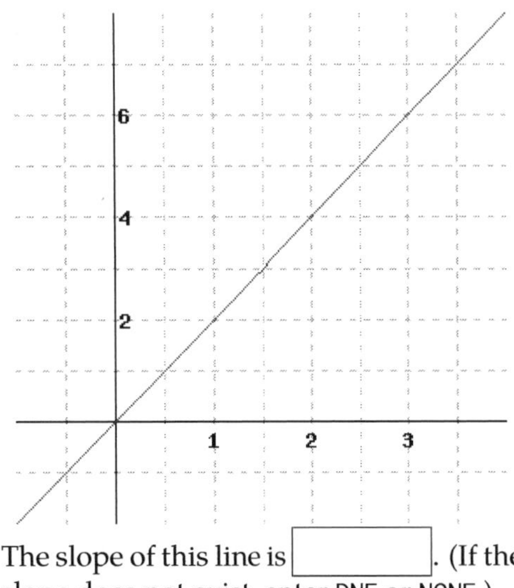

The slope of this line is ______. (If the slope does not exist, enter DNE or NONE.)

2. A line's graph is shown below.

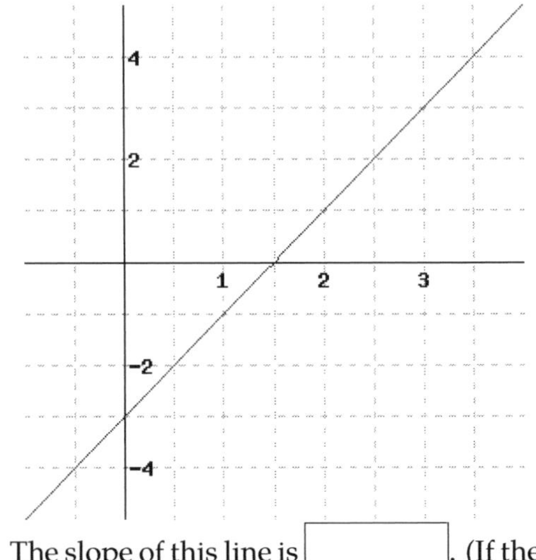

The slope of this line is ______. (If the slope does not exist, enter DNE or NONE.)

3. A line's graph is shown below.

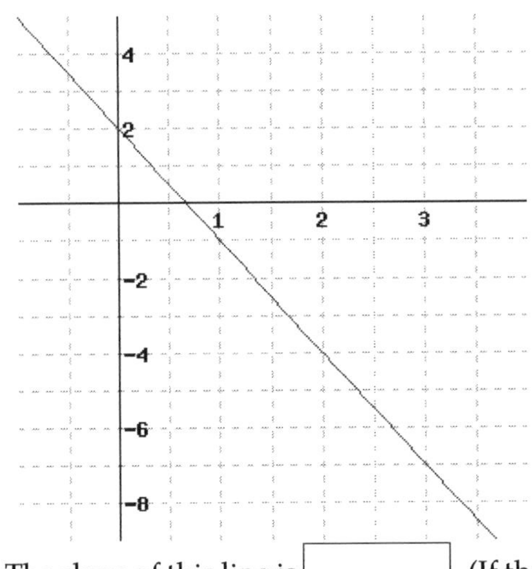

The slope of this line is ______. (If the slope does not exist, enter DNE or NONE.)

4. A line's graph is shown below.

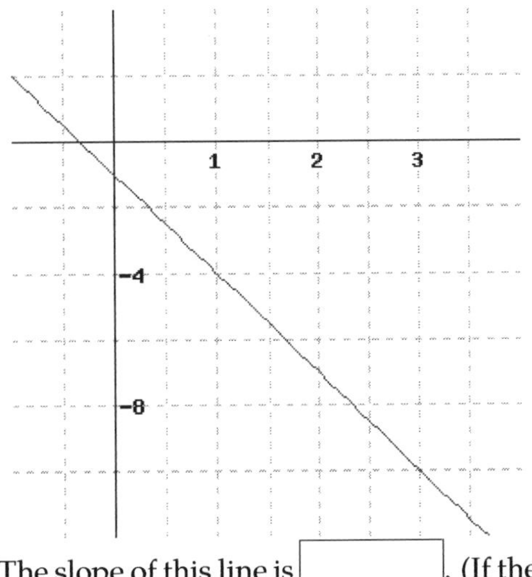

The slope of this line is ______. (If the slope does not exist, enter DNE or NONE.)

5. A line's graph is shown below.

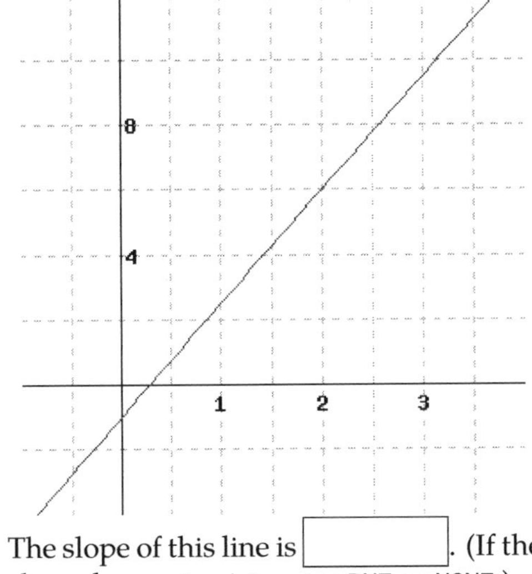

The slope of this line is ______. (If the slope does not exist, enter DNE or NONE.)

6. A line's graph is shown below.

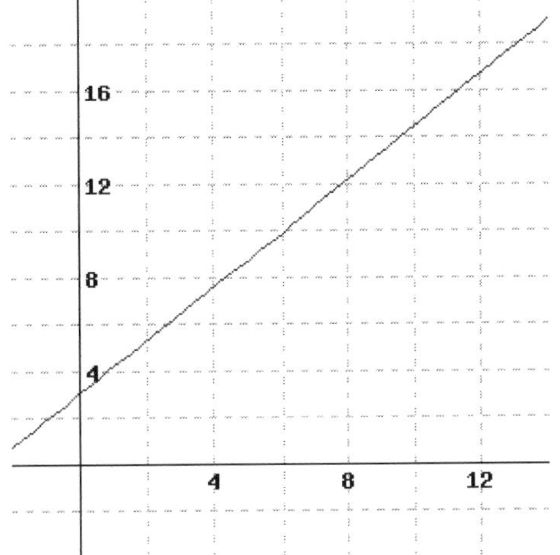

The slope of this line is ______. (If the slope does not exist, enter DNE or NONE.)

7. A line's graph is shown below.

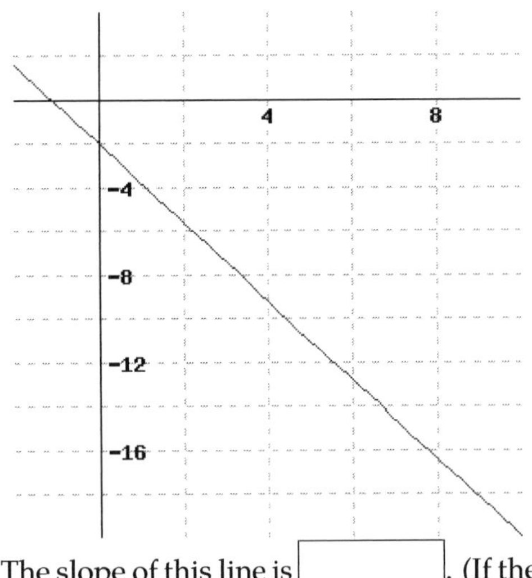

The slope of this line is ______. (If the slope does not exist, enter DNE or NONE.)

8. A line's graph is shown below.

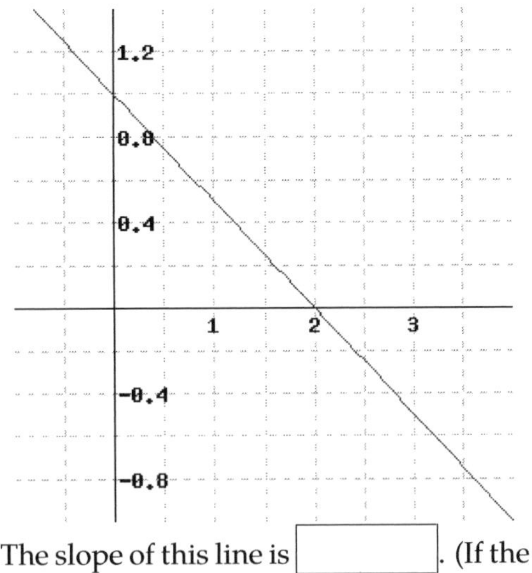

The slope of this line is ______. (If the slope does not exist, enter DNE or NONE.)

9. A line's graph is shown below.

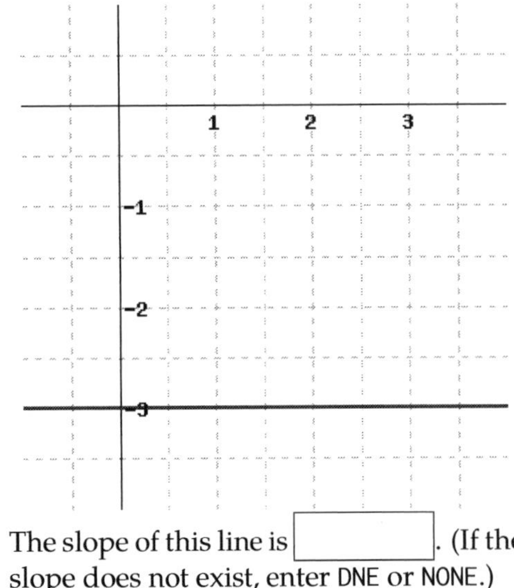

The slope of this line is ______. (If the slope does not exist, enter DNE or NONE.)

10. A line's graph is shown below.

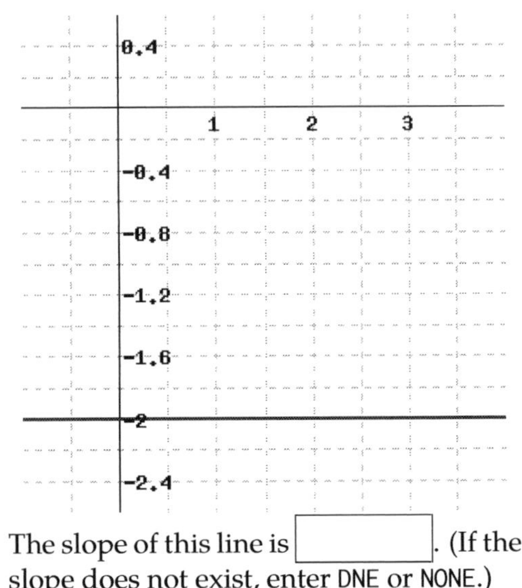

The slope of this line is ______. (If the slope does not exist, enter DNE or NONE.)

11. A line's graph is shown below.

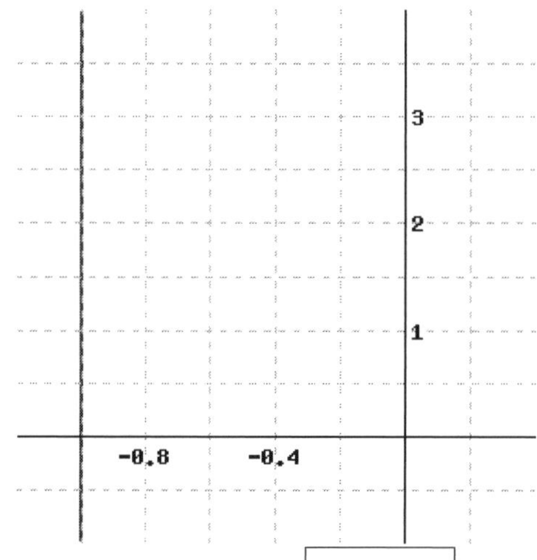

The slope of this line is ______. (If the slope does not exist, enter DNE or NONE.)

12. A line's graph is shown below.

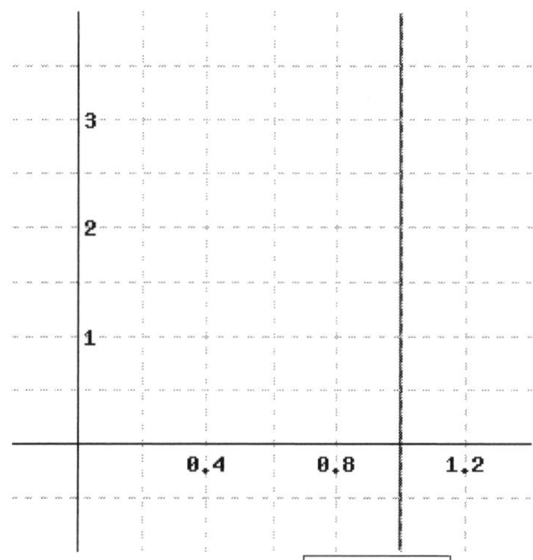

The slope of this line is ______. (If the slope does not exist, enter DNE or NONE.)

Calculate slope using two points.

13. A line passes through the points $(4,8)$ and $(7,17)$. Find this line's slope. If the slope does not exists, you may enter DNE or NONE.

This line's slope is ______.

14. A line passes through the points $(1,13)$ and $(6,33)$. Find this line's slope. If the slope does not exists, you may enter DNE or NONE.

This line's slope is ______.

15. A line passes through the points $(3,-1)$ and $(9,-7)$. Find this line's slope. If the slope does not exists, you may enter DNE or NONE.

This line's slope is ______.

16. A line passes through the points $(5,-10)$ and $(8,-13)$. Find this line's slope. If the slope does not exists, you may enter DNE or NONE.

This line's slope is ______.

17. A line passes through the points $(-3,5)$ and $(-8,0)$. Find this line's slope. If the slope does not exists, you may enter DNE or NONE.

This line's slope is ______.

18. A line passes through the points $(-1,0)$ and $(-5,-4)$. Find this line's slope. If the slope does not exists, you may enter DNE or NONE.

This line's slope is ______.

19. A line passes through the points $(-4, 6)$ and $(3, -15)$. Find this line's slope. If the slope does not exists, you may enter DNE or NONE.

This line's slope is ______.

20. A line passes through the points $(-2, 12)$ and $(2, 4)$. Find this line's slope. If the slope does not exists, you may enter DNE or NONE.

This line's slope is ______.

21. A line passes through the points $(-4, 12)$ and $(-8, 24)$. Find this line's slope. If the slope does not exists, you may enter DNE or NONE.

This line's slope is ______.

22. A line passes through the points $(-3, 2)$ and $(-5, 8)$. Find this line's slope. If the slope does not exists, you may enter DNE or NONE.

This line's slope is ______.

23. A line passes through the points $(7, -4)$ and $(-21, -28)$. Find this line's slope. If the slope does not exists, you may enter DNE or NONE.

This line's slope is ______.

24. A line passes through the points $(5, 4)$ and $(-15, -28)$. Find this line's slope. If the slope does not exists, you may enter DNE or NONE.

This line's slope is ______.

25. A line passes through the points $(-2, 14)$ and $(2, -4)$. Find this line's slope. If the slope does not exists, you may enter DNE or NONE.

This line's slope is ______.

26. A line passes through the points $(-8, -4)$ and $(8, -10)$. Find this line's slope. If the slope does not exists, you may enter DNE or NONE.

This line's slope is ______.

27. A line passes through the points $(3, -8)$ and $(-3, -8)$. Find this line's slope. If the slope does not exists, you may enter DNE or NONE.

This line's slope is ______.

28. A line passes through the points $(1, -5)$ and $(-1, -5)$. Find this line's slope. If the slope does not exists, you may enter DNE or NONE.

This line's slope is ______.

29. A line passes through the points $(-3, -2)$ and $(-3, 3)$. Find this line's slope. If the slope does not exists, you may enter DNE or NONE.

This line's slope is ______.

30. A line passes through the points $(-1, -4)$ and $(-1, 5)$. Find this line's slope. If the slope does not exists, you may enter DNE or NONE.

This line's slope is ______.

4.5 Slope-Intercept Form

In this section, we will explore one of the "standard" ways to write the equation of a line. It's known as **slope-intercept form**.

4.5.1 Slope-Intercept Definition

Recall Example 4.4.5, where Matthew had \$50 in his savings account when the year began, and decided to deposit \$20 each week without withdrawing any money. In that example, we model using x to represent how many weeks have passed. After x weeks, Matthew has added $20x$ dollars. And since he started with \$50, he has

$$y = 20x + 50$$

in his account after x weeks. In this example, there is a constant rate of change of 20 dollars per week, so we call that the **slope** as discussed in Section 4.4. We also saw in Figure 4.4.7 that plotting Matthew's balance over time gives us a straight-line graph.

The graph of Matthew's savings has some things in common with almost every straight-line graph. There is a **slope**, and there is a place where the line crosses the y-axis. Figure 4.5.2 illustrates this in the abstract.

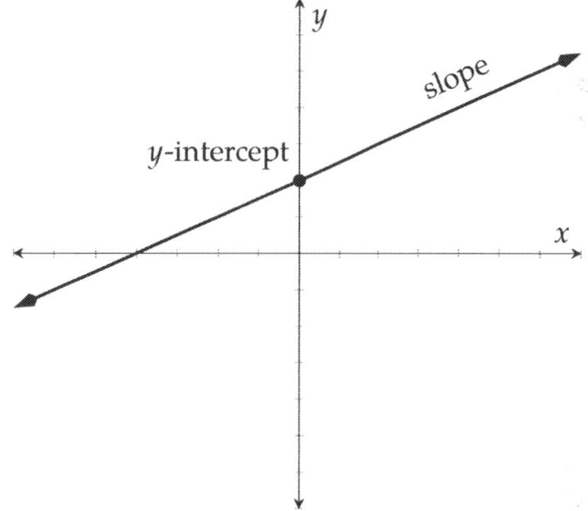

Figure 4.5.2: Generic line with slope and y-intercept

> **What else is there?** Can you think of a type of straight line that does not have a notion of slope? Or that does not cross the y-axis somewhere?

We already have an accepted symbol, m, for the slope of a line. The **y-intercept** is a *point* on the y-axis where the line crosses. Since it's on the y-axis, the x-coordinate of this point is 0. It is standard to call the y-interecept $(0, b)$ where b represents the position of the y-intercept on the y-axis.

Exercise 4.5.3. Use Figure 4.4.7 to answer this question.

What was the value of b in the plot of Matthew's savings?

What is the y-intercept?

Solution. The line crosses the y-axis at $(0, 50)$, so the value of b is 50. And the y-intercept is $(0, 50)$

One way to write the equation for Matthew's savings was

$$y = 20x + 50,$$

where both $m = 20$ and $b = 50$ are immediately visible in the equation. Now we are ready to generalize this.

Definition 4.5.4 (Slope-Intercept Form). When x and y have a linear relationship where m is the slope and $(0, b)$ is the y-intercept, one equation for this relationship is

$$y = mx + b \tag{4.5.1}$$

and this equation is called the **slope-intercept form** of the line. It is called this because the slope and y-intercept are immediately discernible from the numbers in the equation.

Exercise 4.5.5. What are the slope and y-intercept for each of the following line equations?

Equation	Slope	y-intercept
$y = 3.1x + 1.78$	______	______
$y = -17x + 112$	______	______
$y = \frac{3}{7}x - \frac{2}{3}$	______	______
$y = 13 - 8x$	______	______
$y = 1 - \frac{2x}{3}$	______	______
$y = 2x$	______	______
$y = 3$	______	______

Solution. In the first three equations, simply read the slope m according to slope-intercept form (4.5.1). The slopes are 3.1, -17, and $\frac{3}{7}$.

The fourth equation was written with the terms not in the slope-intercept form order. It could be written $y = -8x + 13$, and then it is clear that its slope is -8. In any case, the slope is the coefficient of x.

The fifth equation is also written with the terms not in the slope-intercept form order. Changing the order of the terms, it could be written $y = -\frac{2x}{3} + 1$, but this still does not match the pattern of slope-intercept form. Considering how fraction multiplication works, $\frac{2x}{3} = \frac{2}{3} \cdot \frac{x}{1} = \frac{2}{3}x$. So we can write this equation as $y = -\frac{2}{3}x + 1$, and we see the slope is $-\frac{2}{3}$.

The last two equations could be written $y = 2x + 0$ and $y = 0x + 3$, allowing us to read their slopes as 2 and 0.

For the y-intercepts, remember that we are expected to answer using an ordered pair $(0, b)$, not just a single number b. We can simply read that the first two y-intercepts are $(0, 1.78)$ and $(0, 112)$.

The third equation does not exactly match slope-intercept form (4.5.1), until you view it as $y = \frac{3}{7}x + \left(-\frac{2}{3}\right)$, and then you can see that its y-intercept is $-\frac{2}{3}$.

With the fourth equation, after rewriting it as $y = -8x + 13$, we can see that its y-intercept is $(0, 13)$.

We already explored rewriting the fifth equation as $y = -\frac{2}{3}x + 1$, where we can see that its y-intercept is $(0, 1)$.

The last two equations could be written $y = 2x + 0$ and $y = 0x + 3$, allowing us to read their y-intercepts as $(0, 0)$ and $(0, 3)$.

Alternatively, we know that y-intercepts happen where $x = 0$, and substituting $x = 0$ into each equation gives you the y-value of the y-intercept.

Remark 4.5.6. The number b is the y-value when $x = 0$. Therefore it is common to refer to b as the **initial value** or **starting value** of a linear relationship.

Example 4.5.7. Let's review. With a simple equation like $y = 2x + 3$, we can see that there is a line whose slope is 2 and which has initial value 3. So starting at $y = 3$ when $x = 0$ (that is, on the y-axis), each time you would increase the x-value by 1, the y-value increases by 2. With these basic observations, you may quickly produce a table and/or a graph.

	x	y	
start on y-axis $\longrightarrow$	0	3	$\longleftarrow$ initial value
increase by 1 $\longrightarrow$	1	5	$\longleftarrow$ increase by 2
increase by 1 $\longrightarrow$	2	7	$\longleftarrow$ increase by 2
increase by 1 $\longrightarrow$	3	9	$\longleftarrow$ increase by 2
increase by 1 $\longrightarrow$	4	11	$\longleftarrow$ increase by 2

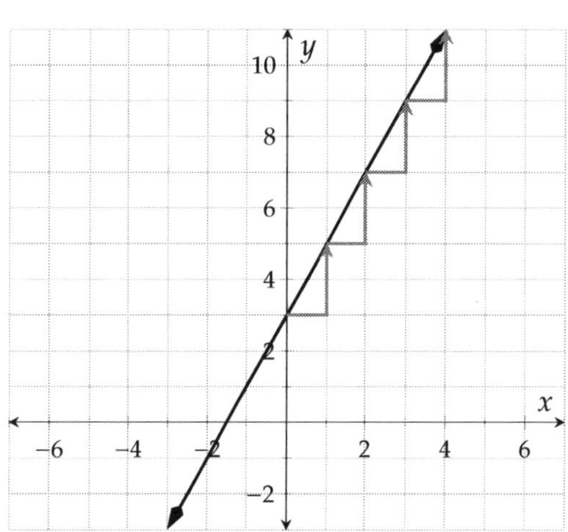

Example 4.5.8. Decide whether data in the table has a linear relationship. If so, write the linear equation in slope-intercept form (4.5.1).

x-values	y-values
0	−4
2	2
5	11
9	23

Solution. To assess whether the relationship is linear, we have to recall from Section 4.3 that we should examine rates of change between data points. Note that the changes in y-values are not consistent. However, the rates of change are calculated thusly:

- When x increases by 2, y increases by 6. The first rate of change is $\frac{6}{2} = 3$.
- When x increases by 3, y increases by 9. The second rate of change is $\frac{9}{3} = 3$.
- When x increases by 4, y increases by 12. The third rate of change is $\frac{12}{4} = 3$.

Since the rates of change are all the same, 3, the relationship is linear and the slope m is 3.

According to the table, when $x = 0$, $y = -4$. So the starting value, b, is -4.

So in slope-intercept form, the line equation is $y = 3x - 4$.

Exercise 4.5.9. Decide whether data in the table has a linear relationship. If so, write the linear equation in slope-intercept form. This may not be as easy as the previous example. Read the solution for a full explanation.

x-values	y-values
3	−2
6	−8
8	−12
11	−18

The data (□ does □ does not) have a linear relationship, because (□ changes in x are not constant □ rates of change between data points are constant □ rates of change between data points are not constant) .

The slope-intercept form of the equation for this line is ______.
(Enter NONE if the data is not linear.)

Solution. To assess whether the relationship is linear, we examine rates of change between data points.

- The first rate of change is $\frac{-6}{3} = -2$.
- The second rate of change is $\frac{-4}{2} = -2$.
- The third rate of change is $\frac{-6}{3} = -2$.

Since the rates of change are all the same, -2, the relationship is linear and the slope m is -2.

So we know that the slope-intercept equation is $y = -2x + b$, but what number is b? The table does not directly tell us what the initial y-value is.

One approach is to use any point that we know the line passes through, and use algebra to solve for b. We know the line passes through $(3, -2)$, so

$$\begin{aligned} y &= -2x + b \\ -2 &= -2(3) + b \\ -2 &= -6 + b \\ 4 &= b \end{aligned}$$

So the equation is $y = -2x + 4$.

4.5.2 Graphing Slope-Intercept Equations

Example 4.5.11. The conversion formula for a Celsius temperature into Fahrenheit is $F = \frac{9}{5}C + 32$. This appears to be in slope-intercept form, except that x and y are replaced with C and F. Suppose you are asked to graph this equation. How will you proceed? You *could* make a table of values as we do in Section 4.2 but that takes time and effort. Since the equation here is in slope-intercept form, there is a nicer way.

Since this equation is for starting with a Celsius temperature and obtaining a Fahrenheit temperature, it makes sense to let C be the horizontal axis variable and F be the vertical axis variable. Note the slope is $\frac{9}{5}$ and the y-intercept is $(0, 32)$.

1. Set up the axes using an appropriate window and labels. Considering the freezing and boiling temperatures of water, it's reasonable to let C run through at least 0 to 100. Similarly it's reasonable to let F run through at least 32 to 212.
2. Plot the y-intercept, which is at $(0, 32)$.
3. Starting at the y-intercept, use slope triangles to reach the next point. Since our slope is $\frac{9}{5}$, that suggests a "run" of 5 and a rise of 9 might work. But as Figure 4.5.12 indicates, such slope triangles are too tiny. Since $\frac{9}{5} = \frac{90}{50}$, we can try a "run" of 50 and a rise of 90.
4. Connect your points, use arrowheads, and label the equation.

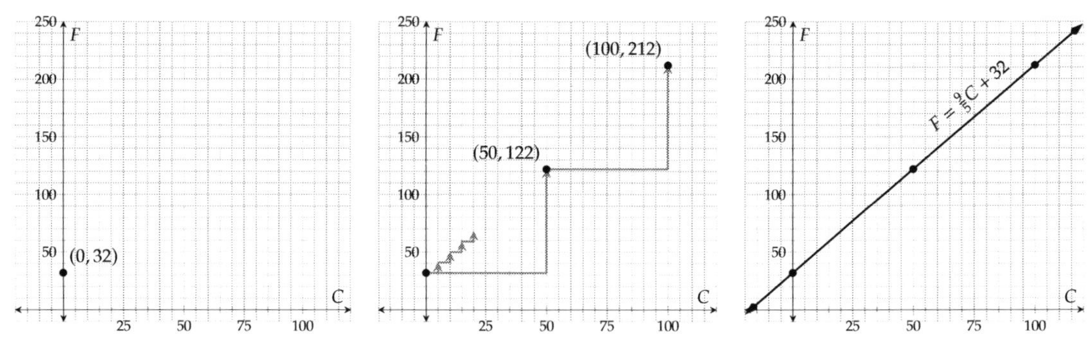

Figure 4.5.12: Graphing $F = \frac{9}{5}C + 32$

Example 4.5.13. Plot $y = -\frac{2}{3}x + 10$ and $y = 3x + 5$. These plots follow the approach form the previous example, but there is no context to the equation.

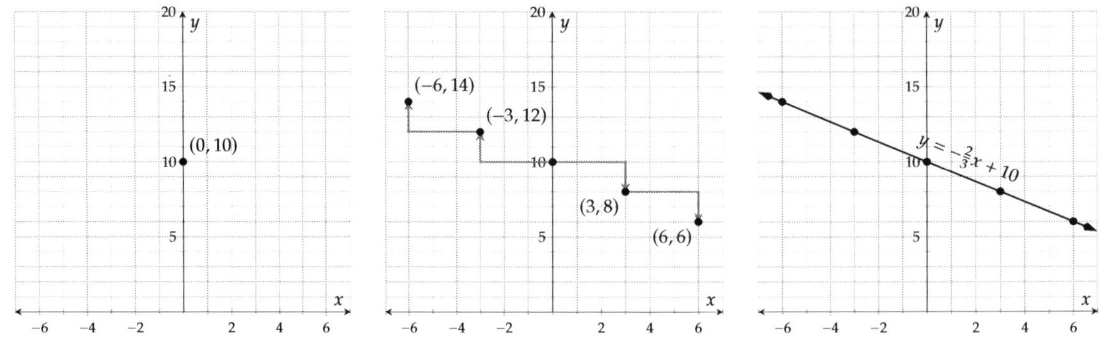

(a) With no context, we only need to make sure that the y-intercept will be visible, and that any "run" and "rise" amounts we wish to use will not make triangles that are too big or too small.

(b) The slope is $-\frac{2}{3} = \frac{-2}{3} = \frac{2}{-3}$. So we can try using a "run" of 3 and a "rise" of −2 or a "run" of −3 and a "rise" of 2.

(c) Arrowheads and labels are encouraged.

Figure 4.5.14: Graphing $y = -\frac{2}{3}x + 10$

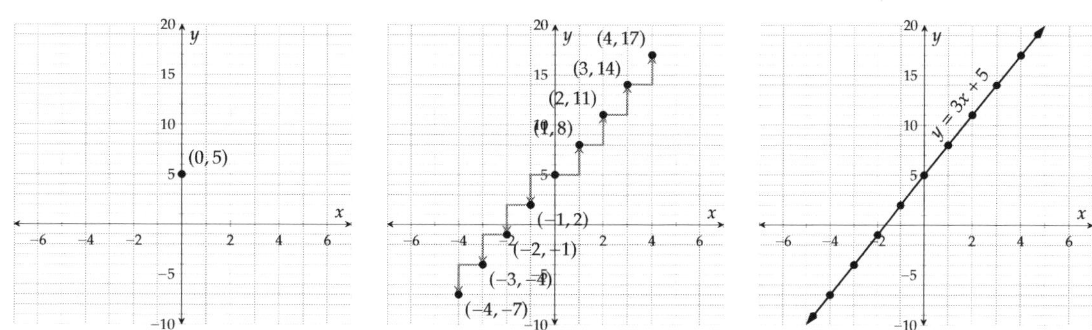

(a) With no context, we only need to make sure that the y-intercept will be visible, and that any "run" and "rise" amounts we wish to use will not make triangles that are too big or too small.

(b) The slope is a whole number 3. Every 1 unit forward causes a change of positive 3 in the y-values.

(c) Arrowheads and labels are encouraged.

Figure 4.5.15: Graphing $y = 3x + 5$

4.5.3 Writing a Slope-Intercept Equation Given a Graph

We can write a linear equation in slope-intercept form based on its graph. We need to be able to calculate the line's slope and see it's y-intercept.

Exercise 4.5.17. Use the graph to write an equation of the line in slope-intercept form.

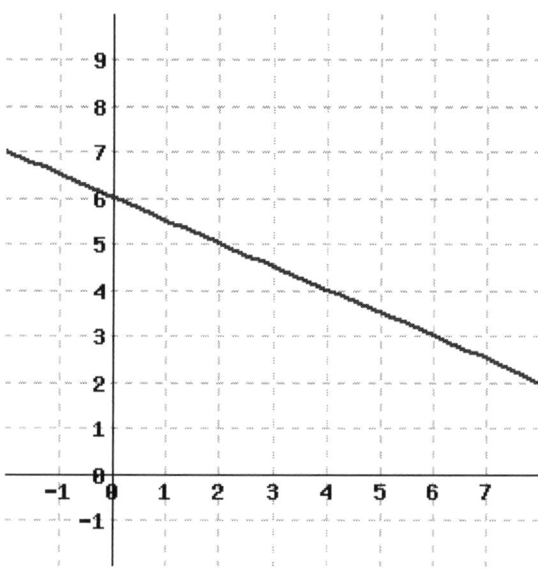

Solution. On the line, pick two points with easy-to-read integer coordinates so that we can calculate slope. It doesn't matter which two points we use; the slope will be the same.

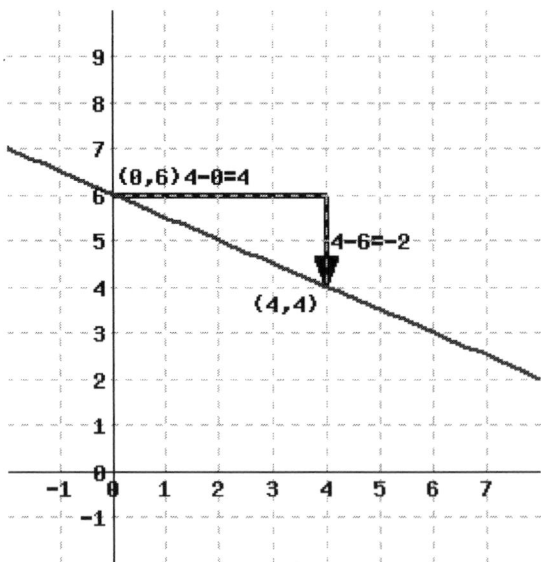

Using the slope triangle, we can calculate the line's slope:

$$\text{slope} = \frac{\Delta y}{\Delta x} = \frac{-2}{4} = -\frac{1}{2}.$$

From the graph, we can see the y-intercept is $(0, 6)$.

With the slope and y-intercept found, we can write the line's equation:

$$y = -\frac{1}{2}x + 6.$$

Exercise 4.5.18. There are seven public four-year colleges in Oregon. The graph plots the annual in-state tuition for each school on the x-axis, and the median income of former students ten years after first enrolling on the y-axis.

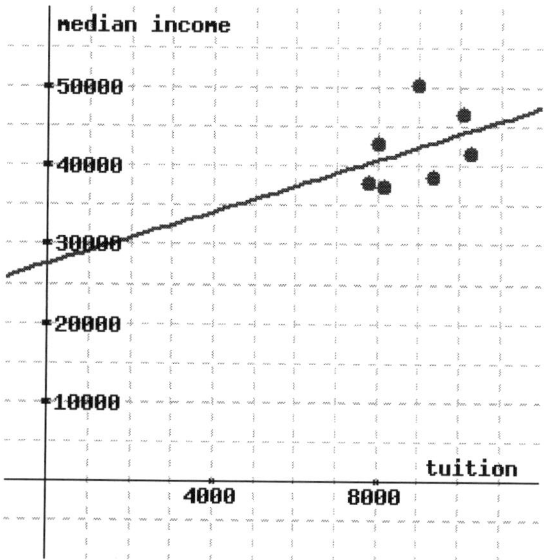

Write an equation for this line in slope-intercept form.

Solution. Do your best to identify two points on the line. We go with (0, 27500) and (8000, 41000).

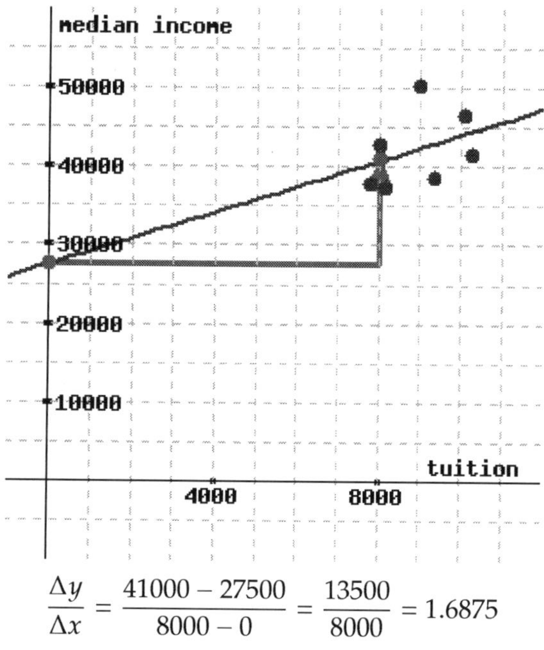

$$\frac{\Delta y}{\Delta x} = \frac{41000 - 27500}{8000 - 0} = \frac{13500}{8000} = 1.6875$$

So the slope is about 1.6875 dollars of median income per dollar of tuition. This is only an estimate since we are not all certain the two points we chose are actually on the line.

Estimating the y-intercept to be at (0, 27500), we have $y = 1.6875x + 27500$.

4.5.4 Writing a Slope-Intercept Equation Given Two Points

The idea that any two points uniquely determine a line has been understood for thousands of years in many cultures around the world. Once you have two specific points, there is a straightforward process to find the slope-intercept form of the equation of the line that connects them.

Example 4.5.19. Find the slope-intercept form of the equation of the line that passes through the points $(0, 5)$ and $(8, -5)$.

Solution. We are trying to write down $y = mx + b$, but with specific numbers for m and b. So the first step is to find the slope, m. To do this, recall the slope formula (4.4.3) from Section 4.4. It says that if a line passes through the points (x_1, y_1) and (x_2, y_2), then the slope is found by the formula $m = \frac{y_2 - y_1}{x_2 - x_1}$.

Applying this to our two points $(\overset{x_1}{0}, \overset{y_1}{5})$ and $(\overset{x_2}{8}, \overset{y_2}{-5})$, we see that the slope is:

$$\begin{aligned} m &= \frac{y_2 - y_1}{x_2 - x_1} \\ &= \frac{-5 - 5}{8 - 0} \\ &= \frac{-10}{8} \\ &= -\frac{5}{4} \end{aligned}$$

We are trying to write $y = mx + b$. Since we already found the slope, we know that we want to write $y = -\frac{5}{4}x + b$ but we need a specific number for b. We *happen* to know that one point on this line is $(0, 5)$, which is on the y-axis because its x-value is 0. So $(0, 5)$ is this line's y-intercept, and therefore $b = 5$. (We're only able to make this conclusion because this point has 0 for its x-coordinate.) So, our equation is

$$y = -\frac{5}{4}x + 5.$$

Example 4.5.20. Find the slope-intercept form of the equation of the line that passes through the points $(-8, 15)$ and $(4, 6)$.

Solution. The first step is always to find the slope between our two points: $(\overset{x_1}{-8}, \overset{y_1}{15})$ and $(\overset{x_2}{4}, \overset{y_2}{6})$. Using the slope formula (4.4.3) again, we have:

$$\begin{aligned} m &= \frac{y_2 - y_1}{x_2 - x_1} \\ &= \frac{6 - 15}{4 - (-8)} \\ &= \frac{-9}{12} \\ &= -\frac{3}{4} \end{aligned}$$

Now that we have the slope, we can write $y = -\frac{3}{4}x + b$. Unlike in Example 4.5.19, we are not given the value of b because neither of our two given points have an x-value of 0. The trick to finding b is to remember that we have two points that we know make the equation true! This means all we have to do is substitute *either* point into the equation for x and y and solve for b. Let's arbitrarily choose $(4, 6)$ to plug in.

$$\begin{aligned} y &= -\frac{3}{4}x + b \\ 6 &= -\frac{3}{4}(4) + b && \text{(Now solve for } b.\text{)} \\ 6 &= -3 + b \\ 6 + 3 &= -3 + b + 3 \\ 9 &= b \end{aligned}$$

In conclusion, the equation for which we were searching is $y = -\frac{3}{4}x + 9$. Don't be tempted to plug in values for x and y at this point. The general equation of a line in any form should have (at least one, and in this case) two variables in the final answer.

Example 4.5.21. Find the slope-intercept form of the equation of the line that passes through the points $(-3, \frac{9}{2})$ and $(4, -\frac{4}{3})$.

Solution.

This example has fractions, but the process is the same: fractions are just numbers after all. First find the slope through our points: $(-3, \frac{9}{2})$ and $(4, -\frac{4}{3})$. For this problem, we choose to do all of our algebra with improper fractions as it often simplifies the process.

$$\begin{aligned} m &= \frac{y_2 - y_1}{x_2 - x_1} \\ &= \frac{-\frac{4}{3} - \frac{9}{2}}{4 - (-3)} \\ &= \frac{-\frac{4}{3} \cdot \frac{2}{2} - \frac{9}{2} \cdot \frac{3}{3}}{7} \\ &= \frac{-\frac{8}{6} - \frac{27}{6}}{7} \\ &= \frac{-\frac{35}{6}}{\frac{7}{1}} \\ &= -\frac{35}{6} \cdot \frac{1}{7} \\ &= -\frac{5}{6} \end{aligned}$$

So far we have $y = -\frac{5}{6}x + b$. Now we need to solve for b since neither of the points given were the vertical intercept. To do this, we will choose one of the two points and plug it into our equation. We choose $\left(-3, \frac{9}{2}\right)$.

$$\begin{aligned} y &= -\frac{5}{6}x + b \\ \frac{9}{2} &= -\frac{5}{6}(-3) + b \\ \frac{9}{2} &= -\frac{15}{6} + b \\ \frac{9}{2} &= -\frac{5}{2} + b \\ \frac{9}{2} + \frac{5}{2} &= -\frac{5}{2} + b + \frac{5}{2} \\ \frac{14}{2} &= b \\ 7 &= b \end{aligned}$$

Lastly, we write our equation.

$$y = -\frac{5}{6}x + 7$$

Exercise 4.5.22. Find the slope-intercept form of the equation of the line that passes through the points $(-3, 150)$ and $(0, 30)$.

Solution. The first step is always to find the slope between our points: $(\overset{x_1}{-3}, \overset{y_1}{150})$ and $(\overset{x_2}{0}, \overset{y_2}{30})$. Using the slope formula (4.4.3), we have:

$$\begin{aligned} m &= \frac{y_2 - y_1}{x_2 - x_1} \\ &= \frac{30 - 150}{0 - (-3)} \\ &= \frac{-120}{3} \\ &= -40 \end{aligned}$$

Now we can write $y = -40x + b$ and to find b we need look no further than one of the given points: $(0, 30)$. Since the x-value is 0, the value of b must be 30. So, the slope-intercept form of the line is

$$y = -40x + 30$$

Exercise 4.5.23. Find the slope-intercept form of the equation of the line that passes through the points $\left(-3, \frac{3}{4}\right)$ and $\left(-6, -\frac{17}{4}\right)$.

Solution. First find the slope through our points: $\left(-3, \frac{3}{4}\right)$ and $\left(-6, -\frac{17}{4}\right)$. For this problem, we choose to do all of our algebra with improper fractions as it often simplifies the process.

$$\begin{aligned} m &= \frac{y_2 - y_1}{x_2 - x_1} \\ &= \frac{-\frac{17}{4} - \frac{3}{4}}{-6 - (-3)} \\ &= \frac{\frac{-20}{4}}{-3} \\ &= \frac{-5}{-3} \\ &= \frac{5}{3} \end{aligned}$$

So far we have $y = \frac{5}{3}x + b$. Now we need to solve for b since neither of the points given were the vertical intercept. Recall that to do this, we will choose one of the two points and plug it into our equation. We shoose $\left(-3, \frac{3}{4}\right)$.

$$\begin{aligned} y &= \frac{5}{3}x + b \\ \frac{3}{4} &= \frac{5}{3}(-3) + b \\ \frac{3}{4} &= -5 + b \\ \frac{3}{4} + 5 &= -5 + b + 5 \\ \frac{3}{4} + \frac{20}{4} &= b \\ \frac{23}{4} &= b \end{aligned}$$

Lastly, we write our equation.

$$y = \frac{5}{3}x + \frac{23}{4}$$

4.5.5 Modeling with Slope-Intercept Form

We can model many relatively simple relationships using slope-intercept form, and then solve related questions using algebra. Here are a few examples.

Example 4.5.25. Uber is a ride-sharing company. Its pricing in Portland factors in how much time and how many miles a trip takes. But if you assume that rides average out at a speed of 30 mph, then their pricing scheme boils down to a base of \$7.35 for the trip, plus \$3.85 per mile. Use a slope-intercept equation and algebra to answer these questions.

1. How much is the fare if a trip is 5.3 miles long?

2. With \$100 available to you, how long a trip can you afford?

Solution. The rate of change (slope) is \$3.85 per mile, and the starting value is \$7.35. So the slope-intercept equation is

$$y = 3.85x + 7.35.$$

In this equation, x stands for the number of miles in a trip, and y stands for the amount of money to be charged.

If a trip is 5 miles long, we substitute $x = 5$ into the equation and we have:

$$\begin{aligned} y &= 3.85x + 7.35 \\ &= 3.85(5) + 7.35 \\ &= 19.25 + 7.35 \\ &= 26.60 \end{aligned}$$

And the 5-mile ride will cost you about \$26.60. (We say "about," because this was all assuming you average 30 mph.)

Next, to find how long of a trip would cost \$100, we substitute $y = 100$ into the equation and solve for x:

$$\begin{aligned} y &= 3.85x + 7.35 \\ 100 &= 3.85x + 7.35 \\ 100 - 7.35 &= 3.85x \\ 92.65 &= 3.85x \\ \frac{92.65}{3.85} &= x \\ 24.06 &\approx x \end{aligned}$$

So with \$100 you could afford a little more than a 24-mile trip.

Exercise 4.5.26. In a certain wildlife reservation in Africa, there are approximately 2400 elephants. Unfortunately, the population has been decreasing by 30 elephants per year. Use a slope-intercept equation and algebra to answer these questions.

1. If the trend continues, what would the elephant population be 15 years from now?

   [ ] elephants

2. If the trend continues, how many years will it be until the elephant population dwindles to 1200?

   [ ] years

Solution. The rate of change (slope) is -30 elephants per year. Notice that since we are losing elephants, the slope is a negative number. The starting value is 2400 elephants. So the slope-intercept equation is

$$y = -30x + 2400.$$

In this equation, x stands for a number of years into the future, and y stands for the elephant population.

To estimate the elephant population 15 years later, we substitute x in the equation with 15, and we have:

$$\begin{aligned} y &= -30x + 2400 \\ &= -30(15) + 2400 \\ &= -450 + 2400 \\ &= 1950 \end{aligned}$$

So if the trend continues, there would be 1950 elephants on this reservation 15 years later.

Next, to find when the elephant population would decrease to 1200, we substitute y in the equation with 1200, and solve for x:

$$\begin{aligned} y &= -30x + 2400 \\ 1200 &= -30x + 2400 \\ 1200 - 2400 &= -30x \\ -1200 &= -30x \\ \frac{-1200}{-30} &= x \\ 40 &= x \end{aligned}$$

So if the trend continues, 40 years later, the elephant population would dwindle to $1,200$.

4.5.6 Exercises

Exercises on Identifying Slope and y-Intercept

For the following exercises: Find the line's slope and y-intercept.

1. A line has equation $y = 6x + 3$.

This line's slope is ______.

This line's y-intercept is ______.

2. A line has equation $y = 7x + 10$.

This line's slope is ______.

This line's y-intercept is ______.

3. A line has equation $y = -4x - 5$.

This line's slope is ______.

This line's y-intercept is ______.

4. A line has equation $y = -3x - 8$.

This line's slope is ______.

This line's y-intercept is ______.

5. A line has equation $y = x + 9$.

This line's slope is $\boxed{}$.

This line's y-intercept is $\boxed{}$.

6. A line has equation $y = x - 10$.

This line's slope is $\boxed{}$.

This line's y-intercept is $\boxed{}$.

7. A line has equation $y = -x - 8$.

This line's slope is $\boxed{}$.

This line's y-intercept is $\boxed{}$.

8. A line has equation $y = -x - 5$.

This line's slope is $\boxed{}$.

This line's y-intercept is $\boxed{}$.

9. A line has equation $y = -\frac{4x}{5} + 9$.

This line's slope is $\boxed{}$.

This line's y-intercept is $\boxed{}$.

10. A line has equation $y = -\frac{4x}{3} - 3$.

This line's slope is $\boxed{}$.

This line's y-intercept is $\boxed{}$.

11. A line has equation $y = \frac{x}{6} + 6$.

This line's slope is $\boxed{}$.

This line's y-intercept is $\boxed{}$.

12. A line has equation $y = \frac{x}{8} - 1$.

This line's slope is $\boxed{}$.

This line's y-intercept is $\boxed{}$.

13. A line has equation $y = -9 + 9x$.

This line's slope is $\boxed{}$.

This line's y-intercept is $\boxed{}$.

14. A line has equation $y = 5 + 10x$.

This line's slope is $\boxed{}$.

This line's y-intercept is $\boxed{}$.

15. A line has equation $y = 1 - x$.

This line's slope is $\boxed{}$.

This line's y-intercept is $\boxed{}$.

16. A line has equation $y = 2 - x$.

This line's slope is $\boxed{}$.

This line's y-intercept is $\boxed{}$.

Exercises on Graphing Lines in Slope-Intercept Form

17. Graph the equation $y = 4x$.

18. Graph the equation $y = 5x$.

19. Graph the equation $y = -3x$.

20. Graph the equation $y = -2x$.

21. Graph the equation $y = \frac{5}{2}x$.

22. Graph the equation $y = \frac{1}{4}x$.

23. Graph the equation $y = -\frac{1}{3}x$.

24. Graph the equation $y = -\frac{5}{4}x$.

25. Graph the equation $y = 5x + 2$.

26. Graph the equation $y = 3x + 6$.

27. Graph the equation $y = -4x + 3$.

28. Graph the equation $y = -2x + 5$.

29. Graph the equation $y = x - 4$.

30. Graph the equation $y = x + 2$.

31. Graph the equation $y = -x + 3$.

32. Graph the equation $y = -x - 5$.

33. Graph the equation $y = \frac{2}{3}x + 4$.

34. Graph the equation $y = \frac{3}{2}x - 5$.

35. Graph the equation $y = -\frac{3}{5}x - 1$.

36. Graph the equation $y = -\frac{1}{5}x + 1$.

Exercises on Writing Slope-Intercept Form Equation from the Graph

For the following exercises: A line's graph is given.

37.

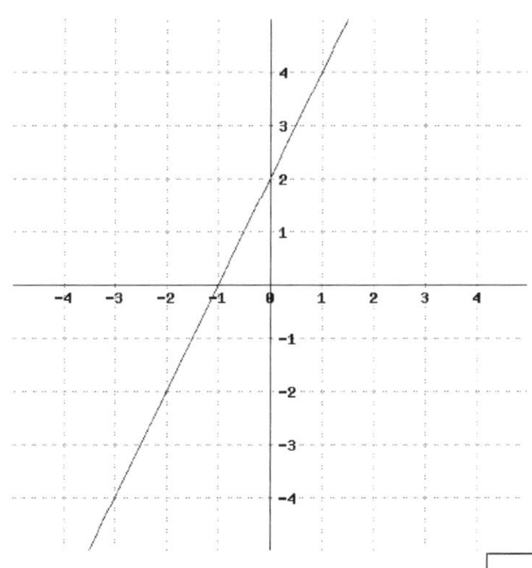

This line's slope-intercept equation is ☐

38.

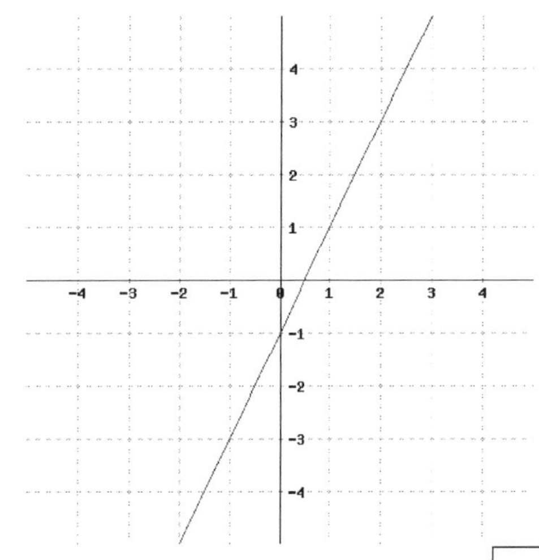

This line's slope-intercept equation is ☐

39.

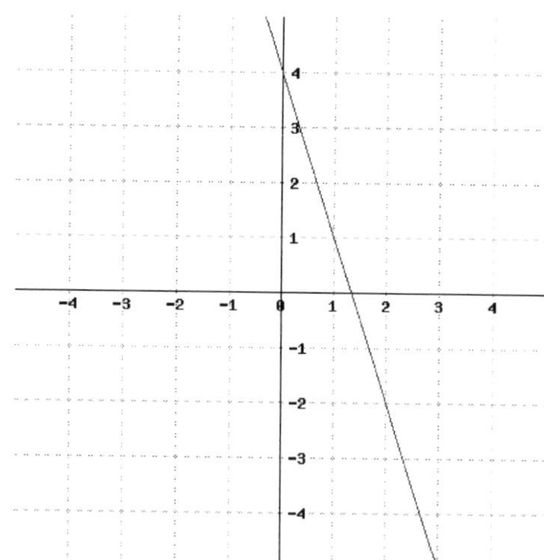

This line's slope-intercept equation is ☐

40.

This line's slope-intercept equation is ☐

41.

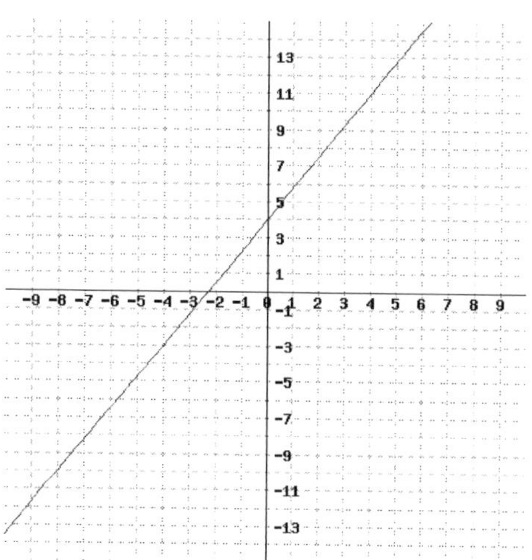

This line's slope-intercept equation is ☐

42.

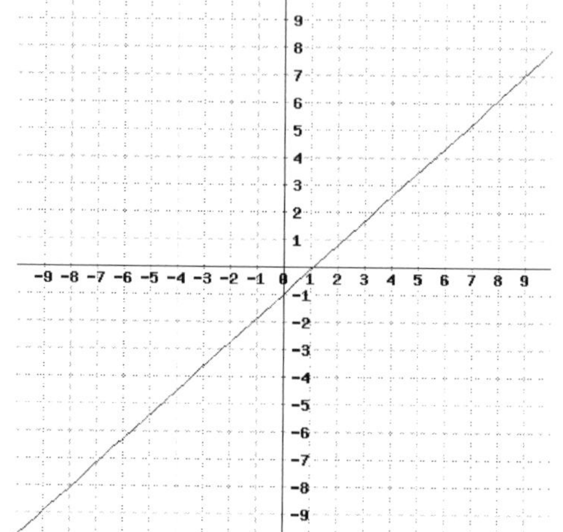

This line's slope-intercept equation is ☐

43.

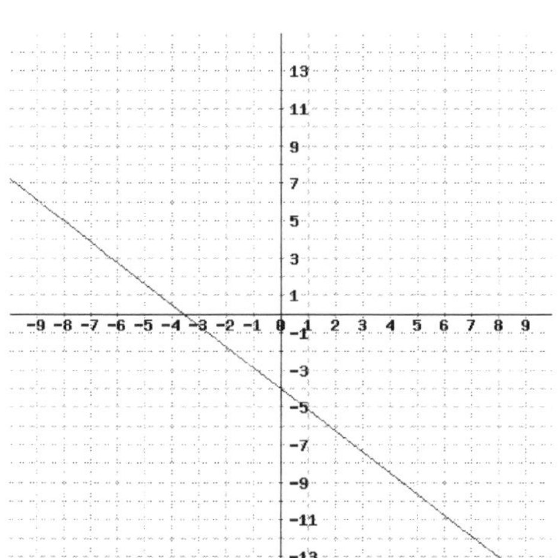

This line's slope-intercept equation is $\boxed{}$

44.

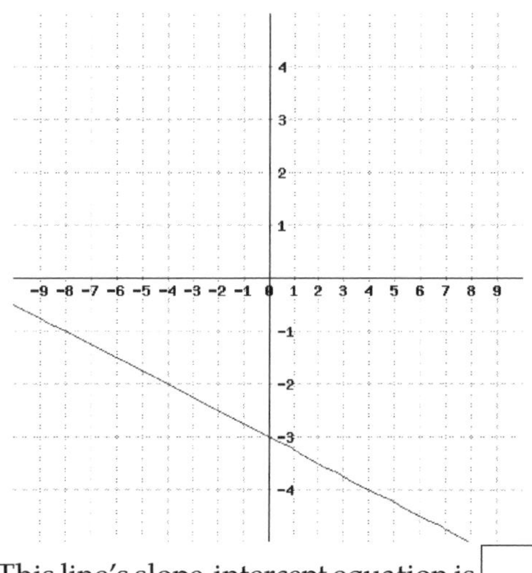

This line's slope-intercept equation is $\boxed{}$

Exercises on Writing Slope-Intercept Form Given Two Points

45. A line passes through the points $(3, 16)$ and $(5, 20)$. Find this line's equation in slope-intercept form.

This line's slope-intercept equation is $\boxed{}$.

46. A line passes through the points $(1, 8)$ and $(3, 12)$. Find this line's equation in slope-intercept form.

This line's slope-intercept equation is $\boxed{}$.

47. A line passes through the points $(-1, -2)$ and $(5, -26)$. Find this line's equation in slope-intercept form.

This line's slope-intercept equation is $\boxed{}$.

48. A line passes through the points $(4, -8)$ and $(2, 0)$. Find this line's equation in slope-intercept form.

This line's slope-intercept equation is $\boxed{}$.

49. A line passes through the points $(-3, 4)$ and $(-2, 3)$. Find this line's equation in slope-intercept form.

This line's slope-intercept equation is $\boxed{}$.

50. A line passes through the points $(-4, 8)$ and $(2, 2)$. Find this line's equation in slope-intercept form.

This line's slope-intercept equation is $\boxed{}$.

51. A line passes through the points $(-5, 1)$ and $(10, 25)$. Find this line's equation in slope-intercept form.

This line's slope-intercept equation is $\boxed{}$.

52. A line passes through the points $(0, 6)$ and $(-5, -3)$. Find this line's equation in slope-intercept form.

This line's slope-intercept equation is $\boxed{}$.

53. A line passes through the points $(-6, 12)$ and $(-3, 11)$. Find this line's equation in slope-intercept form.

This line's slope-intercept equation is ☐.

54. A line passes through the points $(-9, 7)$ and $(9, 3)$. Find this line's equation in slope-intercept form.

This line's slope-intercept equation is ☐.

Applications

55. A gym charges members $25 for a registration fee, and then $28 per month. You became a member some time ago, and now you have paid a total of $501 to the gym. How many months have passed since you joined the gym?

☐ months have passed since you joined the gym.

56. Your cell phone company charges a $17 monthly fee, plus $0.11 per minute of talk time. One month your cell phone bill was $53.30. How many minutes did you spend talking on the phone that month?

You spent ☐ talking on the phone that month.

57. A school purchased a batch of T-shirts from a company. The company charged $6 per T-shirt, and gave the school a $90 rebate. If the school had a net expense of $2,430 from the purchase, how many T-shirts did the school buy?

The school purchased ☐ T-shirts.

58. Holli hired a face-painter for a birthday party. The painter charged a flat fee of $80, and then charged $3.50 per person. In the end, Holli paid a total of $115. How many people used the face-painter's service?

☐ people used the face-painter's service.

59. A certain country has 585.48 million acres of forest. Every year, the country loses 7.14 million acres of forest mainly due to deforestation for farming purposes. If this situation continues at this pace, how many years later will the country have only 321.3 million acres of forest left? (Use an equation to solve this problem.)

After ☐ years, this country would have 321.3 million acres of forest left.

60. Farshad has $86 in his piggy bank. He plans to purchase some Pokemon cards, which costs $2.85 each. He plans to save $57.50 to purchase another toy. At most how many Pokemon cards can he purchase?

Write an equation to solve this problem.

Farshad can purchase at most ☐ Pokemon cards.

61. By your cell phone contract, you pay a monthly fee plus some money for each minute you use the phone during the month. In one month, you spent 230 minutes on the phone, and paid \$28.95. In another month, you spent 310 minutes on the phone, and paid \$34.15.

Let x be the number of minutes you talk over the phone in a month, and let y be your cell phone bill for that month. Use a linear equation to model your monthly bill based on the number of minutes you talk over the phone.

a. This linear model's slope-intercept equation is ______.

b. If you spent 140 minutes over the phone in a month, you would pay ______.

c. If in a month, you paid \$45.85 of cell phone bill, you must have spent ______ minutes on the phone in that month.

62. A company set aside a certain amount of money in the year 2000. The company spent exactly the same amount from that fund each year on perks for its employees. In 2004, there was still \$920,000 left in the fund. In 2007, there was \$860,000 left.

Let x be the number of years since 2000, and let y be the amount of money left in the fund that year. Use a linear equation to model the amount of money left in the fund after so many years.

a. The linear model's slope-intercept equation is ______.

b. In the year 2009, there was ______ left in the fund.

c. In the year ______, the fund will be empty.

63. A biologist has been observing a tree's height. 10 months into the observation, the tree was 17.4 feet tall. 18 months into the observation, the tree was 18.36 feet tall.

Let x be the number of months passed since the observations started, and let y be the tree's height at that time. Use a linear equation to model the tree's height as the number of months pass.

a. This line's slope-intercept equation is [ ].

b. 26 months after the observations started, the tree would be [ ] feet in height.

c. [ ] months after the observation started, the tree would be 23.28 feet tall.

64. Scientists are conducting an experiment with a gas in a sealed container. The mass of the gas is measured, and the scientists realize that the gas is leaking over time in a linear way.

Eight minutes since the experiment started, the gas had a mass of 129.2 grams.

Twelve minutes since the experiment started, the gas had a mass of 114 grams.

Let x be the number of minutes that have passed since the experiment started, and let y be the mass of the gas in grams at that moment. Use a linear equation to model the weight of the gas over time.

a. This line's slope-intercept equation is [ ].

b. 32 minutes after the experiment started, there would be [ ] grams of gas left.

c. If a linear model continues to be accurate, [ ] minutes since the experiment started, all gas in the container will be gone.

65. By your cell phone contract, you pay a monthly fee plus \$0.04 for each minute you spend on the phone. In one month, you spent 300 minutes over the phone, and had a bill totaling \$31.00.

Let x be the number of minutes you spend on the phone in a month, and let y be your total cell phone bill for that month. Use a linear equation to model your monthly bill based on the number of minutes you spend on the phone.

a. This line's slope-intercept equation is ______.

b. If you spend 130 minutes on the phone in a month, you would be billled ______.

c. If your bill was \$38.60 one month, you must have spent ______ minutes on the phone in that month.

66. A company set aside a certain amount of money in the year 2000. The company spent exactly \$33,000 from that fund each year on perks for its employees. In 2003, there was still \$744,000 left in the fund.

Let x be the number of years since 2000, and let y be the amount of money left in the fund that year. Use a linear equation to model the amount of money left in the fund after so many years.

a. The linear model's slope-intercept equation is ______.

b. In the year 2009, there was ______ left in the fund.

c. In the year ______, the fund will be empty.

67. A biologist has been observing a tree's height. This type of tree typically grows by 0.21 feet per month. fourteen months into the observation, the tree was 16.04 feet tall.

Let x be the number of months passed since the observations started, and let y be the tree's height at that time. Use a linear equation to model the tree's height as the number of months pass.

a. This line's slope-intercept equation is [ ].

b. 26 months after the observations started, the tree would be [ ] feet in height.

c. [ ] months after the observation started, the tree would be 25.49 feet tall.

68. Scientists are conducting an experiment with a gas in a sealed container. The mass of the gas is measured, and the scientists realize that the gas is leaking over time in a linear way. Its mass is leaking by 9.2 grams per minute. Six minutes since the experiment started, the remaining gas had a mass of 377.2 grams.

Let x be the number of minutes that have passed since the experiment started, and let y be the mass of the gas in grams at that moment. Use a linear equation to model the weight of the gas over time.

a. This line's slope-intercept equation is [ ].

b. 32 minutes after the experiment started, there would be [ ] grams of gas left.

c. If a linear model continues to be accurate, [ ] minutes since the experiment started, all gas in the container will be gone.

4.6 Point-Slope Form

In Section 4.5, we learned that a linear equation can be written in slope-intercept form, $y = mx + b$. This section covers an alternative that can often be more useful depending on the application: **point-slope form**.

4.6.1 Point-Slope Definition

Sometimes, one problem with slope-intercept form (4.5.1) is that it uses the y-intercept, which might be somewhat meaningless in the context of an application. For example, here is a slope-intercept equation for the population of the United States in year x, where x can be any year from 1990 and beyond, and y is the population measured in millions:

$$y = 2.865x - 5451.$$

What can we say about the two numbers 2.865 and −5451? The slope is 2.865 with units $\frac{y\text{-unit}}{x\text{-unit}} = \frac{\text{million}}{\text{year}}$. OK, so there is meaning there: the population has been growing by 2.865 million people per year.

But what about −5451? This number tells us that the y-intercept is $(0, -5451)$, but what practical use is that? It's nonsense to say that in the year 0, the population of the United States was −5451 million. It doesn't make sense to have a negative population. It doesn't make sense to talk about the Unted States population before there even was a United States. And it doesn't make sense to use this model for years earlier than 1990 because it says clearly that the model is for 1990 and beyond.

Remark 4.6.2. If the x-value 0 is not an appropriate x-value to consider in a linear model, then the "initial value" b from the slope-intercept form is not meaningful in the context of the model. Its only value is to be part of the formula for calculations. It can still be used to mark a y-intercept on the y-axis, but if you are treating the equation as a mathematical model then you shouldn't be thinking too hard about the portion of the line near the y-axis, since the x-values near 0 are not relevant.

Example 4.6.3. Since 1990, the population of the United States has been growing by about 2.865 million per year. Also, back in 1990, the population was 253 million. Since the rate of growth has been roughly constant, a linear model is appropriate. But let's look for a way to write the equation other than slope-intercept form. Here are some things we know:

1. The slope equation is $m = \frac{y_2 - y_1}{x_2 - x_1}$.
2. The slope is $m = 2.865$ (million per year).
3. One point on the line is $(1990, 253)$, since in 1990, the population was 253 million.

If we use the generic (x, y) to represent a point *somewhere* on this line, then the rate of change between (1990, 253) and (x, y) has to be 2.865. So

$$\frac{y - 253}{x - 1990} = 2.865.$$

There is good reason[a] to want to isolate y in this equation:

$$\begin{aligned} \frac{y - 253}{x - 1990} &= 2.865 \\ y - 253 &= 2.865 \cdot (x - 1990) && \text{(could distribute, but not going to)} \\ y &= 2.865(x - 1990) + 253 \end{aligned}$$

This is a good place to stop. We have isolated y, and three *meaningful* numbers appear in the population: the rate of growth, a certain year, and the population in that year. This is a specific example of **point-slope form**.

[a]It will help us to see that y (population) *depends* on x (whatever year it is).

Definition 4.6.4 (Point-Slope Form). When x and y have a linear relationship where m is the slope and (x_0, y_0) is some specific point that the line passes through, one equation for this relationship is

$$y = m(x - x_0) + y_0 \tag{4.6.1}$$

and this equation is called the **point-slope form** of the line. It is called this because the slope and one point on the line are immediately discernable from the numbers in the equation.

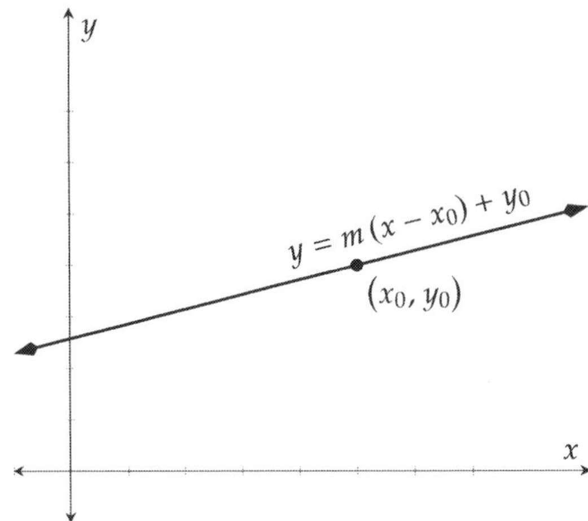

Figure 4.6.5

Alternative Point-Slope Form It is also common to define point-slope form as

$$y - y_0 = m(x - x_0) \tag{4.6.2}$$

by subtracting the y_0 to the right side. Some exercises may appear using this form.

Consider the graph in Figure 4.6.6.

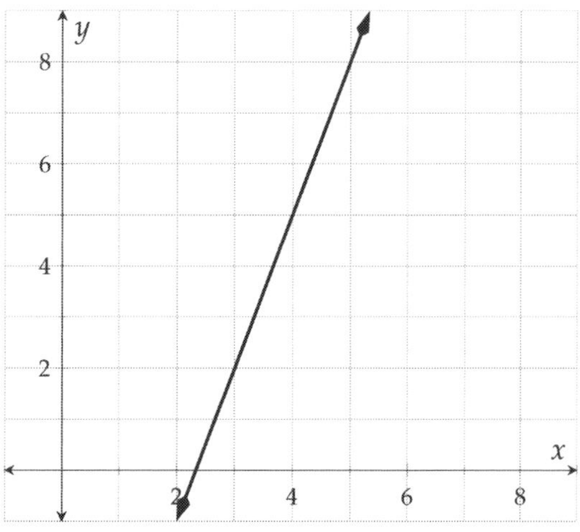

Figure 4.6.6

Exercise 4.6.7.

a. What is the slope of the line in Figure 4.6.6?

b. Identify a point visible on this line that has integer coordinates.

c. Use point-slope form to write an equation for this line, making use of a point with integer coordinates.

Solution.

a. Several slope triangles are visible where the "run" is 1 and the "rise" is 3. So the slope is $\frac{3}{1} = 3$.

b. The visible points with integer coordinates are $(2, -1)$, $(3, 2)$, $(4, 5)$, and $(5, 8)$.

c. Using $(3, 2)$, the point-slope equation is $y = 3(x - 3) + 2$. You can of course use other points, like $(2, -1)$, and have $y = 3(x - 2) + (-1)$ which simplifies to $y = 3(x - 2) - 1$.

In the previous exercise, the solution explains that each of the following are accepatable equations for the same line:

$$y = 3(x - 3) + 2 \qquad\qquad y = 3(x - 2) - 1$$

Are those two equations really equivalent? Let's distribute and simplify each of them to get slope-intercept form (4.5.1).

$$\begin{aligned} y &= 3(x - 3) + 2 \\ y &= 3x - 9 + 2 \end{aligned} \qquad\qquad \begin{aligned} y &= 3(x - 2) - 1 \\ y &= 3x - 6 - 1 \end{aligned}$$

$$y = 3x - 7 \qquad\qquad y = 3x - 7$$

So, yes. It didn't matter which point we focused on in the line in Figure 4.6.6. We get different-looking equations that still represent the same line (which, by the way, has y-intercept at $(0, -7)$).

Point-slope form is preferable when we know a line's slope and a point on it, but we don't know the y-intercept.

Example 4.6.8. A spa chain has been losing customers at a roughly constant rate since the year 2010. In 2013, it had 2975 customers; in 2016, it had 2585 customers. Management estimated that the company will go out of business once its customer base decreases to 1800. If this trend continues, when will the company close?

The given information tells us two points on the line: $(2013, 2975)$ and $(2016, 2585)$. The slope formula (4.4.3) will give us the slope. After labeling those two points as $(\overset{x_1}{2013}, \overset{y_1}{2975})$ and $(\overset{x_2}{2016}, \overset{y_2}{2585})$, we have:

$$\begin{aligned} \text{slope} &= \frac{y_2 - y_1}{x_2 - x_1} \\ &= \frac{2585 - 2975}{2016 - 2013} \\ &= \frac{-390}{3} \\ &= -130 \end{aligned}$$

And considering units, this means they are losing 130 customers per year.

Let's note that we could try to make an equation for this line in slope-intercept form, but then we would need to calculate the y-intercept, which in context would correspond to the number of customers in year 0. We could do it, but we'd be working with numbers that have no real-world meaning in this context.

For point-slope form, since we calculated the slope, we know at least this much:

$$y = -130(x - x_0) + y_0.$$

Now we can pick one of those two given points, say $(2013, 2975)$, and get the equation

$$y = -130(x - 2013) + 2975.$$

Note that all three numbers in this equation have meaning in the context of the spa chain.

We're ready to answer the question about when the chain might go out of business. Substitute y in the equation with 1800 and solve for x, and we will get the answer we seek.

$$\begin{aligned} y &= -130(x - 2013) + 2975 \\ 1800 &= -130(x - 2013) + 2975 \\ 1800 - 2975 &= -130(x - 2013) \end{aligned}$$

$$
\begin{aligned}
-1175 - 130(x - 2013) \\
\frac{-1175}{-130} &= x - 2013 \\
9.038 &\approx x - 2013 \\
9.038 + 2013 &\approx x \\
2022 &\approx x
\end{aligned}
$$

And so we find that at this rate, the company is headed toward a collapse in 2022.

Exercise 4.6.9. If we go state by state and compare the Republican candidate's 2012 vote share (x) to the Republican candidate's 2016 vote share (y), we get the following graph where a trendline has been superimposed. 

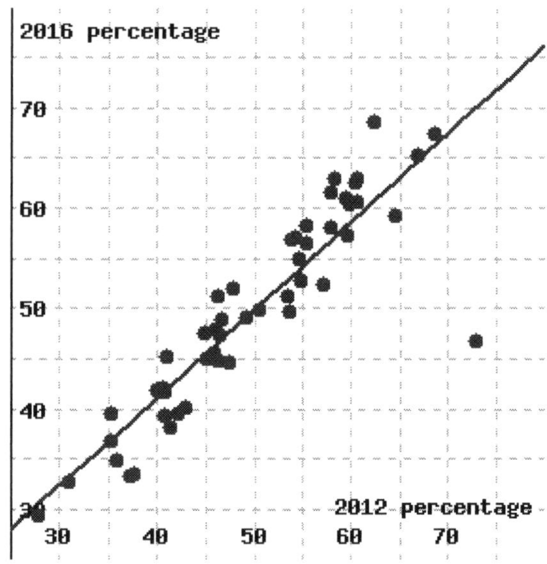

Find a point-slope equation for this line. (Note that a slope-intercept equation would use the y-intercept cooridnate b, and that would not be meaningful in context, since no state had anywhere near zero percent Republican vote.)

Solution. We need to calculate slope first. And for that, we need to identify two points on the line. conveniently, the line apears to pass right through $(50, 50)$. We have to take a second point somewhere, and $(75, 72)$ seems like a reasonable roughly accurate choice. The slope equation gives us that

$$m = \frac{72 - 50}{75 - 50} = \frac{22}{25} = 0.88.$$

Using $(50, 50)$ as the point, the point-slope equation would then be

$$y = 0.88(x - 50) + 50.$$

4.6.2 Using Two Points to Build a Linear Equation

Since two points can determine a line's location, we can calculate a line's equation using just the coordinates from any two points it passes through.

Example 4.6.10. A line passes through $(-6, 0)$ and $(9, -10)$. Find this line's equation in both point-slope and slope-intercept form.

Solution. We will use the slope formula (4.4.3) to find the slope first. After labeling those two points as $(\overset{x_1}{-6}, \overset{y_1}{0})$ and $(\overset{x_2}{9}, \overset{y_2}{-10})$, we have:

$$\begin{aligned} \text{slope} &= \frac{y_2 - y_1}{x_2 - x_1} \\ &= \frac{-10 - 0}{9 - (-6)} \\ &= \frac{-10}{15} \\ &= -\frac{2}{3} \end{aligned}$$

Now the point-slope equation looks like $y = -\frac{2}{3}(x - x_0) + y_0$. Next, we will use $(9, -10)$ and substitute x_0 with 9 and y_0 with -10, and we have:

$$\begin{aligned} y &= -\frac{2}{3}(x - x_0) + y_0 \\ y &= -\frac{2}{3}(x - 9) + (-10) \\ y &= -\frac{2}{3}(x - 9) - 10 \end{aligned}$$

Next, we will change the point-slope equation into slope-intercept form:

$$\begin{aligned} y &= -\frac{2}{3}(x - 9) - 10 \\ y &= -\frac{2}{3}x + 6 - 10 \\ y &= -\frac{2}{3}x - 4 \end{aligned}$$

Exercise 4.6.11. A line passes through $(13, -108)$ and $(-42, 23)$. Find equations for this line using both point-slope and slope-intercept form.

A point-slope equation: [ ]

A point-slope equation: [ ]

4.6.3 Exercises

Point-Slope Form Basics

For the following exercises: A line's equation is given in point-slope form:

1. $y = 2(x - 5) + 14$

This line's slope is ______.

A point on this line that is apparent from the given equation is ______.

2. $y = 2(x - 2)$

This line's slope is ______.

A point on this line that is apparent from the given equation is ______.

3. $y = 3(x + 3) - 7$

This line's slope is ______.

A point on this line that is apparent from the given equation is ______.

4. $y = -3(x + 5) + 11$

This line's slope is ______.

A point on this line that is apparent from the given equation is ______.

5. $y = \frac{5}{3}(x + 9) - 11$

This line's slope is ______.

A point on this line that is apparent from the given equation is ______.

6. $y = \left(-\frac{6}{7}\right)(x + 14) + 14$

This line's slope is ______.

A point on this line that is apparent from the given equation is ______.

Use given information to find a line's point-slope form equation.

7. A line passes through the points $(2, 15)$ and $(1, 10)$. Find this line's equation in point-slope form.

Using the point $(2, 15)$, this line's point-slope form equation is ______.

Using the point $(1, 10)$, this line's point-slope form equation is ______.

8. A line passes through the points $(4, 21)$ and $(1, 6)$. Find this line's equation in point-slope form.

Using the point $(4, 21)$, this line's point-slope form equation is ______.

Using the point $(1, 6)$, this line's point-slope form equation is ______.

9. A line passes through the points $(-4, 28)$ and $(3, -7)$. Find this line's equation in point-slope form.

Using the point $(-4, 28)$, this line's point-slope form equation is ______.

Using the point $(3, -7)$, this line's point-slope form equation is ______.

10. A line passes through the points $(1, 2)$ and $(0, 4)$. Find this line's equation in point-slope form.

Using the point $(1, 2)$, this line's point-slope form equation is ______.

Using the point $(0, 4)$, this line's point-slope form equation is ______.

11. A line passes through the points $(-3,-12)$ and $(-9,-16)$. Find this line's equation in point-slope form.

Using the point $(-3,-12)$, this line's point-slope form equation is [ ].

Using the point $(-9,-16)$, this line's point-slope form equation is [ ].

12. A line passes through the points $(21,8)$ and $(-21,-10)$. Find this line's equation in point-slope form.

Using the point $(21,8)$, this line's point-slope form equation is [ ].

Using the point $(-21,-10)$, this line's point-slope form equation is [ ].

13. A line's slope is 3. The line passes through the point $(5,17)$. Find an equation for this line in both point-slope and slope-intercept form.

An equation for this line in point-slope form is: [ ].

An equation for this line in slope-intercept form is: [ ].

14. A line's slope is 4. The line passes through the point $(2,9)$. Find an equation for this line in both point-slope and slope-intercept form.

An equation for this line in point-slope form is: [ ].

An equation for this line in slope-intercept form is: [ ].

15. A line's slope is -3. The line passes through the point $(2,-4)$. Find an equation for this line in both point-slope and slope-intercept form.

An equation for this line in point-slope form is: [ ].

An equation for this line in slope-intercept form is: [ ].

16. A line's slope is -2. The line passes through the point $(-4,6)$. Find an equation for this line in both point-slope and slope-intercept form.

An equation for this line in point-slope form is: [ ].

An equation for this line in slope-intercept form is: [ ].

17. A line's slope is 1. The line passes through the point $(5,9)$. Find an equation for this line in both point-slope and slope-intercept form.

An equation for this line in point-slope form is: [ ].

An equation for this line in slope-intercept form is: [ ].

18. A line's slope is 1. The line passes through the point $(1,6)$. Find an equation for this line in both point-slope and slope-intercept form.

An equation for this line in point-slope form is: [ ].

An equation for this line in slope-intercept form is: [ ].

19. A line's slope is -1. The line passes through the point $(-3, -1)$. Find an equation for this line in both point-slope and slope-intercept form.

An equation for this line in point-slope form is: ________.

An equation for this line in slope-intercept form is: ________.

20. A line's slope is -1. The line passes through the point $(5, -8)$. Find an equation for this line in both point-slope and slope-intercept form.

An equation for this line in point-slope form is: ________.

An equation for this line in slope-intercept form is: ________.

21. A line's slope is $\frac{3}{5}$. The line passes through the point $(-10, -2)$. Find an equation for this line in both point-slope and slope-intercept form.

An equation for this line in point-slope form is: ________.

An equation for this line in slope-intercept form is: ________.

22. A line's slope is $\frac{4}{3}$. The line passes through the point $(-6, -7)$. Find an equation for this line in both point-slope and slope-intercept form.

An equation for this line in point-slope form is: ________.

An equation for this line in slope-intercept form is: ________.

23. A line's slope is $-\frac{5}{8}$. The line passes through the point $(24, -20)$. Find an equation for this line in both point-slope and slope-intercept form.

An equation for this line in point-slope form is: ________.

An equation for this line in slope-intercept form is: ________.

24. A line's slope is $-\frac{6}{5}$. The line passes through the point $(15, -17)$. Find an equation for this line in both point-slope and slope-intercept form.

An equation for this line in point-slope form is: ________.

An equation for this line in slope-intercept form is: ________.

Convert to slope-intercept form.

For the following exercises: Change this equation from point-slope form to slope-intercept form.

25. $y = 5(x - 4) + 16$

In slope-intercept form, this line's equation would be ________.

26. $y = 5(x + 3) - 11$

In slope-intercept form, this line's equation would be ________.

27. $y = -2(x - 3) - 4$

In slope-intercept form, this line's equation would be ________.

28. $y = -5(x + 4) + 16$

In slope-intercept form, this line's equation would be ________.

29. $y = \frac{2}{3}(x + 6) - 1$

In slope-intercept form, this line's equation would be ______.

30. $y = \frac{3}{5}(x - 15) + 14$

In slope-intercept form, this line's equation would be ______.

31. $y = -\frac{4}{7}(x + 14) + 11$

In slope-intercept form, this line's equation would be ______.

32. $y = -\frac{5}{7}(x + 7) + 8$

In slope-intercept form, this line's equation would be ______.

Applications

33. By your cell phone contract, you pay a monthly fee plus some money for each minute you use the phone during the month. In one month, you spent 220 minutes on the phone, and paid \$23.90. In another month, you spent 320 minutes on the phone, and paid \$28.40.

Let x be the number of minutes you talk over the phone in a month, and let y be your cell phone bill for that month. Use a linear equation to model your monthly bill based on the number of minutes you talk over the phone.

a. This linear model's slope-intercept equation is ______.

b. If you spent 180 minutes over the phone in a month, you would pay ______.

c. If in a month, you paid \$33.35 of cell phone bill, you must have spent ______ minutes on the phone in that month.

34. A company set aside a certain amount of money in the year 2000. The company spent exactly the same amount from that fund each year on perks for its employees. In 2003, there was still \$490,000 left in the fund. In 2007, there was \$318,000 left.

Let x be the number of years since 2000, and let y be the amount of money left in the fund that year. Use a linear equation to model the amount of money left in the fund after so many years.

a. The linear model's slope-intercept equation is ______.

b. In the year 2011, there was ______ left in the fund.

c. In the year ______, the fund will be empty.

35. A biologist has been observing a tree's height. 10 months into the observation, the tree was 19.4 feet tall. 18 months into the observation, the tree was 21.64 feet tall.

Let x be the number of months passed since the observations started, and let y be the tree's height at that time. Use a linear equation to model the tree's height as the number of months pass.

a. This line's slope-intercept equation is [ ].

b. 29 months after the observations started, the tree would be [ ] feet in height.

c. [ ] months after the observation started, the tree would be 31.16 feet tall.

36. Scientists are conducting an experiment with a gas in a sealed container. The mass of the gas is measured, and the scientists realize that the gas is leaking over time in a linear way.

Seven minutes since the experiment started, the gas had a mass of 180.6 grams.

Thirteen minutes since the experiment started, the gas had a mass of 155.4 grams.

Let x be the number of minutes that have passed since the experiment started, and let y be the mass of the gas in grams at that moment. Use a linear equation to model the weight of the gas over time.

a. This line's slope-intercept equation is [ ].

b. 37 minutes after the experiment started, there would be [ ] grams of gas left.

c. If a linear model continues to be accurate, [ ] minutes since the experiment started, all gas in the container will be gone.

37. By your cell phone contract, you pay a monthly fee plus \$0.03 for each minute you spend on the phone. In one month, you spent 290 minutes over the phone, and had a bill totaling \$19.70.

Let x be the number of minutes you spend on the phone in a month, and let y be your total cell phone bill for that month. Use a linear equation to model your monthly bill based on the number of minutes you spend on the phone.

a. This line's slope-intercept equation is [ ].

b. If you spend 170 minutes on the phone in a month, you would be billed [ ].

c. If your bill was \$23.90 one month, you must have spent [ ] minutes on the phone in that month.

38. A company set aside a certain amount of money in the year 2000. The company spent exactly \$26,000 from that fund each year on perks for its employees. In 2002, there was still \$812,000 left in the fund.

Let x be the number of years since 2000, and let y be the amount of money left in the fund that year. Use a linear equation to model the amount of money left in the fund after so many years.

a. The linear model's slope-intercept equation is [ ].

b. In the year 2011, there was [ ] left in the fund.

c. In the year [ ], the fund will be empty.

39. A biologist has been observing a tree's height. This type of tree typically grows by 0.16 feet per month. fourteen months into the observation, the tree was 15.74 feet tall.

Let x be the number of months passed since the observations started, and let y be the tree's height at that time. Use a linear equation to model the tree's height as the number of months pass.

a. This line's slope-intercept equation is ______.

b. 29 months after the observations started, the tree would be ______ feet in height.

c. ______ months after the observation started, the tree would be 21.82 feet tall.

40. Scientists are conducting an experiment with a gas in a sealed container. The mass of the gas is measured, and the scientists realize that the gas is leaking over time in a linear way. Its mass is leaking by 9.6 grams per minute. Five minutes since the experiment started, the remaining gas had a mass of 374.4 grams.

Let x be the number of minutes that have passed since the experiment started, and let y be the mass of the gas in grams at that moment. Use a linear equation to model the weight of the gas over time.

a. This line's slope-intercept equation is ______.

b. 37 minutes after the experiment started, there would be ______ grams of gas left.

c. If a linear model continues to be accurate, ______ minutes since the experiment started, all gas in the container will be gone.

4.7 Standard Form

We've seen that a linear relationship can be expressed with an equation in slope-intercept form (4.5.1) or with an equation in point-slope-form (4.6.1). There is a third standard form that you can use to write line equations. It's so "standard" that it's actually known as **standard form**.

4.7.1 Standard Form Definition

Imagine trying to gather donations to pay for a \$10,000 medical procedure you cannot afford. Oversimplifying the mathematics a bit, suppose that there were only two types of donors in the world: those who will donate \$20 and those who will donate \$100. How many of each, or what combination, do you need to reach the funding goal? As in, if x people donate \$20 and y people donate \$100, what numbers could x and y be? The donors of the first type have collectively donated $20x$ dollars, and the donors of the second type have collectively donated $100y$. So altogether you'd need

$$20x + 100y = 10000.$$

This is an example of a line equation in **standard form**.

Definition 4.7.2 (Standard Form). It is always possible to write an equation for a line in the form

$$Ax + By = C \tag{4.7.1}$$

where A, B, and C are three numbers (each of which might be 0, although at least one of B and C must be nonzero). This form of a line equation is called **standard form**. In the context of an application, the meaning of A, B, and C depends on that context. This equation is called **standard form** perhaps because *any* line can be written this way, even vertical lines which cannot be written using the two previous forms we've studied.

Exercise 4.7.3. For each of the following equations, identify what form they are in.

$2.7x + 3.4y = -82$	(□ slope-intercept	□ point-slope	□ standard	□ other linear	□ not linear)
$y = \frac{2}{7}(x - 3) + \frac{1}{10}$	(□ slope-intercept	□ point-slope	□ standard	□ other linear	□ not linear)
$12x - 3 = y + 2$	(□ slope-intercept	□ point-slope	□ standard	□ other linear	□ not linear)
$y = x^2 + 5$	(□ slope-intercept	□ point-slope	□ standard	□ other linear	□ not linear)
$x - y = 10$	(□ slope-intercept	□ point-slope	□ standard	□ other linear	□ not linear)
$y = 4x + 1$	(□ slope-intercept	□ point-slope	□ standard	□ other linear	□ not linear)

Solution. $2.7x + 3.4y = -82$ is in standard form, with $A = 2.7$, $B = 3.4$, and $C = -82$.

$y = \frac{2}{7}(x - 3) + \frac{1}{10}$ is in point-slope form, with slope $\frac{2}{7}$, and passing through $\left(3, \frac{1}{10}\right)$.

$12x - 3 = y + 2$ is linear, but not in any of the forms we have studied. Using algebra, you can rearrange it to read $y = 12x - 5$.

$y = x^2 + 5$ is not linear. The exponent on x is a dead giveaway.

$x - y = 10$ is in standard form, with $A = 1$, $B = -1$, and $C = 10$.

$y = 4x + 1$ is in slope-intercept form, with slope 4 and y-intercept at $(0, 1)$.

Returning to the example with donations for the medical procedure, let's examine the equation

$$20x + 100y = 10000.$$

What units are attached to all of the parts of this equation? Both x and y are numbers of people. The 10000 is in dollars. Both the 20 and the 100 are in dollars per person. Note how both sides of the equation are in dollars. On the right, that fact is clear. On the left, $20x$ is in dollars since 20 is in dollars per person, and x is in people. The same is true for $100y$, and the two dollar amounts $20x$ and $100y$ add to a dollar amount.

What is the slope of the linear relationship? It's not immediately visible since m is not part of the standard form equation. But we can use algebra to isolate y:

$$\begin{aligned} 20x + 100y &= 10000 \\ 100y &= -20x + 10000 \\ y &= \frac{-20x + 10000}{100} \\ y &= \frac{-20x}{100} + \frac{10000}{100} \\ y &= -\frac{1}{5}x + 100. \end{aligned}$$

And we see that the slope is $-\frac{1}{5}$. OK, what units are on that slope? As always, the units on slope are $\frac{y\text{-unit}}{x\text{-unit}}$. In this case that's $\frac{\text{person}}{\text{person}}$, which sounds a little weird and seems like it should be simplified away to unitless. But this slope of $-\frac{1}{5}\frac{\text{person}}{\text{person}}$ is saying that for every one extra person who donates \$20, you only need $\frac{1}{5}$ fewer people donating \$100 to still reach your goal.

What is the y-intercept? Since we've already converted the equation into slope-intercept form, we can see that it is at $(0, 100)$. This tells us that if 0 people donate \$20, then you will need 100 people to each donate \$100.

What does a graph for this line look like? We've already converted into slope-intercept form, and we could use that to make the graph. But when given a line in standard form, there is another approach that is often used. Returning to

$$20x + 100y = 10000,$$

let's calculate the y-intercept and the x-intercept. Recall that these are *points* where the line crosses the y-axis and x-axis. To be on the y-axis means that $x = 0$, and to be on the x-axis means that $y = 0$. All these zeros make the resulting algebra easy to solve:

$$\begin{aligned} 20x + 100y &= 10000 \\ 20(0) + 100y &= 10000 \\ 100y &= 10000 \\ y &= \frac{10000}{100} \\ y &= 100 \end{aligned} \qquad \begin{aligned} 20x + 100y &= 10000 \\ 20x + 100(0) &= 10000 \\ 20x &= 10000 \\ x &= \frac{10000}{20} \\ x &= 500 \end{aligned}$$

So we have a y-intercept at $(0, 100)$ and an x-interept at $(500, 0)$. If we plot these, we get to mark especially relevant points given the context, and then drawing a straight line between them gives us Figure 4.7.4.

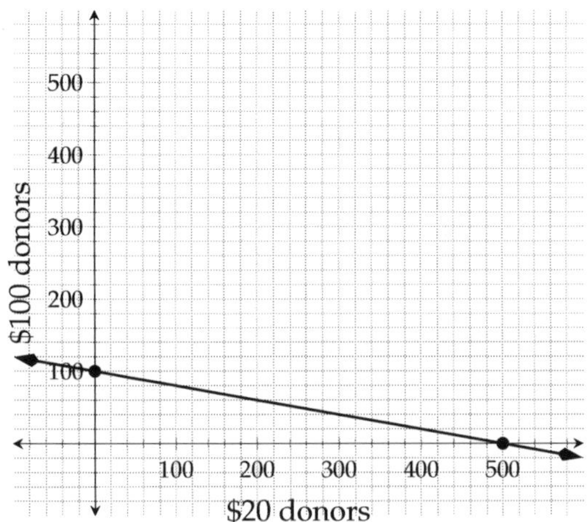

Figure 4.7.4

4.7.2 The x- and y-Intercepts

With a linear relationship (and other types of equations), we are often interested in the x-intercept and y-intercept because they are important in the context. For example, in Figure 4.7.4, the x-intercept implies that if *no one* donates \$100, you need 500 people to donate \$20 to get us to \$10,000. And the y-intercept implies if *no one* donates \$20, you need 100 people to donate \$100. Let's look at another example.

Example 4.7.5. A restaurant purchased 1200 lb of flour. The restaurant uses about 32 lb of flour every day. Model the amount of flour that remains x days later with a linear equation, and interpret the meaning of its x-intercept and y-intercept.

Since the rate of change is constant (-32 lb every day), and we know the initial value, we can model the amount of flour at this restaurant with a slope-intercept equation (4.5.1):

$$y = -32x + 1200$$

where x represents the number of days passed since the initial purchase, and y represents the amount of flour left (in lb.)

A line's x-intercept is in the form of $(x, 0)$, since to be on the x-axis, the y-coordinate must be 0. To find this line's x-intercept, we substitute y in the equation with 0, and solve for x:

$$\begin{aligned} y &= -32x + 1200 \\ 0 &= -32x + 1200 \end{aligned}$$

$$\begin{aligned} 0 - 1200 &= -32x \\ -1200 &= -32x \\ \frac{-1200}{-32} &= x \\ 37.5 &= x \end{aligned}$$

So the line's x-intercept is at $(37.5, 0)$. In context this means that 37.5 days after the 1200 lb of flour was purchased, all the flour would be used up.

A line's y-intercept is in the form of $(0, y)$. This line equation is already in slope-intercept form, so we can just see that its y-intercept is at $(0, 1200)$. In general though, we would substitute x in the equation with 0, and we have:

$$\begin{aligned} y &= -32x + 1200 \\ y &= -32(0) + 1200 \\ y &= 1200 \end{aligned}$$

So yes, the line's y-intercept is at $(0, 1200)$. This means that on the day the flour was purchased, there was 1200 lb of it. In other words, the y-intercept tells us one of the original pieces of information: in the beginning, the restaurant purchased 1200 lb of flour.

If a line is in standard form, it's often easiest to graph it using its two intercepts.

Example 4.7.7. Graph $2x - 3y = -6$ using its intercepts. And then use the intercepts to calculate the line's slope.

Solution. To graph a line by its x-intercept and y-intercept, it might help to first set up a table like Table 4.7.8:

	x-value	y-value	Intercepts
x-intercept		0	
y-intercept	0		

Table 4.7.8: Intercepts of $2x - 3y = -6$

A table like this might help you stay focused on the fact that we are searching for *two* points. As we've noted earlier, an x-intercept is on the x-axis, and so its y-coordinate must be 0. This is worth taking special note of: to find an x-intercept, y must be 0. This is why we put 0 in the y-value cell of the x-intercept. Similarly, a line's y-intercept has $x = 0$, and we put 0 into the x-value cell of the y-intercept.

Next, we calculate the line's x-intercept by substituting $y = 0$ into the equation

$$\begin{aligned} 2x - 3y &= -6 \\ 2x - 3(0) &= -6 \\ 2x &= -6 \\ x &= -3 \end{aligned}$$

So the line's x-intercept is $(-3, 0)$.

Similarly, we substitute $x = 0$ into the equation to calculate the y-intercept:

$$\begin{aligned} 2x - 3y &= -6 \\ 2(0) - 3y &= -6 \\ -3y &= -6 \\ y &= 2 \end{aligned}$$

So the line's y-intercept is $(0, 2)$.

Now we can complete the table:

	x-value	y-value	Intercepts
x-intercept	-3	0	$(-3, 0)$
y-intercept	0	2	$(0, 2)$

Table 4.7.9: Intercepts of $2x - 3y = -6$

With both intercepts' coordinates, we can graph the line:

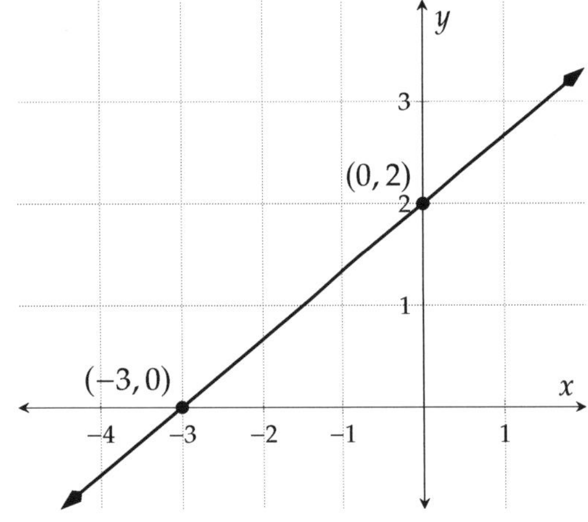

Figure 4.7.10: Graph of $2x - 3y = -6$

There is a slope triangle from the x-intercept to the origin up to the y-intercept. It tells us that the slope is

$$m = \frac{\Delta y}{\Delta x} = \frac{2}{3}.$$

This last example generalizes to a fact worth noting.

Fact 4.7.11. *If a line's x-intercept is at $(r, 0)$ and its y-intercept is at $(0, b)$, then the slope of the line is $-\frac{b}{r}$. (Unless the line passes through the origin. Then both r and b equal 0, and then this fraction is undefined. And the slope of the line could be anything.)*

Exercise 4.7.12. Consider the line with equation $2x + 4.3y = \frac{1000}{99}$.

a. What is its x-intercept? [ ]

b. What is its y-intercept? [ ]

c. What is its slope? [ ]

Solution.

a. To find the x-intercept:

$$\begin{aligned} 2x + 4.3y &= \frac{1000}{99} \\ 2x + 4.3(0) &= \frac{1000}{99} \\ 2x &= \frac{1000}{99} \\ x &= \frac{500}{99} \end{aligned}$$

So the x-intercept is at $\left(\frac{500}{99}, 0\right)$.

b. To find the y-intercept:

$$\begin{aligned} 2x + 4.3y &= \frac{1000}{99} \\ 2(0) + 4.3y &= \frac{1000}{99} \\ 4.3y &= \frac{1000}{99} \\ y &= \frac{1}{4.3} \cdot \frac{1000}{99} \\ y &\approx 2.349\ldots \end{aligned}$$

So the y-intercept is at about $(0, 2.349)$.

c. Since we have the x- and y-intercepts, we can calulate the slope:

$$m \approx -\frac{2.349}{\frac{500}{99}} = -\frac{2.349 \cdot 99}{500} \approx -0.4561.$$

4.7.3 Transforming between Standard Form and Slope-Intercept Form

Sometimes a linear equation arises in standard form (4.7.1), but it would be useful to see that equation in slope-intercept form (4.5.1). Or perhaps, vice versa.

A linear equation in slope-intercept form (4.5.1) tells us important information about the line: its slope m and y-intercept $(0, b)$. However, a line's standard form does not show those two important values. As a result, we often need to change a line's equation from standard form to slope-intercept form. Let's look at some examples.

Example 4.7.14. Change $2x - 3y = -6$ to slope-intercept form, and then graph it.

Solution. Since a line in slope-intercept form looks like $y = \ldots$, we will solve for y in $2x-3y = -6$:

$$\begin{aligned} 2x - 3y &= -6 \\ -3y &= -6 - 2x \\ -3y &= -2x - 6 \\ y &= \frac{-2x - 6}{-3} \\ y &= \frac{-2x}{-3} - \frac{6}{-3} \\ y &= \frac{2}{3}x + 2 \end{aligned}$$

In the third line, we wrote $-2x-6$ on the right side, instead of $-6 - 2x$. The only reason we did this is becasue we are headed to slope-intercept form, where the x-term is traditionally written first.

Now we can see that the slope is $\frac{2}{3}$ and the y-intercept is at $(0, 2)$. With these things found, we can graph the line using slope triangles.

Compare this graphing method with the Graphing by Intercepts method in Example 4.7.7. We have more points in this graph, thus we can graph the line more accurately.

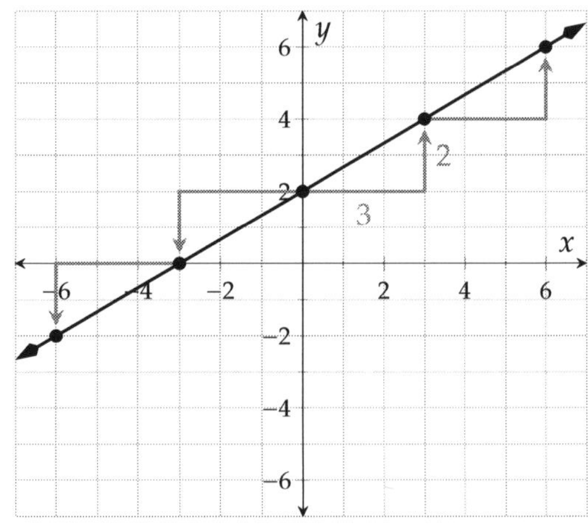

Figure 4.7.15: Graphing $2x - 3y = -6$ with Slope Triangles

Example 4.7.16. Graph $2x - 3y = 0$.

Solution. First We will try (and fail) to graph this line using its x- and y-intercepts.

Trying to find the x-intercept:

$$\begin{aligned} 2x - 3y &= 0 \\ 2x - 3(0) &= 0 \end{aligned}$$

$$\begin{aligned} 2x &= 0 \\ x &= 0 \end{aligned}$$

So the line's x-intercept is at $(0, 0)$, at the origin.

Huh, that is *also* on the y-axis...

Trying to find the y-intercept:

$$\begin{aligned} 2x - 3y &= 0 \\ 2(0) - 3y &= 0 \\ -3y &= 0 \\ y &= 0 \end{aligned}$$

So the line's y-intercept is also at $(0, 0)$.

Since both intercepts are the same point, there is no way to use the intercepts alone to graph this line. So what can be done?

Several approaches are out there, but one is to convert the line equation into slope-intercept form:

$$\begin{aligned} 2x - 3y &= 0 \\ -3y &= 0 - 2x \\ -3y &= -2x \\ y &= \frac{-2x}{-3} \\ y &= \frac{2}{3}x \end{aligned}$$

So the line's slope is $\frac{2}{3}$, and we can graph the line using slope triangles and the intercept at $(0, 0)$, as in Figure 4.7.17.

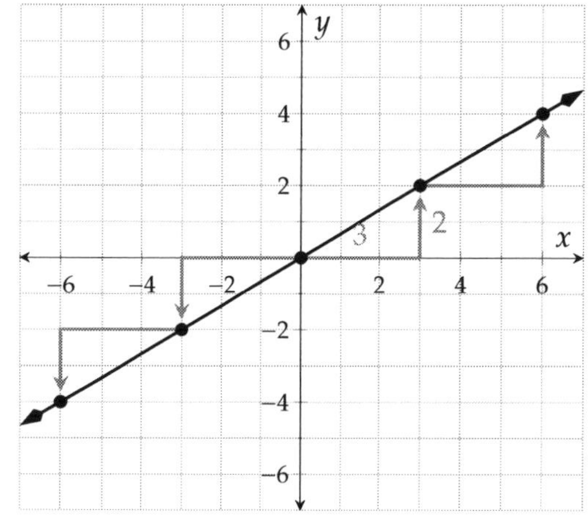

Figure 4.7.17: Graphing $2x - 3y = 0$ with Slope Triangles

In summary, if $C = 0$ in a standard form equation (4.7.1), it's convenient to graph it by first converting the equation to slope-intercept form (4.5.1).

Example 4.7.18. Write the equation $y = \frac{2}{3}x + 2$ in standard form.

Solution. Once we subtract $\frac{2}{3}x$ on both sides of the equation, we have

$$-\frac{2}{3}x + y = 2$$

Technically, this equation is already in standard form $Ax + By = C$. However, you might like to end up with an equation that has no fractions, and so you can take some extra steps.

$$\begin{aligned} y &= \frac{2}{3}x + 2 \\ y - \frac{2}{3}x &= 2 \\ -\frac{2}{3}x + y &= 2 \end{aligned}$$

This is in standard form, but we keep going to clear away the fraction.

$$\begin{aligned} 3 \cdot \left(-\frac{2}{3}x + y\right) &= 3 \cdot 2 \\ -2x + 3y &= 6 \end{aligned}$$

4.7.4 Exercises

Find Slope and y-intercept.

For the following exercises: Find the line's slope and y-intercept.

1. A line has equation $-8x + y = 9$.

This line's slope is ______.

This line's y-intercept is ______.

2. A line has equation $-x - y = 8$.

This line's slope is ______.

This line's y-intercept is ______.

3. A line has equation $10x + 5y = -15$.

This line's slope is ______.

This line's y-intercept is ______.

4. A line has equation $10x - 2y = 2$.

This line's slope is ______.

This line's y-intercept is ______.

5. A line has equation $x + 2y = -10$.

This line's slope is ______.

This line's y-intercept is ______.

6. A line has equation $3x + 4y = -4$.

This line's slope is ______.

This line's y-intercept is ______.

7. A line has equation $7x - 4y = 16$.

This line's slope is ______.

This line's y-intercept is ______.

8. A line has equation $12x + 9y = 9$.

This line's slope is ______.

This line's y-intercept is ______.

9. A line has equation $8x - 10y = 0$.

This line's slope is ______.

This line's y-intercept is ______.

10. A line has equation $20x - 8y = 0$.

This line's slope is ______.

This line's y-intercept is ______.

11. A line has equation $12x + 10y = 3$.

This line's slope is ______.

This line's y-intercept is ______.

12. A line has equation $24x + 20y = 3$.

This line's slope is ______.

This line's y-intercept is ______.

Plot a line given in standard form.

13. Make a graph of the line $x + y = 2$.

14. Make a graph of the line $-5x - y = -3$.

15. Make a graph of the line $x + 5y = 5$.

16. Make a graph of the line $x - 2y = 2$.

17. Make a graph of the line $20x - 4y = 8$.

18. Make a graph of the line $3x + 5y = 10$.

19. Make a graph of the line $-3x + 2y = 6$.

20. Make a graph of the line $-4x - 5y = 10$.

21. Make a graph of the line $4x - 5y = 0$.

22. Make a graph of the line $5x + 7y = 0$.

Convert to standard form.

23. Rewrite $y = 2x + 5$ in standard form.

24. Rewrite $y = 3x - 2$ in standard form.

25. Rewrite $y = \frac{4}{3}x - 1$ in standard form.

26. Rewrite $y = -\frac{5}{7}x + 2$ in standard form.

Exercises on Graphing Lines by Intercepts

27. Find the y-intercept and x-intercept of the line given by the equation

$$5x + 4y = 20$$

If a particular intercept does not exist, enter *none* into all the answer blanks for that row.

	x-value	y-value	Location
y-intercept	________	________	________________
x-intercept	________	________	________________

28. Find the y-intercept and x-intercept of the line given by the equation

$$6x + 3y = -36$$

If a particular intercept does not exist, enter *none* into all the answer blanks for that row.

	x-value	y-value	Location
y-intercept	________	________	________________
x-intercept	________	________	________________

29. Find the y-intercept and x-intercept of the line given by the equation

$$8x - 5y = -120$$

If a particular intercept does not exist, enter *none* into all the answer blanks for that row.

	x-value	y-value	Location
y-intercept	________	________	________________
x-intercept	________	________	________________

30. Find the y-intercept and x-intercept of the line given by the equation

$$x - 7y = -7$$

If a particular intercept does not exist, enter *none* into all the answer blanks for that row.

	x-value	y-value	Location
y-intercept	________	________	________________
x-intercept	________	________	________________

31. Find the x- and y-intercepts of the line with equation $4x + 6y = 24$. Then use the intercepts to plot the line.

32. Find the x- and y-intercepts of the line with equation $4x+5y = -40$. Then use the intercepts to plot the line.

33. Find the x- and y-intercepts of the line with equation $5x - 2y = 10$. Then use the intercepts to plot the line.

34. Find the x- and y-intercepts of the line with equation $5x-6y = -90$. Then use the intercepts to plot the line.

35. Find the x- and y-intercepts of the line with equation $x + 5y = -15$. Then use the intercepts to plot the line.

36. Find the x- and y-intercepts of the line with equation $6x + y = -18$. Then use the intercepts to plot the line.

Interpreting Intercepts in Context

37. Kim is buying some tea bags and some sugar bags. Each tea bag costs 6 cents, and each sugar bag costs 10 cents. She can spend a total of $1.80.

Assume Kim will purchase x tea bags and y sugar bags. Use a linear equation to model the number of tea bags and sugar bags she can purchase.

Find this line's x-intercept, and interpret its meaning in this context.

- ○ *A*. The x-intercept is (18,0). It implies Kim can purchase 18 tea bags with no sugar bags.
- ○ *B*. The x-intercept is (0,30). It implies Kim can purchase 30 sugar bags with no tea bags.
- ○ C. The x-intercept is (0, 18). It implies Kim can purchase 18 sugar bags with no tea bags.
- ○ *D*. The x-intercept is (30,0). It implies Kim can purchase 30 tea bags with no sugar bags.

38. Sharell is buying some tea bags and some sugar bags. Each tea bag costs 6 cents, and each sugar bag costs 5 cents. She can spend a total of \$2.70.

Assume Sharell will purchase x tea bags and y sugar bags. Use a linear equation to model the number of tea bags and sugar bags she can purchase.

Find this line's y-intercept, and interpret its meaning in this context.

- ○ *A*. The y-intercept is (45,0). It implies Sharell can purchase 45 tea bags with no sugar bags.
- ○ *B*. The y-intercept is (54,0). It implies Sharell can purchase 54 tea bags with no sugar bags.
- ○ *C*. The y-intercept is (0, 54). It implies Sharell can purchase 54 sugar bags with no tea bags.
- ○ *D*. The y-intercept is (0,45). It implies Sharell can purchase 45 sugar bags with no tea bags.

39. An engine's tank can hold 75 gallons of gasoline. It was refilled with a full tank, and has been running without breaks, consuming 3 gallons of gas per hour.

Assume the engine has been running for x hours since its tank was refilled, and assume there are y gallons of gas left in the tank. Use a linear equation to model the amount of gas in the tank as time passes.

Find this line's x-intercept, and interpret its meaning in this context.

- ○ *A*. The x-intercept is (0,25). It implies the engine started with 25 gallons of gas in its tank.
- ○ *B*. The x-intercept is (25,0). It implies the engine will run out of gas 25 hours after its tank was refilled.
- ○ *C*. The x-intercept is (0,75). It implies the engine started with 75 gallons of gas in its tank.
- ○ *D*. The x-intercept is (75,0). It implies the engine will run out of gas 75 hours after its tank was refilled.

40. An engine's tank can hold 200 gallons of gasoline. It was refilled with a full tank, and has been running without breaks, consuming 4 gallons of gas per hour.

Assume the engine has been running for x hours since its tank was refilled, and assume there are y gallons of gas left in the tank. Use a linear equation to model the amount of gas in the tank as time passes.

Find this line's y-intercept, and interpret its meaning in this context.

○ *A.* The y-intercept is (200,0). It implies the engine will run out of gas 200 hours after its tank was refilled.

○ *B.* The y-intercept is (0,50). It implies the engine started with 50 gallons of gas in its tank.

○ *C.* The y-intercept is (0,200). It implies the engine started with 200 gallons of gas in its tank.

○ *D.* The y-intercept is (50,0). It implies the engine will run out of gas 50 hours after its tank was refilled.

41. A new car of a certain model costs $46,200.00. According to Blue Book, its value decreases by $2,200.00 every year.

Assume x years since its purchase, the car's value is y dollars. Use a linear equation to model the car's value.

Find this line's x-intercept, and interpret its meaning in this context.

○ *A.* The x-intercept is (0,21). It implies the car would have no more value 21 years since its purchase.

○ *B.* The x-intercept is (0,46200). It implies the car's initial value was 46200.

○ *C.* The x-intercept is (21,0). It implies the car would have no more value 21 years since its purchase.

○ *D.* The x-intercept is (46200,0). It implies the car's initial value was 46200.

42. A new car of a certain model costs $40,800.00. According to Blue Book, its value decreases by $2,400.00 every year.

Assume x years since its purchase, the car's value is y dollars. Use a linear equation to model the car's value.

Find this line's y-intercept, and interpret its meaning in this context.

- ○ *A*. The y-intercept is (40800,0). It implies the car's initial value was 40800.
- ○ *B*. The y-intercept is (17,0). It implies the car would have no more value 17 years since its purchase.
- ○ C. The y-intercept is (0,40800). It implies the car's initial value was 40800.
- ○ *D*. The y-intercept is (0,17). It implies the car would have no more value 17 years since its purchase.

4.8 Horizontal, Vertical, Parallel, and Perpendicular Lines

Horizontal and vertical lines have some special features worth our attention. Also if a pair of lines are parallel or perpendicular to each other, we have some interesting things to say about them. This section looks at these geometric features that lines may have.

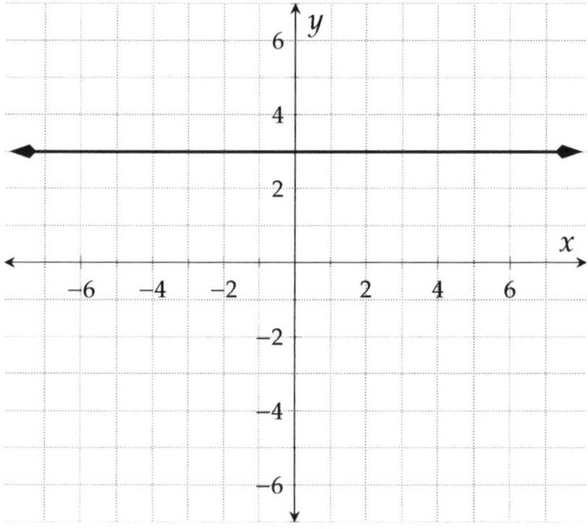

Figure 4.8.1: Horizontal Line

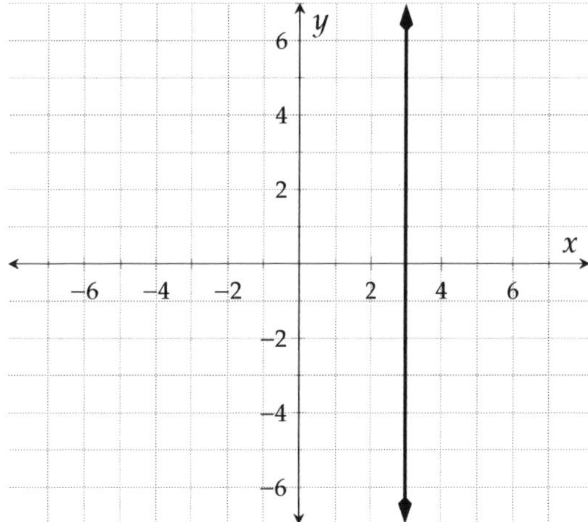

Figure 4.8.2: Vertical Line

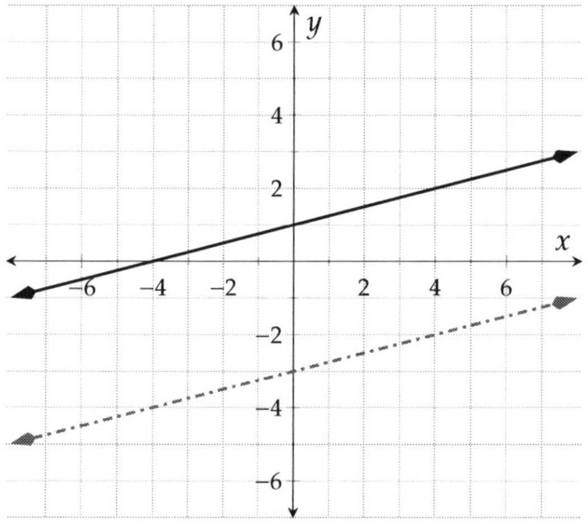

Figure 4.8.3: Parallel Lines

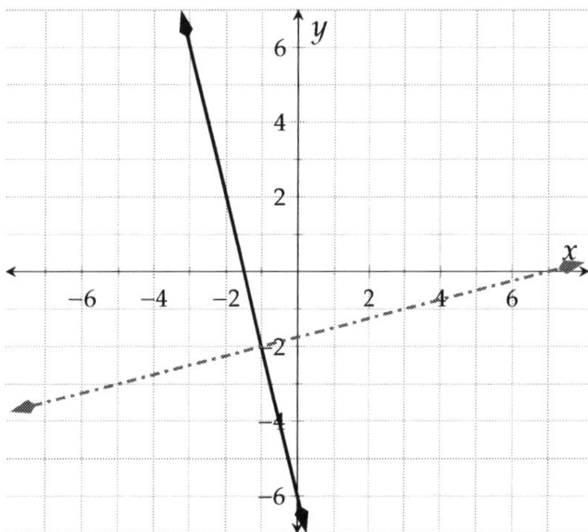

Figure 4.8.4: Perpendicular Lines

4.8.1 Horizontal Lines and Vertical Lines

We learned in Section 4.7 that all lines can be written in standard form (4.7.1). When either A or B equal 0, we end up with a horizontal or vertical line, as we will soon see. Let's take the standard form (4.7.1) line equation, let $A = 0$ and $B = 0$ one at a time and simplify each equation.

$$\begin{aligned} Ax + By &= C \\ 0x + By &= C \\ By &= C \\ y &= \frac{C}{B} \\ y &= k \end{aligned} \qquad \begin{aligned} Ax + By &= C \\ Ax + 0y &= C \\ Ax &= C \\ x &= \frac{C}{A} \\ x &= h \end{aligned}$$

At the end we just renamed the constant numbers $\frac{C}{B}$ and $\frac{C}{A}$ to k and h because of tradition. What is important, is that you view h and k (as well as A, B, and C) as constants: numbers that have some specific value and don't change in the context of one problem.

Think about just one of these last equations: $y = k$. It says that the y-value is the same no matter where you are on the line. If you wanted to plot points on this line, you are free to move far to the left or far to the right on the x-axis, but then you always move up (or down) to make the y-value equal k. What does such a line look like?

Example 4.8.6. Let's plot the line with equation $y = 3$. (Note that this is the same as $0x + 1y = 3$.)

To plot some points, it doesn't matter what x-values we use. All that matters is that y is *always* 3.

A line like this is **horizontal**, parallel to the horizontal axis. All lines with an eqaution in the form

$$y = k$$

(or, in standard form, $0x + By = C$) are **horizontal**.

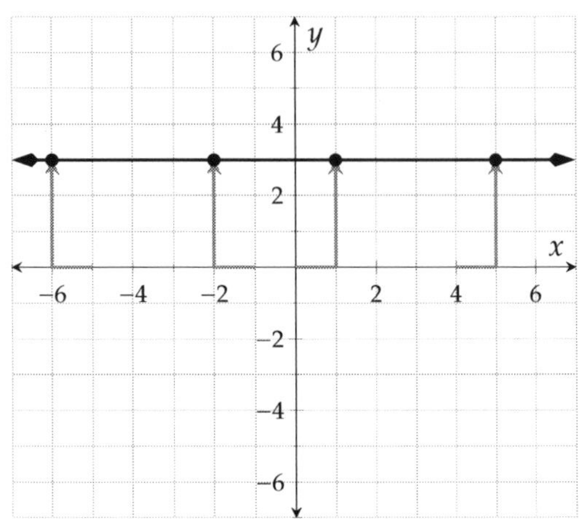

Figure 4.8.7: $y = 3$

Example 4.8.8. Let's plot the line with equation $x = 5$. (Note that this is the same as $1x + 0y = 5$.)

The line has $x = 5$, so to plot points, we are *required* to move over to $x = 5$. From there, we have complete freedom to move however far we like up or down.

A line like this is **vertical**, parallel to the vertical axis. All lines with an equation in the form

$$x = h$$

(or, in standard form, $Ax + 0y = C$) are vertical.

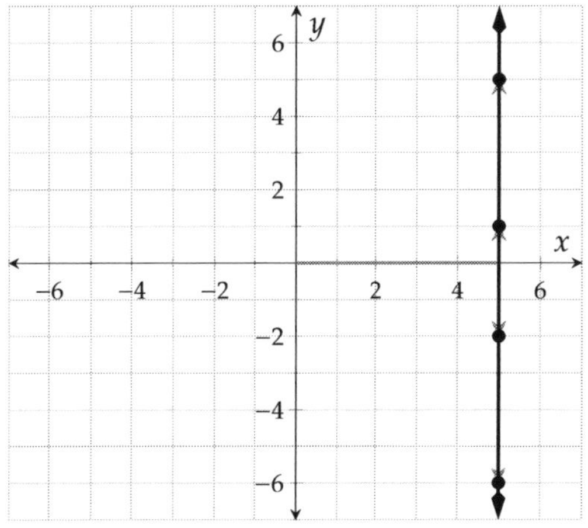

Figure 4.8.9: $x = 5$

Example 4.8.10 (Zero Slope). In Exercise 4.4.22, we learned that a horizontal line's slope is 0, because the distance doesn't change as time moves on. So the numerator in the slope formula (4.4.3) is 0. Now, if we know a line's slope and its y-intercept, we can use slope-intercept form (4.5.1) to write its equation:

$$y = mx + b$$
$$y = 0x + b$$
$$y = b$$

This provides us with an alternative way to think about equations of horizontal lines. They have a certain y-intercept b, and they have slope 0.

We use horizontal lines to model scenarios where there is no change in y-values, like when Tammy stopped for 12 hours (she deserved a rest!)

Exercise 4.8.11 (Plotting Points). Suppose you need to plot the equation $y = -4.25$. Since the equation is in "$y =$" form, you decide to make a table of points. Fill out some points for this table.

x	y

Solution. We can use whatever values for x that we like, as long as they are all different. The equation tells us the y-value has to be -4.25 each time.

x	y
-2	-4.25
-1	-4.25
0	-4.25
1	-4.25
2	-4.25

The reason we made a table was to help with plotting the line.

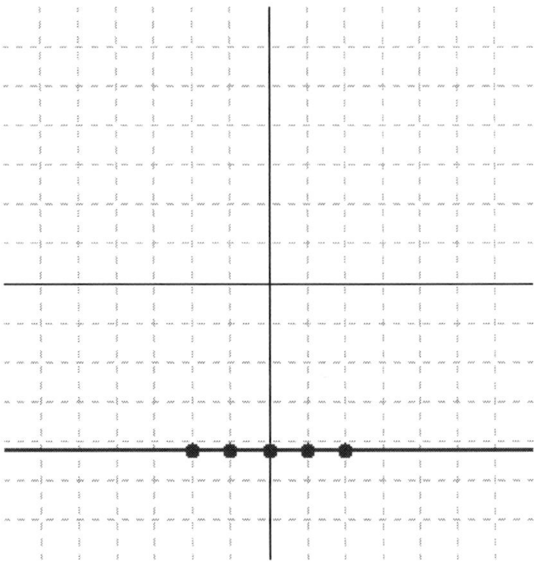

Example 4.8.12 (Undefined Slope). What is the slope of a vertical line? Figure 4.8.13 shows three lines passing through the origin, each steeper than the last. In each graph, you can see a slope triangle that uses a "rise" of 4 each time.

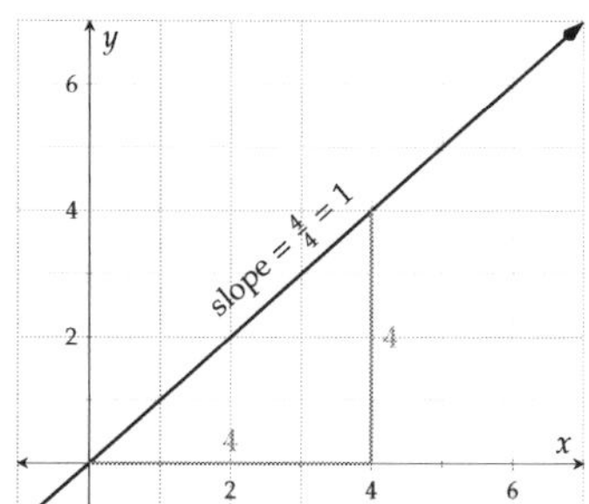

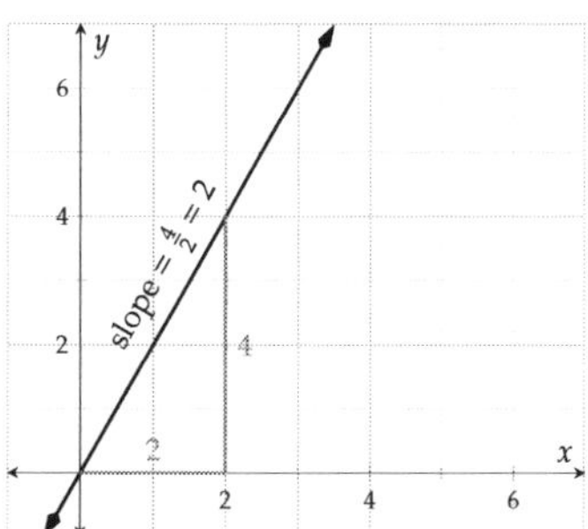

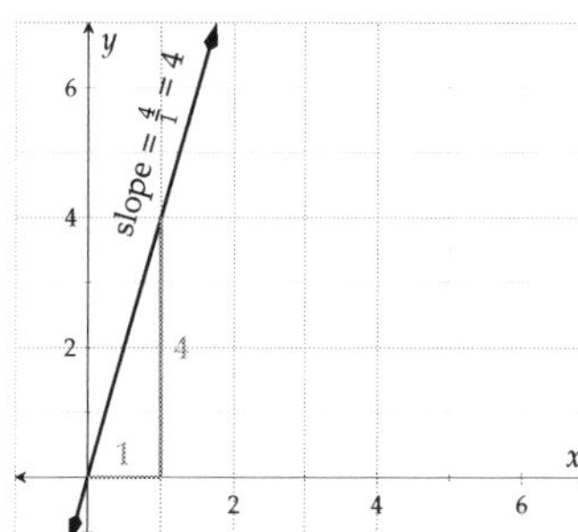

Figure 4.8.13

If we continued making the line steeper and steeper until it was vertical, the slope triangle would still have a "rise" of 4, but the "run" would become smaller and smaller, closer to 0. And then the slope would be $m = \frac{4}{\text{very small}} = \text{very large}$. So the slope of a vertical line can be thought of as "infinitely large."

If we actually try to compute the slope using the slope triangle when the run is 0, we would have $\frac{4}{0}$, which is undefined. So we also say that the slope of a vertical line is *undefined*. Some people say that a vertical line *has no slope*.

Remark 4.8.14. Be careful not to mix up "no slope" (which means "its slope is undefined") with "has slope 0." If a line has slope 0, it *does* have a slope.

Exercise 4.8.15 (Plotting Points)**.** Suppose you need to plot the equation $x = 3.14$. You decide to try making a table of points. Fill out some points for this table.

x	y
______	______
______	______
______	______
______	______
______	______

Solution. Since the equation says x is always the number 3.14, we have to use this for the x value in all the points. This is different from how we would plot a "$y =$" equation, where we would use several different x-values. We can use whatever values for y that we like, as long as they are all different.

x	y
3.14	−2
3.14	−1
3.14	0
3.14	1
3.14	2

The reason we made a table was to help with plotting the line.

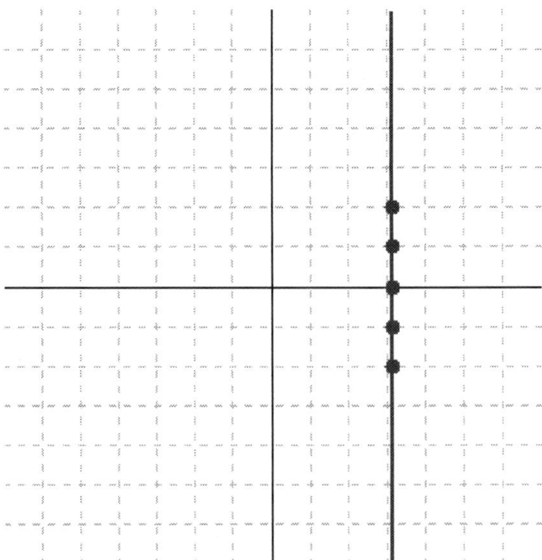

Example 4.8.16. Let x represent the price of a new 60-inch television at Target on Black Friday (which was \$650), and let y be the number of hours you will watch something on this TV over its lifetime. What is the relationship between x and y?

Well, there is no getting around the fact that $x = 650$. As for y, without any extra information about your viewing habits, it could theoretically be as low as 0 or it could be anything larger than that. If we graph this scenario, we have to graph the equation $x = 650$ which we now know to give a vertical line, and we get Figure 4.8.17.

Figure 4.8.17: New TV: hours watched versus purchase price; negative y-values omitted since they make no sense in context

Theorem 4.8.18 (Summary of Horizontal and Vertical Line Equations).

Horizontal Lines	*Vertical Lines*
*A line is **horizontal** if and only if its equation can be written* $$y = k$$ *for some constant k.*	*A line is **vertical** if and only if its equation can be written* $$x = h$$ *for some constant h.*
In standard form (4.7.1), any line with equation $$0x + By = C$$ *is horizontal.*	*In standard form (4.7.1), any line with equation* $$Ax + 0y = C$$ *is vertical.*
If the line with equation $y = k$ is horizontal, it has a y-intercept at $(0, k)$ and has slope 0.	*If the line with equation $x = h$ is vertical, it has an x-intercept at $(h, 0)$ and its slope is undefined. Some say it has no slope, and some say the slope is infinitely large.*
In slope-intercept form (4.5.1), any line with equation $$y = 0x + b$$ *is horizontal.*	*It's impossible to write the equation of a vertical line in slope-intercept form (4.5.1), because to do so would require having a slope.*

4.8.2 Parallel Lines

Example 4.8.19.

Two trees were planted in the same year, and their growth over time is modeled by the two lines in Figure 4.8.20. Use linear equations to model each tree's growth, and interpret their meanings in this context.

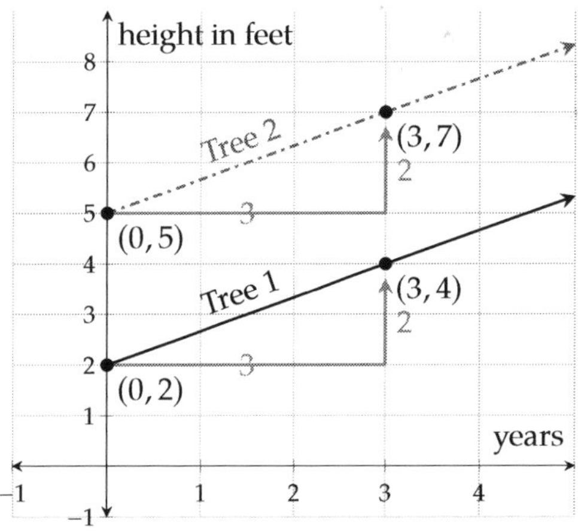

Figure 4.8.20: Two Trees' Growth Chart

We can see Tree 1's equation is $y = \frac{2}{3}x + 2$, and Tree 2's equation is $y = \frac{2}{3} + 5$. Tree 1 was 2 feet tall when it was planted, and Tree 2 was 5 feet tall when it was planted. Both trees have been growing at the same rate, $\frac{2}{3}$ feet per year, or 2 feet every 3 years.

An important observation right now is that those two lines are parallel. Why? For lines with positive slopes, the bigger a line's slope, the steeper the line is slanted. As a result, if two lines have the same slope, they are slanted at the same angle, thus they are parallel.

Fact 4.8.21. *Of course, any two vertical lines are parallel to each other. For two nonvertical lines, they are parallel if and only if they have the same slope.*

Exercise 4.8.22. A line ℓ is parallel to the line with equation $y = 17.2x - 340.9$, but ℓ has y-intercept at $(0, 128.2)$. What is an equation for ℓ?

Solution. Parallel lines have the same slope, and the slope of $y = 17.2x - 340.9$ is 17.2. So ℓ has slope 17.2. And we have been given that ℓ's y-intercept is at $(0, 128.2)$. So we can use slope-intercept form to write its equation as

$$y = 17.2x + 128.2.$$

Exercise 4.8.23. A line κ is parallel to the line with equation $y = -3.5x + 17$, but κ passes through the point $(-12, 23)$. What is an equation for κ?

Solution. Parallel lines have the same slope, and the slope of $y = -3.5x + 17$ is -3.5. So κ has slope -3.5. And we know a point that κ passes through, so we can use point-slope form to write its equation as

$$y = -3.5(x + 12) + 23.$$

4.8.3 Perpendicular Lines

The slopes of two perpendiular lines have a special relationship too.

Figure 4.8.26 walks you through an explanation of this realationship.

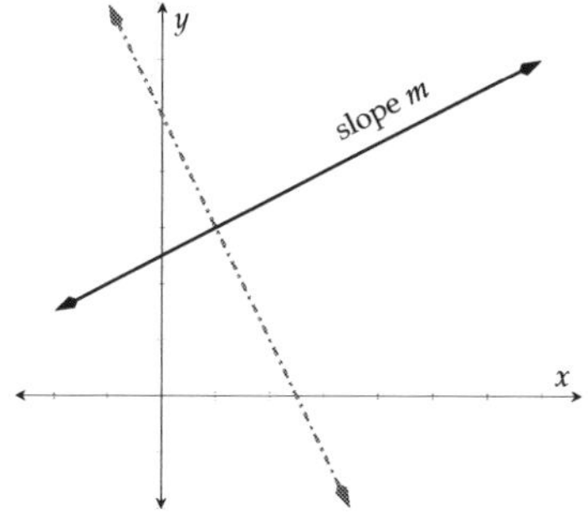

(a) Two generic perpendicular lines, where one has slope m.

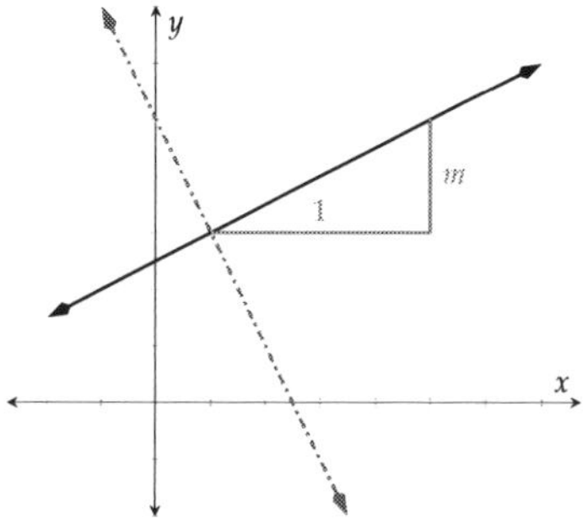

(b) Since the one slope is m, we can draw a slope triangle with "run" 1 and "rise" m.

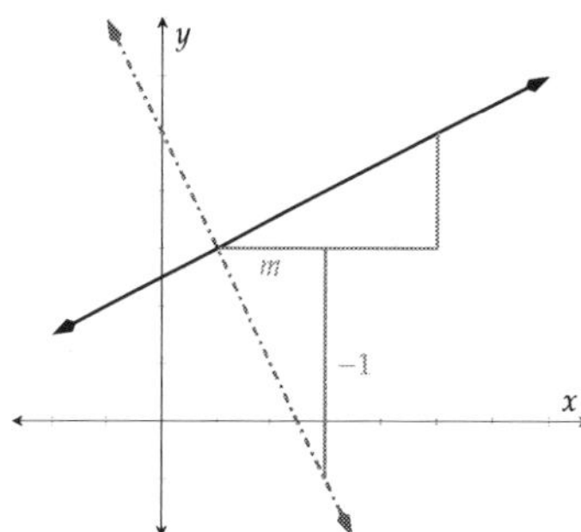

(c) A *congruent* slope triangle can be drawn for the perpendicular line. It's legs have the same lengths, but in different positions, and one is negative.

Figure 4.8.26: The relationship between slopes of perpendicular lines

The second line in Figure 4.8.26 has slope

$$\frac{\Delta y}{\Delta x} = \frac{-1}{m} = -\frac{1}{m}.$$

Fact 4.8.27. *Of course, a vertical line and a horizontal line are perpendicular. For lines that are neither vertical nor horizontal, they are perpendiclar if an only if the slope of one is the negative reciprocal of the slope of the other. That is, if one has slope m, the other has slope $-\frac{1}{m}$.*

Another way to say this is that the product of the slopes of two perpendicular lines is -1 (assuming both of the lines have a slope in the first place).

Not convinced? Here are three pairs of perpendicular lines where we can see if the pattern holds.

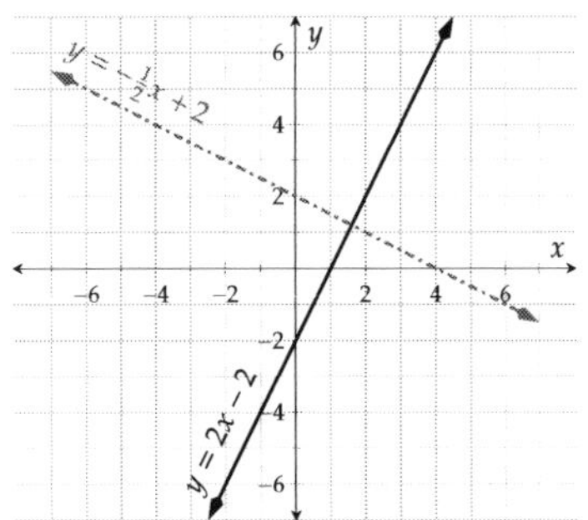

Figure 4.8.28: Graphing $y = 2x - 2$ and $y = -\frac{1}{2}x + 2$. Note the relationship between their slopes: $2 = -\frac{1}{-1/2}$

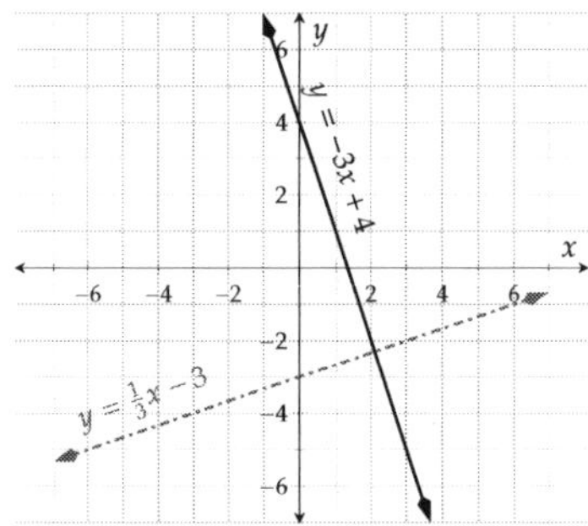

Figure 4.8.29: Graphing $y = -3x + 4$ and $y = \frac{1}{3}x - 3$. Note the relationship between their slopes: $-3 = -\frac{1}{1/3}$

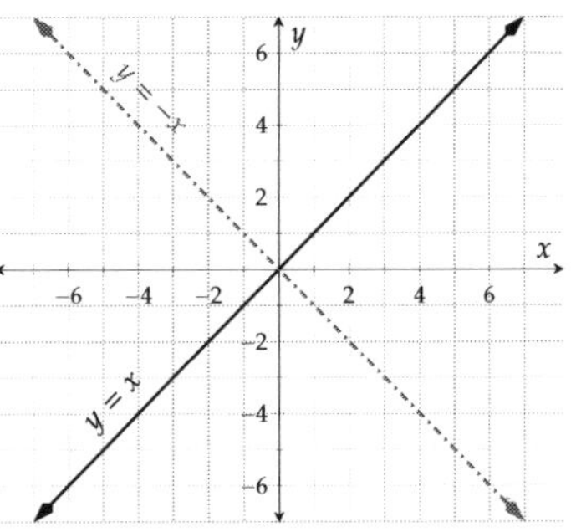

Figure 4.8.30: Graphing $y = x$ and $y = -x$. Note the relationship between their slopes: $1 = -\frac{1}{-1}$

Example 4.8.31. Line A passes through $(-2, 10)$ and $(3, -10)$. Line B passes through $(-4, -4)$ and $(8, -1)$. Determine whether these two lines are parallel, perpendicular or neither.

Solution. We will use the slope formula to find both lines' slopes:

$$\begin{aligned} \text{Line } A\text{'s slope} &= \frac{y_2 - y_1}{x_2 - x_1} \\ &= \frac{-10 - 10}{3 - (-2)} \\ &= \frac{-20}{5} \\ &= -4 \end{aligned} \qquad \begin{aligned} \text{Line } B\text{'s slope} &= \frac{y_2 - y_1}{x_2 - x_1} \\ &= \frac{-1 - (-4)}{8 - (-4)} \\ &= \frac{3}{12} \\ &= \frac{1}{4} \end{aligned}$$

Their slopes are not the same, so those two lines are not parallel.

The product of their slopes is $(-4) \cdot \frac{1}{4} = -1$, which means the two lines are perpendicular.

Exercise 4.8.32. Line A and Line B are perpendicular. Line A's equation is $2x + 3y = 12$. Line B passes through the point $(4, -3)$. Find an equation for Line B.

Solution. First, we will find Line A's slope by rewriting its equation from standard form to slope-intercept form:

$$\begin{aligned} 2x + 3y &= 12 \\ 3y &= 12 - 2x \\ 3y &= -2x + 12 \\ y &= \frac{-2x + 12}{3} \\ y &= -\frac{2}{3}x + 4 \end{aligned}$$

So Line A's slope is $-\frac{2}{3}$. Since Line B is perpendicular to Line A, its slope is $-\frac{1}{-\frac{2}{3}} = \frac{3}{2}$.

It's also given that Line B passes through (4, −3), so we can write Line B's point-slope form equation:

$$y = m(x - x_0) + y_0$$
$$y = \frac{3}{2}(x - 4) - 3$$

4.8.4 Exercises

Creating tables for horizontal and vertical lines.

1. Fill out this table for the equation $y = 2$. The first row is an example.

x	y	Points
−3	2	(−3, 2)
−2	______	__________
−1	______	__________
0	______	__________
1	______	__________
2	______	__________

2. Fill out this table for the equation $y = 4$. The first row is an example.

x	y	Points
−3	4	(−3, 4)
−2	______	__________
−1	______	__________
0	______	__________
1	______	__________
2	______	__________

3. Fill out this table for the equation $x = -6$. The first row is an example.

x	y	Points
−6	−3	(−6, −3)
______	−2	__________
______	−1	__________
______	0	__________
______	1	__________
______	2	__________

4. Fill out this table for the equation $x = -5$. The first row is an example.

x	y	Points
−5	−3	(−5, −3)
______	−2	__________
______	−1	__________
______	0	__________
______	1	__________
______	2	__________

Determining whether a point is on a horizontal or vertical line.

5. Consider the equation $y = 1$.

Which of the following ordered pairs are solutions to the given equation? There may be more than one correct answer.

□ $(-5, 1)$ □ $(7, 1)$ □ $(1, 5)$
□ $(0, 10)$

6. Consider the equation $y = 1$.

Which of the following ordered pairs are solutions to the given equation? There may be more than one correct answer.

□ $(-7, 1)$ □ $(1, 4)$ □ $(8, 1)$
□ $(0, 8)$

7. Consider the equation $x + 1 = 0$.

Which of the following ordered pairs are solutions to the given equation? There may be more than one correct answer.

□ $(-1, 0)$ □ $(1, -1)$ □ $(-1, 5)$
□ $(0, -10)$

8. Consider the equation $x + 1 = 0$.

Which of the following ordered pairs are solutions to the given equation? There may be more than one correct answer.

□ $(-1, 0)$ □ $(1, -1)$ □ $(0, -6)$
□ $(-1, 5)$

Given two points, find the equations of the line connecting them.

9. A line passes through the points $(-2, -9)$ and $(1, -9)$. Find an equation for this line.

An equation for this line is ______.

10. A line passes through the points $(-5, -6)$ and $(-3, -6)$. Find an equation for this line.

An equation for this line is ______.

11. A line passes through the points $(-4, 2)$ and $(-4, -1)$. Find an equation for this line.

An equation for this line is ______.

12. A line passes through the points $(-2, -2)$ and $(-2, 4)$. Find an equation for this line.

An equation for this line is ______.

Given linear graphs, find the equations of horizontal and vertical lines.

For the following exercises: A line's graph is given.

13.

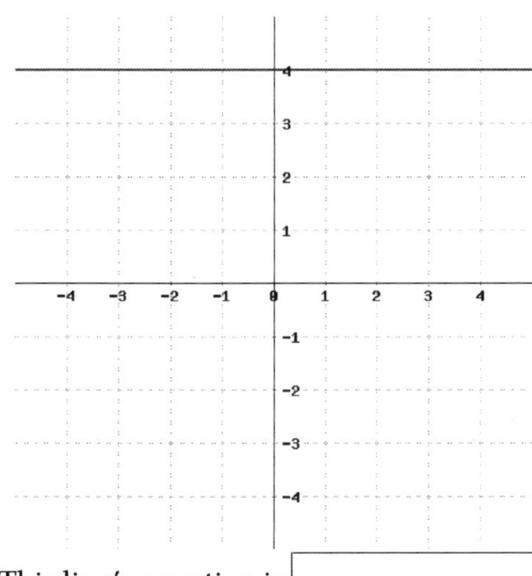

This line's equation is [ ]

14.

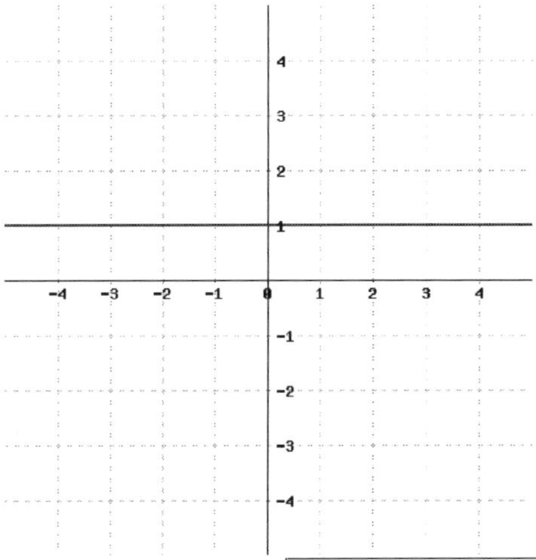

This line's equation is [ ]

15.

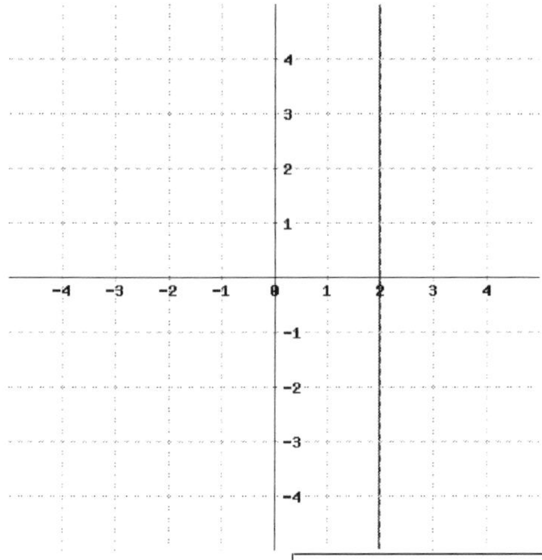

This line's equation is [ ]

16.

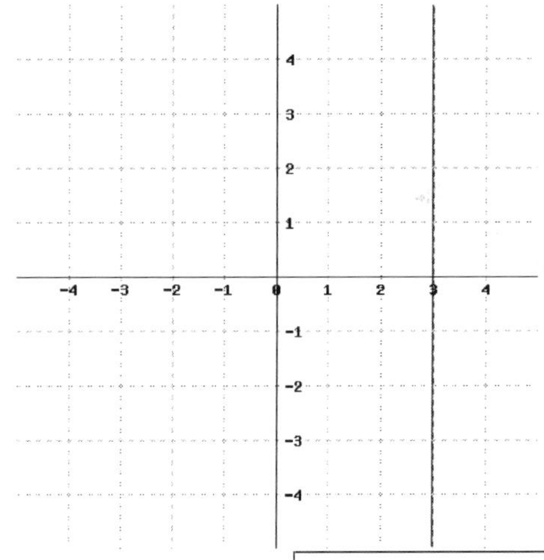

This line's equation is [ ]

Finding the intercepts of horizontal and vertical lines.

17. Find the y-intercept and x-intercept of the line given by the equation

$$x = 10$$

If a particular intercept does not exist, enter *none* into all the answer blanks for that row.

	x-value	y-value	Location
y-intercept	______	______	____________
x-intercept	______	______	____________

18. Find the y-intercept and x-intercept of the line given by the equation

$$x = -9$$

If a particular intercept does not exist, enter *none* into all the answer blanks for that row.

	x-value	y-value	Location
y-intercept	______	______	____________
x-intercept	______	______	____________

19. Find the y-intercept and x-intercept of the line given by the equation

$$y = -7$$

If a particular intercept does not exist, enter *none* into all the answer blanks for that row.

	x-value	y-value	Location
y-intercept	______	______	____________
x-intercept	______	______	____________

20. Find the y-intercept and x-intercept of the line given by the equation

$$y = -4$$

If a particular intercept does not exist, enter *none* into all the answer blanks for that row.

	x-value	y-value	Location
y-intercept	______	______	____________
x-intercept	______	______	____________

Given their equation, graph horizontal and vertical lines.

21. Graph the line $y = 1$.

22. Graph the line $y + 5 = 0$.

23. Graph the line $x = 2$.

24. Graph the line $x - 3 = 0$.

Finding equations of parallel lines.

25. A line passes through the point $(2, -7)$, and it's parallel to the line $y = -2$. Find an equation for this line.

An equation for this line is [ ].

26. A line passes through the point $(-10, 7)$, and it's parallel to the line $y = 0$. Find an equation for this line.

An equation for this line is [ ].

27. A line passes through the point $(0, -1)$, and it's parallel to the line $x = 3$. Find an equation for this line.

An equation for this line is [ ].

28. A line passes through the point $(-7, 8)$, and it's parallel to the line $x = 5$. Find an equation for this line.

An equation for this line is [ ].

29. Line k has the equation $y = 4x - 8$.

Line ℓ is parallel to line k, but passes through the point $(-2, -5)$.

Find an equation for line ℓ in both slope-intercept form and point-slope form.

An equation for ℓ in slope-intercept form is: [ ].

An equation for ℓ in point-slope form is: [ ].

30. Line k has the equation $y = 5x + 7$.

Line ℓ is parallel to line k, but passes through the point $(2, 9)$.

Find an equation for line ℓ in both slope-intercept form and point-slope form.

An equation for ℓ in slope-intercept form is: [ ].

An equation for ℓ in point-slope form is: [ ].

31. Line k has the equation $y = -\frac{1}{2}x + 1$.

Line ℓ is parallel to line k, but passes through the point $(6, -7)$.

Find an equation for line ℓ in both slope-intercept form and point-slope form.

An equation for ℓ in slope-intercept form is: [ ].

An equation for ℓ in point-slope form is: [ ].

32. Line k has the equation $y = -\frac{2}{9}x - 2$.

Line ℓ is parallel to line k, but passes through the point $(9, -1)$.

Find an equation for line ℓ in both slope-intercept form and point-slope form.

An equation for ℓ in slope-intercept form is: [ ].

An equation for ℓ in point-slope form is: [ ].

Determining whether Two Lines Are Parallel or Perpendicular

33. Line m passes points $(5, 8)$ and $(-5, -12)$.

Line n passes points $(-4, -1)$ and $(-2, 3)$.

Determine how the two lines are related.

These two lines are

- ○ parallel
- ○ perpendicular
- ○ neither parallel nor perpendicular

34. Line m passes points $(6, -11)$ and $(-9, 9)$.

Line n passes points $(6, -13)$ and $(-12, 11)$.

Determine how the two lines are related.

These two lines are

- ○ parallel
- ○ perpendicular
- ○ neither parallel nor perpendicular

35. Line m passes points $(8, 2)$ and $(-4, 5)$.

Line n passes points $(-1, -12)$ and $(4, 8)$.

Determine how the two lines are related.

These two lines are

- ○ parallel
- ○ perpendicular
- ○ neither parallel nor perpendicular

36. Line m passes points $(-10, 6)$ and $(10, -18)$.

Line n passes points $(6, -5)$ and $(-6, -15)$.

Determine how the two lines are related.

These two lines are

- ○ parallel
- ○ perpendicular
- ○ neither parallel nor perpendicular

37. Line m passes points $(1, -12)$ and $(4, -18)$.

Line n passes points $(5, 21)$ and $(-5, -19)$.

Determine how the two lines are related.

These two lines are

- ○ parallel
- ○ perpendicular
- ○ neither parallel nor perpendicular

38. Line m passes points $(4, 7)$ and $(-10, 7)$.

Line n passes points $(6, 2)$ and $(-10, 2)$.

Determine how the two lines are related.

These two lines are

- ○ parallel
- ○ perpendicular
- ○ neither parallel nor perpendicular

39. Line m passes points $(10, -4)$ and $(10, -1)$.

Line n passes points $(-3, 6)$ and $(-3, 0)$.

Determine how the two lines are related.

These two lines are

- ○ parallel
- ○ perpendicular
- ○ neither parallel nor perpendicular

40. Line m passes points $(-9, 10)$ and $(-9, 8)$.

Line n passes points $(5, -10)$ and $(5, 6)$.

Determine how the two lines are related.

These two lines are

- ○ parallel
- ○ perpendicular
- ○ neither parallel nor perpendicular

Find a Line's Equation Perpendicular to a Given Equation

41. Line k has the equation $y = -x + 3$.

Line ℓ is perpendicular to line k, and passes through the point $(3, 1)$.

Find an equation for line ℓ in both slope-intercept form and point-slope form.

An equation for ℓ in slope-intercept form is: ______.

An equation for ℓ in point-slope form is: ______.

42. Line k has the equation $y = -3x + 2$.

Line ℓ is perpendicular to line k and passes through the point $(-9, 0)$.

Find an equation for ℓ in both slope-intercept form and point-slope forms.

An equation for ℓ in slope-intercept form is: ______.

An equation for ℓ in point-slope form is: ______.

43. Line k's equation is $y = -\frac{5}{9}x - 3$.

Line ℓ is perpendicular to line k and passes through the point $(-15, -25)$.

Find an equation for line ℓ in both slope-intercept form and point-slope forms.

An equation for ℓ in slope-intercept form is: ______.

An equation for ℓ in point-slope form is: ______.

44. Line k has the equation $x - 6y = -30$.

Line ℓ is perpendicular to line k and passes through the point $(-5, 32)$.

Find line ℓ's equation in both slope-intercept form and point-slope form.

An equation for ℓ in slope-intercept form is: ______.

An equation for ℓ in point-slope form is: ______.

4.9 Summary of Graphing Lines

The previous several sections have demonstrated several methods for plotting a graph of a linear equation. In this section, we review these methods.

We have learned three forms to write a linear equation:

- slope-intercept form

  $y = mx + b$

- standard form

  $Ax + By = C$

- point-slope form

  $y = m(x - x_0) + y_0$

We have studied two special types of line:

- horizontal line: $y = k$
- vertical line: $x = h$

We have practiced three ways to graph a line:

- building a table of x- and y-values
- plotting *one* point (often the y-intercept) and drawing slope triangles
- plotting its x-intercept and y-intercept

4.9.1 Graphing Lines in Slope-Intercept Form

In the following examples we will graph $y = -2x + 1$, which is in slope-intercept form (4.5.1), with different methods and compare them.

Example 4.9.2 (Building a Table of x- and y-values). First, we will graph $y = -2x+1$ by building a table of values. In theory this method can be used for any type of equation, linear or not. Every student must feel comfortable with building a table of values based on an equation.

x-value	y-value	Point
-2	$y = -2(-2) + 1 = 5$	$(-2, 5)$
-1	$y = -2(-1) + 1 = 3$	$(-1, 3)$
0	$y = -2(0) + 1 = 1$	$(0, 1)$
1	$y = -2(1) + 1 = -1$	$(1, -1)$
2	$y = -2(2) + 1 = -3$	$(2, -3)$

Table 4.9.3: Table for $y = -2x + 1$

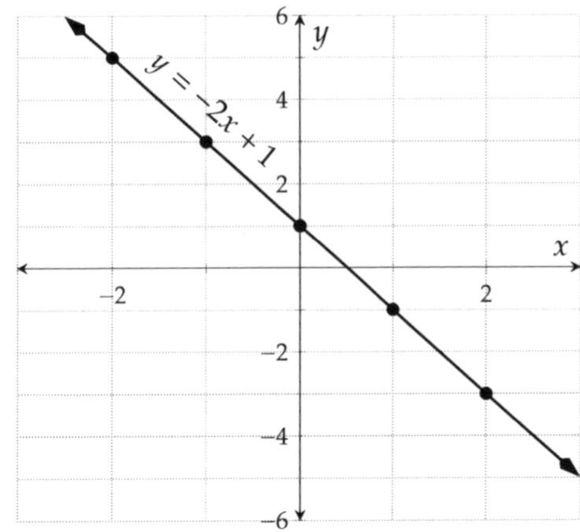

Figure 4.9.4: Graphing $y = -2x + 1$ by Building a Table of Values

Example 4.9.5 (Using Slope Triangles). Although making a table is straightforward, the slope triangle method is both faster and reinforces the true meaning of slope. In the slope triangle method, we first identify some point on the line. With a line in slope-intercept form (4.5.1), we know the y-intercept, which is $(0, 1)$. Then, we can draw slope triangles in both directions to find more points.

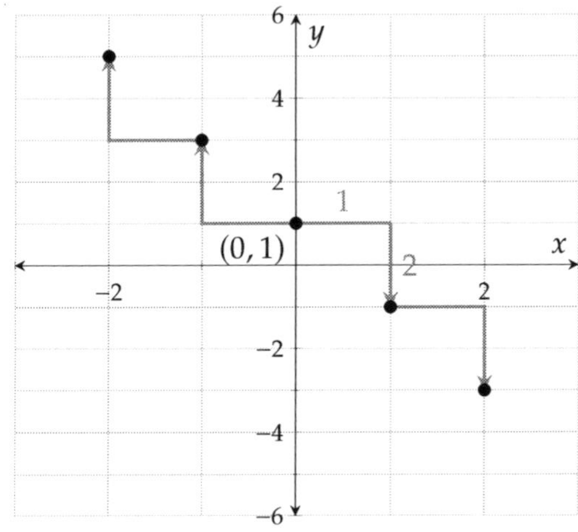

Figure 4.9.6: Marking a point and some slope triangles

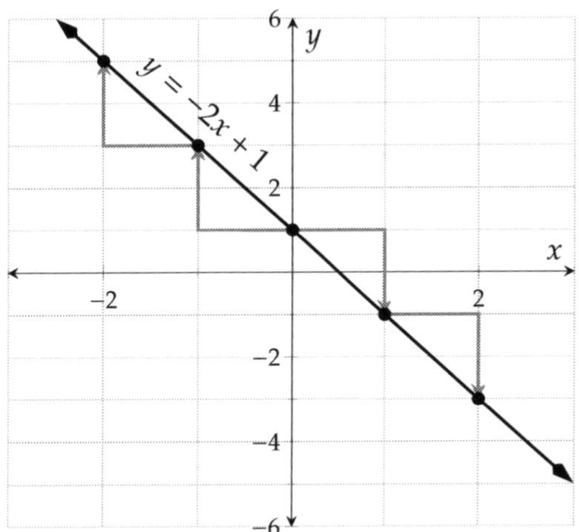

Figure 4.9.7: Graphing $y = -2x + 1$ by slope triangles

Compared to the table method, the slope triangle method:

- is less straightforward
- doesn't take the time and space to make a table
- doesn't involve lots of calculations where you might make a human error
- shows slope triangles, which reinforces the meaning of slope

Example 4.9.8 (Using intercepts). If we use the x- and y-intercepts to plot $y = -2x + 1$, we have some calculation to do. While it is apparent that the y-intercept is at $(0, 1)$, where is the x-intercept? Here are two methods to find it.

Set $y = 0$.

$$\begin{aligned} y &= -2x + 1 \\ 0 &= -2x + 1 \\ 0 - 1 &= -2x \\ -1 &= -2x \\ \frac{-1}{-2} &= x \\ \frac{1}{2} &= x \end{aligned}$$

Factor out the coefficient of x.

$$\begin{aligned} y &= -2x + 1 \\ y &= -2x + (-2)\left(-\frac{1}{2}\right)1 \\ y &= -2\left(x + \left(-\frac{1}{2}\right)1\right) \\ y &= -2\left(x - \frac{1}{2}\right) \end{aligned}$$

And now it is easy to see that substituting $x = \frac{1}{2}$ would make $y = 0$.

So the x-intercept is at $\left(\frac{1}{2}, 0\right)$. Plotting both intercepts:

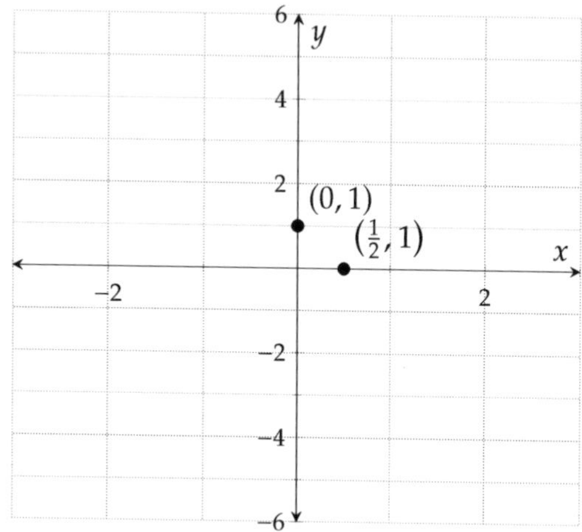

Figure 4.9.9: Marking intercepts

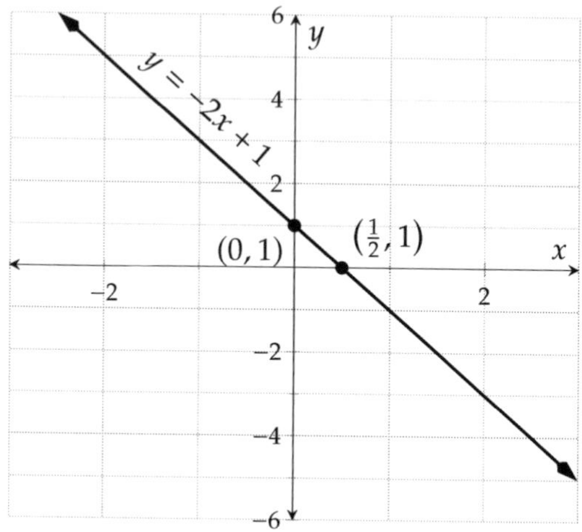

Figure 4.9.10: Graphing $y = -2x + 1$ by slope triangles

This worked, but here are some observations about why this method is not the greatest.

- We had to plot a point with fractional coordinates.
- We only plotted two points and they turned out very close to each other, so even the slightest inaccuracy in our drawing skills could result in a line that is way off.

When a line is presented in slope-intercept form (4.5.1), our opinion is that the slope triangle method is the best choice for making its graph.

4.9.2 Graphing Lines in Standard Form

In the following examples we will graph $3x + 4y = 12$, which is in standard form (4.7.1), with different methods and compare them.

Example 4.9.11 (Building a Table of x- and y-values). To make a table, we could substitute x for various numbers and use algebra to find the corresponding y-values. Let's start with $x = -2$, planning to move on to $x = -1, 0, 1, 2$.

$$\begin{aligned} 3x + 4y &= 12 \\ 3(-2) + 4y &= 12 \\ -6 + 4y &= 12 \\ 4y &= 12 + 6 \\ 4y &= 18 \\ y &= \frac{18}{4} \\ y &= \frac{9}{2} \end{aligned}$$

The first point we found is $\left(-2, \frac{9}{2}\right)$. This has been a lot of calculation, and we ended up with a fraction we will have to plot. *And* we have to repeat this process a few more times to get more points for the table. The table method is generally not a preferred way to graph a line in standard form (4.7.1). Let's look at other options.

Example 4.9.12 (Using intercepts). Next, we will try graphing $3x + 4y = 12$ using intercepts. We set up a small table to record the two intercepts:

	x-value	y-value	Intercept
x-intercept		0	
y-intercept	0		

We have to calculate the line's x-intercept by substituting $y = 0$ into the equation:

$$\begin{aligned} 3x + 4y &= 12 \\ 3x + 4(0) &= 12 \\ 3x &= 12 \\ x &= \frac{12}{3} \\ x &= 4 \end{aligned}$$

And similarly for the y-intercept:

$$\begin{aligned} 3x + 4y &= 12 \\ 3(0) + 4y &= 12 \\ 4y &= 12 \\ y &= \frac{12}{4} \\ y &= 3 \end{aligned}$$

So the line's x-intercept is at $(4, 0)$ and its y-intercept is at $(0, 3)$. Now we can complete the table and then graph the line:

	x-value	y-value	Intercepts
x-intercept	4	0	$(4, 0)$
y-intercept	0	3	$(0, 3)$

Table 4.9.13: Intercepts of $3x + 4y = 12$

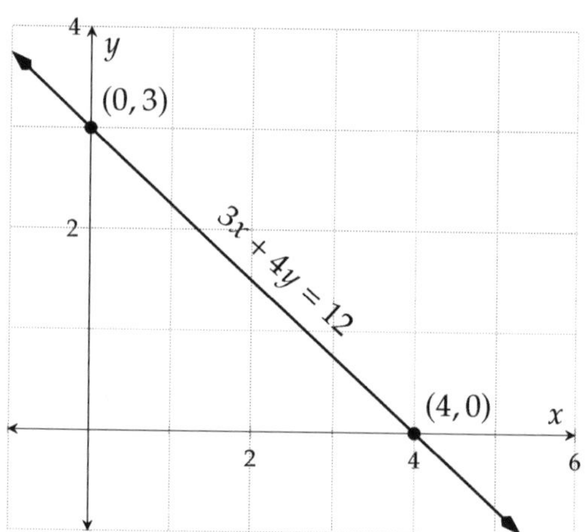

Figure 4.9.14: Graph of $3x + 4y = 12$

Example 4.9.15.

We can always rearrange $3x + 4y = 12$ into slope-intercept form (4.5.1), and then graph it with the slope triangle method:

$$\begin{aligned}3x + 4y &= 12\\ 4y &= 12 - 3x\\ 4y &= -3x + 12\\ y &= \frac{-3x + 12}{4}\\ y &= -\frac{3}{4}x + 3\end{aligned}$$

With the y-intercept at $(0, 3)$ and slope $-\frac{3}{4}$, we can graph the line using slope triangles:

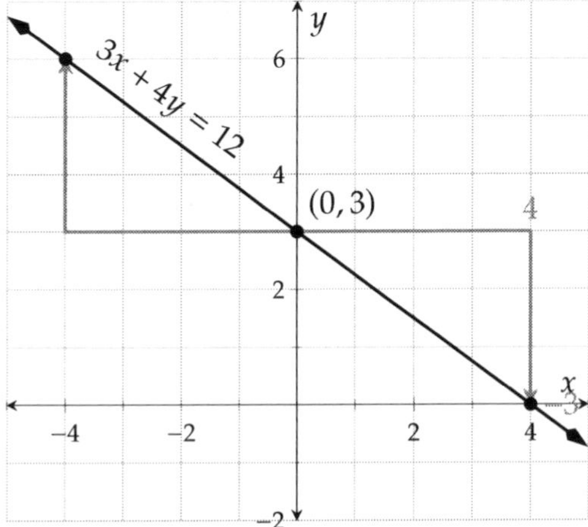

Figure 4.9.16: Graphing $3x + 4y = 12$ using slope triangles

Compared with the intercepts method, the slope triangle method takes more time, but shows more points with slope triangles, and thus a more accurate graph. Also sometimes (as with Example 4.7.16) when we graph a standard form equation like $2x - 3y = 0$, the intercepts method doesn't work because both intercepts are actually at the same point, and we have to resort to something else like slope triangles anyway.

Here are some observations about graphing a line equation that is in standard form (4.7.1):

- The intercepts method might be the quickest approach.
- The intercepts method only tells us two intercepts of the line. When we need to know more information, like the line's slope, and get a more accurate graph, we should spend more time and use the slope triangle method.
- When $C = 0$ in a standard form equation (4.7.1) we have to use something else like slope triangles anyway.

4.9.3 Graphing Lines in Point-Slope Form

When we graph a line in point-slope form (4.6.1) like $y = \frac{2}{3}(x + 1) + 3$, the slope triangle method is the obvious choice. We can see a point on the line, $(-1, 3)$, and the slope is apparent: $\frac{2}{3}$. Here is the graph:

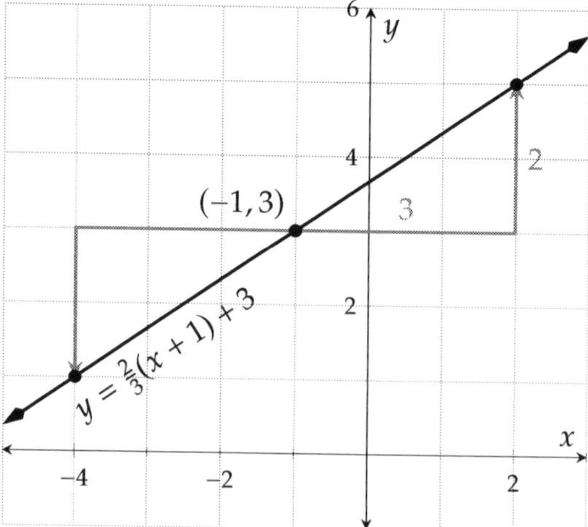

Figure 4.9.17: Graphing $y - 3 = \frac{2}{3}(x + 1)$ using slope triangles

Other graphing methods would take more work and miss the purpose of point-slope form (4.6.1). To graph a line in point-slope form (4.6.1), we recommend always using slope triangles.

4.9.4 Graphing Horizontal and Vertical Lines

We learned in Section 4.8 that equations in the form $x = h$ and $y = k$ make vertical and horizontal lines. But perhaps you will one day find yourself not remembering which is which. Making a table and plotting points can quickly remind you which type of equation makes which type of line. Let's build a table for $y = 2$ and another one for $x = -3$:

x-value	y-value	Point
0	2	$(0, 2)$
1	2	$(1, 2)$

Table 4.9.18: Table of Data for $y = 2$

x-value	y-value	Point
-3	0	$(-3, 0)$
-3	1	$(-3, 1)$

Table 4.9.19: Table of Data for $x = -3$

With two points on each line, we can graph them:

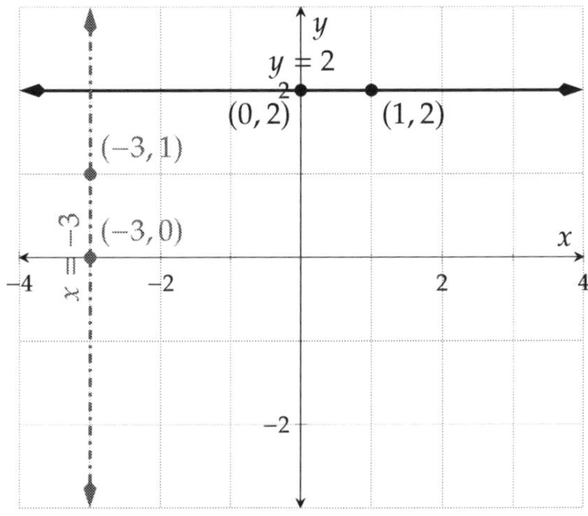

Figure 4.9.20: Graphing $y = 2$ and $x = -3$

4.9.5 Exercises

Graphing by Table

1. Use a table to make a plot of $y = 4x + 3$.

2. Use a table to make a plot of $y = -5x - 1$.

3. Use a table to make a plot of $y = -\frac{3}{4}x - 1$.

4. Use a table to make a plot of $y = \frac{5}{3}x + 3$.

Graphing Standard Form Equations

5. First find the x- and y-intercepts of the line with equation $6x + 5y = -90$. Then use your results to graph the line.

6. First find the x- and y-intercepts of the line with equation $2x - 3y = -6$. Then use your results to graph the line.

7. First find the x- and y-intercepts of the line with equation $3x + y = -9$. Then use your results to graph the line.

8. First find the x- and y-intercepts of the line with equation $-15x + 3y = -3$. Then use your results to graph the line.

9. First find the x- and y-intercepts of the line with equation $4x + 3y = -3$. Then use your results to graph the line.

10. First find the x- and y-intercepts of the line with equation $-4x - 5y = 5$. Then use your results to graph the line.

11. First find the x- and y-intercepts of the line with equation $5x - 3y = 0$. Then use your results to graph the line.

12. First find the x- and y-intercepts of the line with equation $2x + 9y = 0$. Then use your results to graph the line.

Graphing Slope-Intercept Equations

13. Use the slope and y-intercept from the line $y = -5x$ to plot the line. Use slope triangles.

14. Use the slope and y-intercept from the line $y = 3x - 6$ to plot the line. Use slope triangles.

15. Use the slope and y-intercept from the line $y = -\frac{2}{5}x + 2$ to plot the line. Use slope triangles.

16. Use the slope and y-intercept from the line $y = \frac{10}{3}x - 3$ to plot the line. Use slope triangles.

Graphing Horizontal and Vertical Lines

17. Plot the line $y = 1$.

18. Plot the line $x = -8$.

Choosing the Best Method to Graph Lines

19. Use whatever method you think best to plot $y = 2x + 2$.

20. Use whatever method you think best to plot $y = -\frac{3}{4}x - 1$.

21. Use whatever method you think best to plot $y = -\frac{3}{4}(x - 5) + 2$.

22. Use whatever method you think best to plot $3x + 2y = 6$.

23. Use whatever method you think best to plot $3x - 4y = 0$.

24. Use whatever method you think best to plot $x = -3$.

4.10 Linear Inequalities in Two Variables

We have learned how to graph lines like $y = 2x + 1$. In this section, we will learn how to graph linear inequalities like $y > 2x + 1$.

Example 4.10.2 (Office Supplies). Michael has a budget of \$133.00 to purchase some staplers and markers for the office supply closet. Each stapler costs \$19.00, and each marker costs \$1.75. If we use variable names so that he will purchase x staplers and y markers. Write and plot a linear inequality to model the relationship between the number of staplers and markers Michael can purchase. Keep in mind that Michael imight not spend all of the \$133.00.

The cost of buying x staplers would be $19x$ dollars. Similarly, the cost of buying y markers would be $1.75y$ dollars. Since whatever Michael spends needs to be no more than 133 dollars, we have the inequality

$$19x + 1.75y \leq 133.$$

This is a standard-form inequality, similar to Equation (4.7.1). Next, let's graph it.

The first method to graph the inequality is to graph the corresponding equation, $19x + 1.75y = 133$. Its x- and y-intercepts can be found this way:

$$\begin{aligned} 19x + 1.75y &= 133 \\ 19x + 1.75(0) &= 133 \\ 19x &= 133 \\ \frac{19x}{19} &= \frac{133}{19} \\ x &= 7 \end{aligned} \qquad \begin{aligned} 19x + 1.75y &= 133 \\ 19(0) + 1.75y &= 133 \\ 1.75y &= 133 \\ \frac{1.75y}{1.75} &= \frac{133}{1.75} \\ y &= 76 \end{aligned}$$

So the intercepts are $(7, 0)$ and $(0, 76)$, and we can plot the line in Figure 4.10.3.

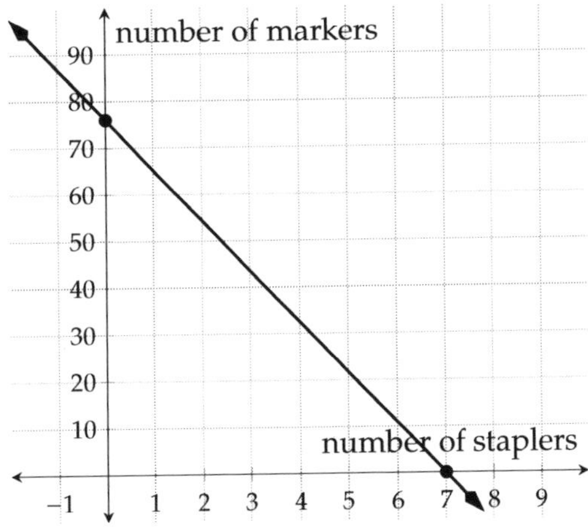

Figure 4.10.3: $19x + 1.75y = 133$

The points *on* this line represent ways in which Michael can spend exactly all of the \$133. But what does a point like $(2, 40)$ in Figure 4.10.4, which is not on the line, mean in this context? That would mean Michael bought 2 staplers and 40 markers, spending $19 \cdot 2 + 1.75 \cdot 40 = 108$ dollars. That is within Michael's budget.

In fact, any point on the lower left side of this line represents a total purchase within Michael's budget. The shading in Figure 4.10.5 captures *all* solutions to $19x + 1.75y \leq 133$. Some of those solutions have negative x- and y-values, which make no sense in context. So in Figure 4.10.6, we restrict the shading to solutions which make physical sense.

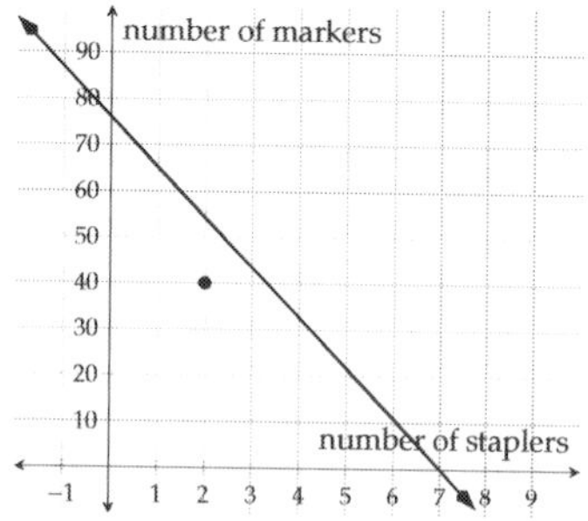

Figure 4.10.4: The line $19x + 1.75y = 133$ with a point identified that is within Michael's budget.

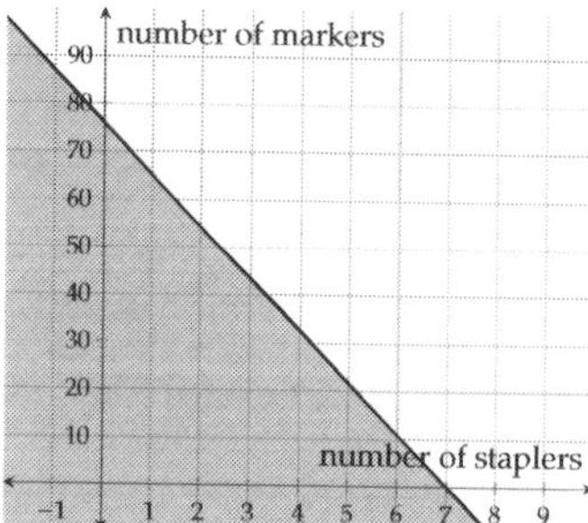

Figure 4.10.5: Shading all points that solve the inequality.

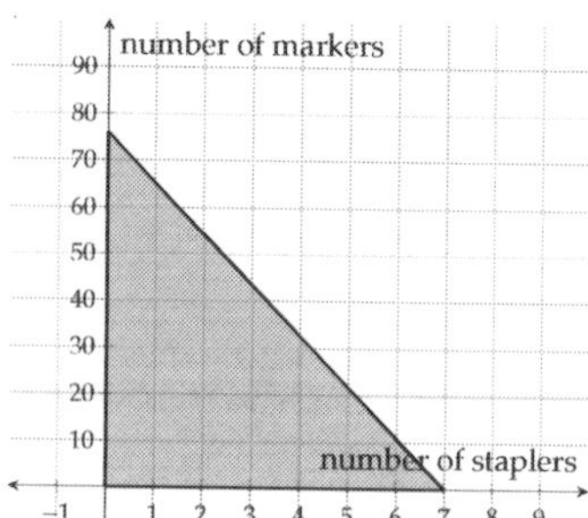

Figure 4.10.6: Shading restricted to points that make physical sense in context.

Let's look at some more examples of graphing linear inequalities in two variables.

Example 4.10.7. Is the point $(1, 2)$ a solution of $y > 2x + 1$?

In the inequality $y > 2x + 1$, substitute x with 1 and y with 2, and we will see whether the inequality is true:

$$\begin{aligned} y &> 2x + 1 \\ 2 &\stackrel{?}{>} 2(2) + 1 \\ 2 &\stackrel{\text{no}}{>} 5 \end{aligned}$$

Since $2 > 5$ is not true, $(1, 2)$ is not a solution of $y > 2x + 1$.

Example 4.10.8. Graph $y > 2x + 1$.

There are two steps to graphing this linear inequality in two variables.

1. Graph the line $y = 2x + 1$. Because the inequality symbol is $>$ (instead of $\geq$), the line should be dashed (instead of solid).

2. Next, we need to decide whether to shade the region above $y = 2x + 1$ or below it. We will choose a point to test whether $y > 2x + 1$ is true. As long as the line doesn't cross $(0, 0)$, we will use $(0, 0)$ to test because the number 0 is the easiest number for calculation.

$$y > 2x + 1$$
$$0 \overset{?}{>} 2(0) + 1$$
$$0 \overset{no}{>} 1$$

Because $0 > 1$ is not true, the point $(0, 0)$ is not a solution and should not be shaded. As a result, we shade the region *without* $(0, 0)$.

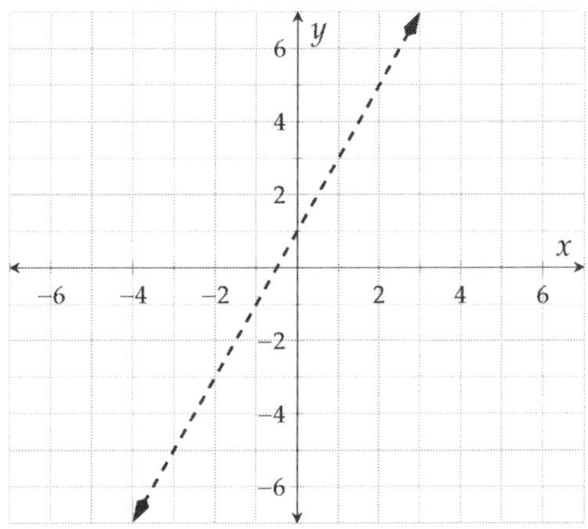

Figure 4.10.9: Step 1 of graphing $y > 2x+1$

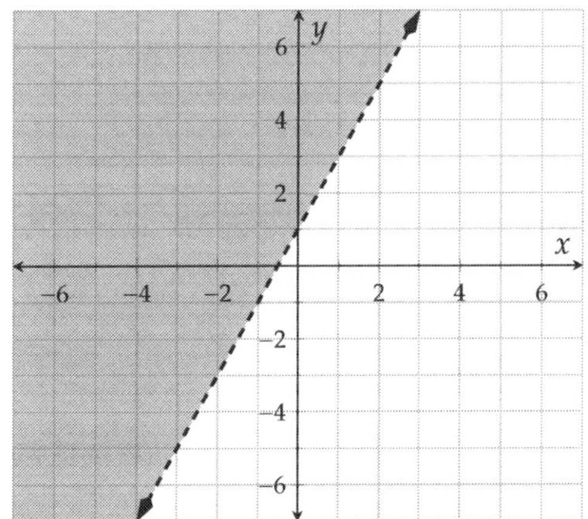

Figure 4.10.10: Complete graph of $y > 2x+1$

Example 4.10.11. Graph $y \leq -\frac{5}{3}x + 2$.

There are two steps to graphing this linear inequality in two variables.

1. Graph the line $y = -\frac{5}{3}x + 2$. Because the inequality symbol is $\leq$ (instead of $<$), the line should be solid.

2. Next, we need to decide whether to shade the region above $y = -\frac{5}{3}x + 2$ or below it. We will choose a point to test whether $y \leq -\frac{5}{3}x + 2$ is true there. Using $(0, 0)$ as a test point:

$$y \leq -\frac{5}{3}x + 2$$
$$0 \overset{?}{\leq} -\frac{5}{3}(0) + 2$$

$$0 \overset{\checkmark}{\leq} 2$$

Because $0 \leq 2$ is true, the point $(0, 0)$ is a solution. As a result, we shade the region *with* $(0, 0)$.

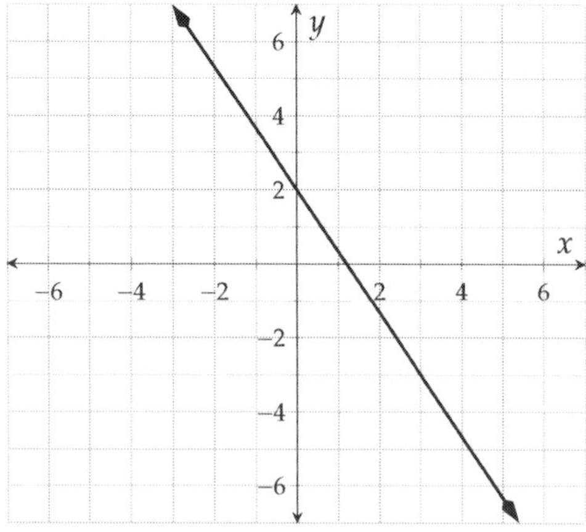

Figure 4.10.12: Step 1 of graphing $y \leq -\frac{5}{3}x + 2$

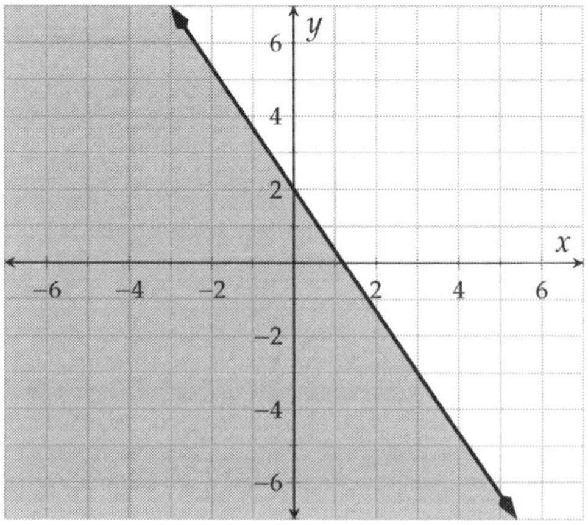

Figure 4.10.13: Complete graph of $y \leq -\frac{5}{3}x + 2$

4.10.1 Exercises

Exercises on Graphing Two-Variable Inequalities

1. Graph the linear inequality $y \geq -4x$.

2. Graph the linear inequality $y \leq -\frac{1}{2}x - 3$.

3. Graph the linear inequality $y < 3x + 5$.

4. Graph the linear inequality $y > \frac{4}{3}x + 1$.

5. Graph the linear inequality $2x + y \geq 3$.

6. Graph the linear inequality $3x + 2y < -6$.

7. Graph the linear inequality $y \geq 3$.

8. Graph the linear inequality $x < -1$.

Application Exercises on Graphing Two-Variable Inequalities

9. You fed your grandpa's cat while he was on vacation. When he was back, he took out a huge bank of coins, including quarters and dimes. He said you can take as many coins as you want, but the total value must be less than \$30.00.

(a) Write an inequality to model this situation, with q representing the number of quarters you will take, and d representing the number of dimes.

(b) Graph this linear inequality.

10. A couple is planning their wedding. They want the cost of the reception and the ceremony to be no more than \$8,000.

(a) Write an inequality to model this situation, with r as the cost of the reception (in dollars) and c as the cost of the ceremony (in dollars).

(b) Graph this linear inequality.

4.11 Graphing Lines Chapter Review

4.11.1 Review of Cartesian Coordinates

Cartesian Coordinate System The Cartesian coordinate system identifies the location of every point in a plane with an ordered pair.

Example 4.11.1. On paper, sketch a Cartesian coordinate system with units, and then plot the following points: $(3, 2), (-5, -1), (0, -3), (4, 0)$.

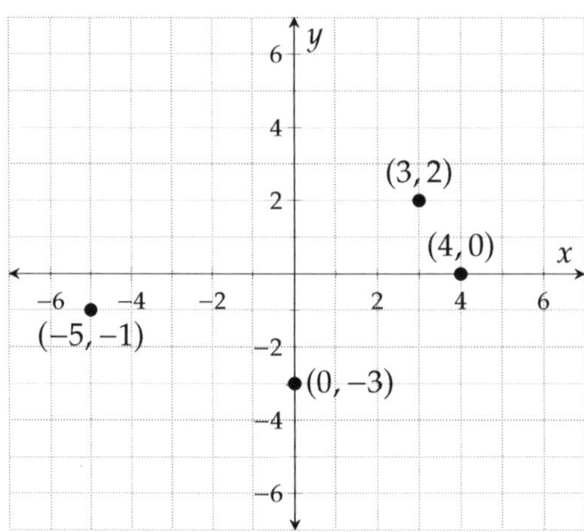

4.11.2 Review of Graphing Equations

Graphing Equations To graph an equation with two variables x and y, we can choose some reasonable x-values, then calculate the corresponding y-values, and then plot the (x, y)-pairs as points. For many (not-so-complicated) algebraic equations, connecting those points with a smooth curve will produce an excellent graph.

Example 4.11.2. Graph the equation $y = -2x + 5$.

x	$y = -2x + 5$	Point
-2		
-1		
0		
1		
2		

(a) Set up the table

x	$y = -2x + 5$	Point
-2	$-2(-2) + 5 = 9$	$(-2, 9)$
-1	$-2(-1) + 5 = 7$	$(-1, 7)$
0	$-2(0) + 5 = 5$	$(0, 5)$
1	$-2(1) + 5 = 3$	$(1, 3)$
2	$-2(2) + 5 = 1$	$(2, 1)$

(b) Complete the table

Figure 4.11.3: Making a table for $y = -2x + 5$

We use points from the table to graph the equation. First, plot each point carefully. Then, connect the points with a smooth curve. Here, the curve is a straight line. Lastly, we can communicate that the graph extends further by sketching arrows on both ends of the line.

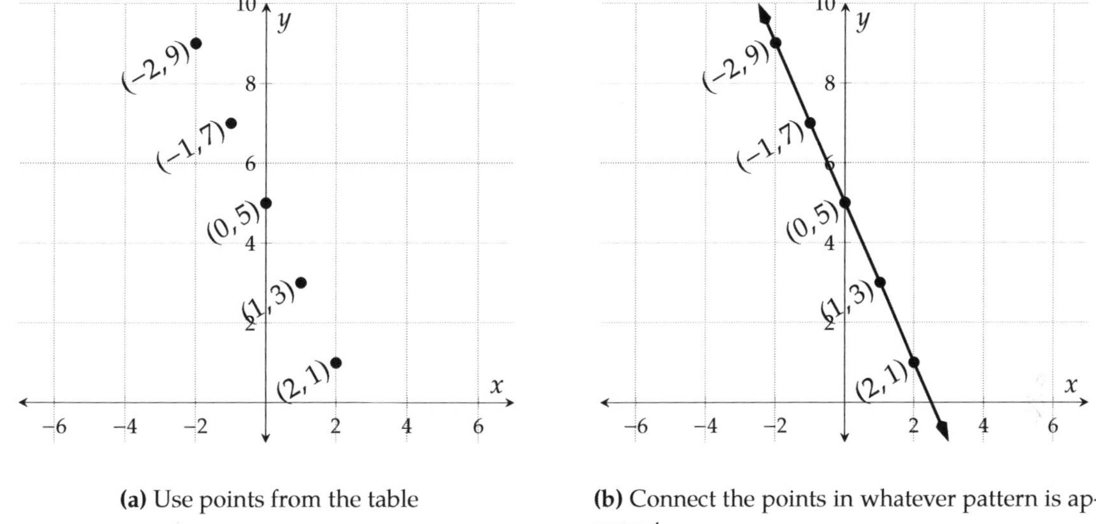

(a) Use points from the table

(b) Connect the points in whatever pattern is apparent

Figure 4.11.4: Graphing the Equation $y = -2x + 5$

4.11.3 Review of Two-Variable Data and Rate of Change

Exploring Two-Variable Data and Rate of Change For a linear relationship, by its data in a table, we can see the rate of change (slope) and the line's y-intercept, thus writing the equation.

Example 4.11.5. Write an equation in the form $y = \ldots$ suggested by the pattern in the table.

x	y
0	−4
1	−6
2	−8
3	−10

We consider how the values change from one row to the next. From row to row, the x-value increases by 1. Also, the y-value decreases by 2 from row to row.

	x	y	
	0	−4	
$+1 \rightarrow$	1	−6	$\leftarrow -2$
$+1 \rightarrow$	2	−8	$\leftarrow -2$
$+1 \rightarrow$	3	−10	$\leftarrow -2$

Since row-to-row change is always 1 for x and is always -2 for y, the rate of change from one row to another row is always the same: -2 units of y for every 1 unit of x.

We know that the output for $x = 0$ is $y = -4$. And our observation about the constant rate of change tells us that if we increase the input by x units from 0, the ouput should decrease by $\overbrace{(-2) + (-2) + \cdots + (-2)}^{x \text{ times}}$, which is $-2x$. So the output would be $-4 - 2x$.

So the equation is $y = -2x - 4$.

4.11.4 Review of Slope

Slope When x and y are two variables where the rate of change between any two points is always the same, we call this common rate of change the **slope**. Since having a constant rate of change means the graph will be a straight line, it's also called the **slope of the line**.

We can find a line's slope by drawing a slope-triangle on the line's graph, and then using the formula

$$m = \frac{\text{change in } y}{\text{change in } x} = \frac{\Delta y}{\Delta x} \tag{4.11.1}$$

Example 4.11.6. Find the slope of the line in the following graph.

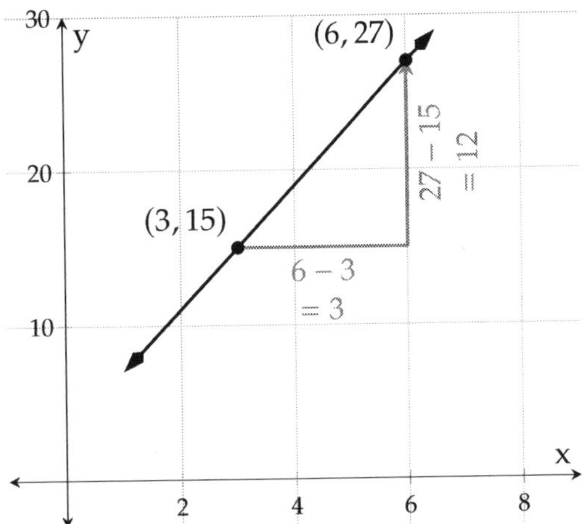

We picked two points on the line, and then drew a slope triangle. Next, we will do:

$$\text{slope} = \frac{12}{3} = 4$$

The line's slope is 4.

Finding a Line's Slope by the Slope Formula

If we know two points on a line, we can find its slope without graphing and, instead, using the slope formula slope $= \frac{y_2 - y_1}{x_2 - x_1}$

Example

A line passes the points $(-5, 25)$ and $(4, -2)$. Find this line's slope.

$$\begin{aligned}\text{slope} &= \frac{y_2 - y_1}{x_2 - x_1} \\ &= \frac{-2 - (25)}{4 - (-5)} \\ &= \frac{-27}{9} \\ &= -3\end{aligned}$$

The line's slope is -3.

4.11.5 Review of Slope-Intercept Form

Graphing a Line in Slope-Intercept Form A line's equation in slope-intercept form looks like $y = mx + b$, where m is the line's slope, and b is the line's y-intercept.

We can use a line's y-intercept and slope triangles to graph it.

Example 4.11.7. Graph the line $y = -\frac{2}{3}x + 10$.

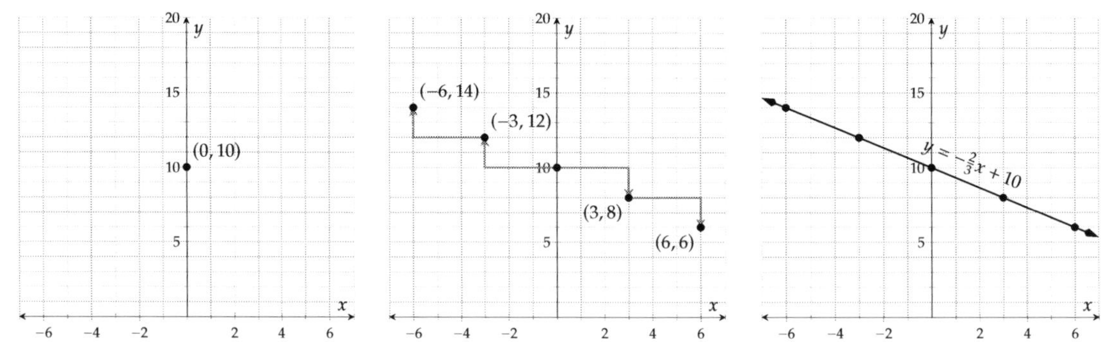

(a) First, plot the line's y-intercept, $(0, 10)$.

(b) The slope is $-\frac{2}{3} = \frac{-2}{3} = \frac{2}{-3}$. So we can try using a "run" of 3 and a "rise" of -2 or a "run" of -3 and a "rise" of 2.

(c) Arrowheads and labels are encouraged.

Figure 4.11.8: Graphing $y = -\frac{2}{3}x + 10$

Writing a Line's Equation in Slope-Intercept Form Based on Graph Given a line's graph, we can identify its y-intercept, and then find its slope by a slope triangle. With a line's slope and y-intercept, we can write its equation in the form of $y = mx + b$.

Example 4.11.9. Find the equation of the line in the graph.

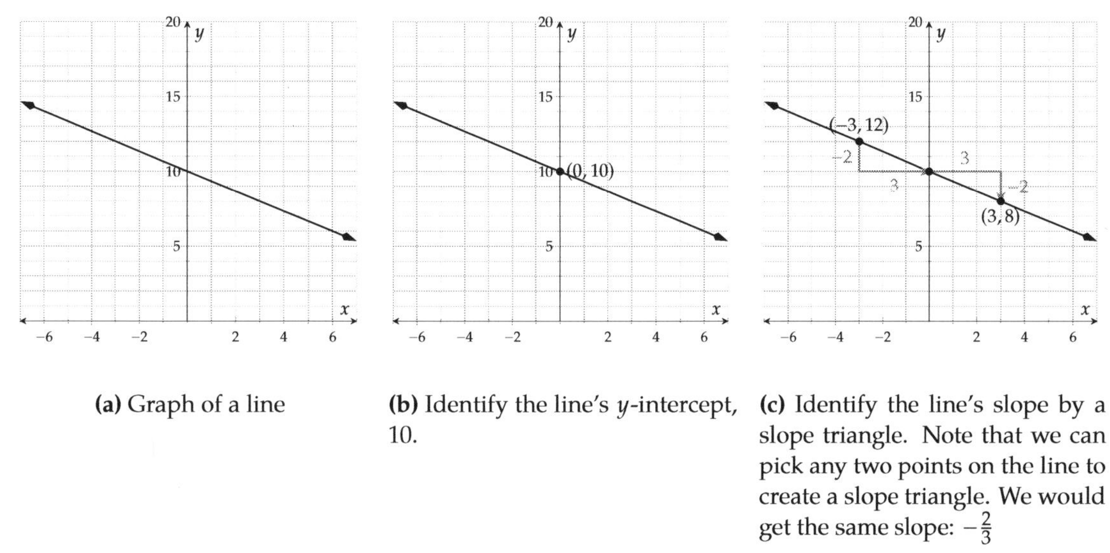

(a) Graph of a line

(b) Identify the line's y-intercept, 10.

(c) Identify the line's slope by a slope triangle. Note that we can pick any two points on the line to create a slope triangle. We would get the same slope: $-\frac{2}{3}$

Figure 4.11.10: Find the equation of the line in the graph.

With the line's slope $-\frac{2}{3}$ and y-intercept 10, we can write the line's equation in slope-intercept form: $y = -\frac{2}{3}x + 10$.

4.11.6 Review of Point-Slope Form

Point-Slope Form A line's point-slope form equation is in the form of $y = m(x - x_0) + y_0$, where m is the line's slope, and (x_0, y_0) is a point on the line.

Example 4.11.11. A line passes through $(-6, 0)$ and $(9, -10)$. Find this line's equation in both point-slope and slope-intercept form.

Solution. We will use the slope formula (4.4.3) to find the slope first. After labeling those two points as $(\overset{x_1}{-6}, \overset{y_1}{0})$ and $(\overset{x_2}{9}, \overset{y_2}{-10})$, we have:

$$\begin{aligned} \text{slope} &= \frac{y_2 - y_1}{x_2 - x_1} \\ &= \frac{-10 - 0}{9 - (-6)} \\ &= \frac{-10}{15} \\ &= -\frac{2}{3} \end{aligned}$$

Now the point-slope equation looks like $y = -\frac{2}{3}(x - x_0) + y_0$. Next, we will use $(9, -10)$ and substitute x_0 with 9 and y_0 with -10, and we have:

$$\begin{aligned} y &= -\frac{2}{3}(x - x_0) + y_0 \\ y &= -\frac{2}{3}(x - 9) + (-10) \\ y &= -\frac{2}{3}(x - 9) - 10 \end{aligned}$$

Next, we will change the point-slope equation into slope-intercept form:

$$\begin{aligned} y &= -\frac{2}{3}(x - 9) - 10 \\ y &= -\frac{2}{3}x + 6 - 10 \\ y &= -\frac{2}{3}x - 4 \end{aligned}$$

4.11.7 Review of Standard Form

Standard Form A line's equation in standard form looks like $Ax + By = C$. We need to convert a line's equation from standard form to slope-intercept form, and vice versa.

Examples

Converting from Standard Form to Slope-Intercept Form

Convert $2x + 3y = 6$ into slope-intercept form.

$$\begin{aligned} 2x + 3y &= 6 \\ 2x + 3y - 2x &= 6 - 2x \\ 3y &= -2x + 6 \\ \frac{3y}{3} &= \frac{-2x + 6}{3} \\ y &= \frac{-2x}{3} + \frac{6}{3} \\ y &= -\frac{2}{3}x + 2 \end{aligned}$$

The line's equation in slope-intercept form is $y = -\frac{2}{3}x + 2$.

Converting from Slope-Intercept Form to Standard Form

Convert $y = -\frac{2}{3}x + 2$ into standard form.

$$\begin{aligned} y &= -\frac{2}{3}x + 2 \\ 3 \cdot y &= 3 \cdot (-\frac{2}{3}x + 2) \\ 3y &= 3 \cdot (-\frac{2}{3}x) + 3 \cdot 2 \\ 3y &= -2x + 6 \\ 3y + 2x &= -2x + 6 + 2x \\ 2x + 3y &= 6 \end{aligned}$$

The line's equation in standard form is $2x + 3y = 6$.

To graph a line in standard form, we could first change it to slope-intercept form, and then graph the line by its y-intercept and slope triangles. A second method is to graph the line by its x-intercept and y-intercept.

Example 4.11.12. Graph $2x - 3y = -6$ using its intercepts. And then use the intercepts to calculate the line's slope.

Solution. We calculate the line's x-intercept by substituting $y = 0$ into the equation

$$\begin{aligned} 2x - 3y &= -6 \\ 2x - 3(0) &= -6 \\ 2x &= -6 \\ x &= -3 \end{aligned}$$

So the line's x-intercept is $(-3, 0)$.

Similarly, we substitute $x = 0$ into the equation to calculate the y-intercept:

$$\begin{aligned} 2x - 3y &= -6 \\ 2(0) - 3y &= -6 \\ -3y &= -6 \\ y &= 2 \end{aligned}$$

So the line's y-intercept is $(0, 2)$.

With both intercepts' coordinates, we can graph the line:

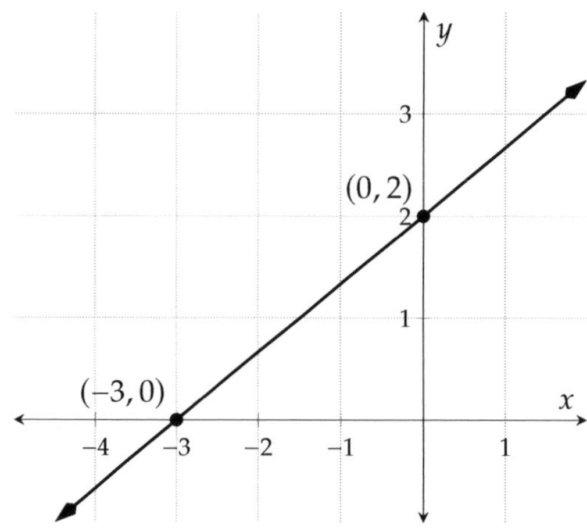

Figure 4.11.13: Graph of $2x - 3y = -6$

4.11.8 Review of Horizontal, Vertical, Parallel, and Perpendicular Lines

Horizontal, Vertical, Parallel, and Perpendicular Lines A horizontal line's equation looks like $y = k$, while a vertical line's equation looks like $x = h$. The following figure has graphs of a horizontal line and a vertical line.

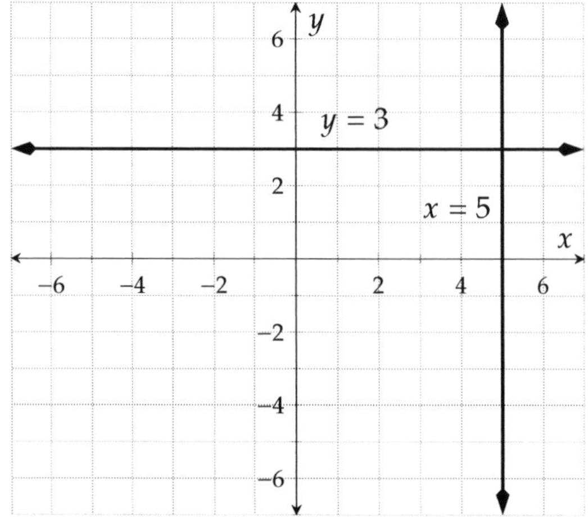

Figure 4.11.14: Graphs of $y = 3$ and $x = 5$

Parallel and Perpendicular Lines

Two lines are parallel if and only if they have the same slope.

Examples

Line m's equation is $y = -2x + 20$. Line n is parallel to m, and line n also passes the point $(4, -3)$. Find line n's equation.

Since parallel lines have the same slope, line n's slope is also -2. Since line n also passes the point $(4, -3)$, we can write line n's equation in point-slope form:

$$y = m(x - x_1) + y_1$$
$$y = -2(x - 4) + (-3)$$
$$y = -2(x - 4) - 3$$

We can also easily get the line's equation in slope-intercept form:

$$y = -2(x - 4) - 3$$
$$y = -2x + 8 - 3$$
$$y = -2x + 5$$

Two lines are perpendicular if and only if the product of their slopes is -1.

Line m's equation is $y = -2x + 20$. Line n is perpendicular to m, and line n also passes the point $(4, -3)$. Find line n's equation.

Since line m and n are perpendicular, the product of their slopes is -1. Because line m's slope is given as -2, we can find line n's slope is $\frac{1}{2}$.

Since line n also passes the point $(4, -3)$, we can write line n's equation in point-slope form:

$$y = m(x - x_1) + y_1$$
$$y = \frac{1}{2}(x - 4) + (-3)$$
$$y = \frac{1}{2}(x - 4) - 3$$

We can also easily get the line's equation in slope-intercept form:

$$y = \frac{1}{2}(x - 4) - 3$$
$$y = \frac{1}{2}x - 2 - 4$$
$$y = \frac{1}{2}x - 6$$

4.11.9 Review of Linear Inequalities in Two Variables

Linear Inequalities in Two Variables When we graph lines like $y = 2x + 1$, we are graphing points which satisfy the relationship $y = 2x + 1$, like $(0, 1), (1, 3), (2, 5)$, etc. Similarly, we can graph linear inequalities like $y > 2x + 1$ by plotting all points which satisfies the inequality, like $(0, 2), (0, 3), (1, 4), (1, 5)$, etc. All these points form a region, instead of a line.

Example 4.11.15. Graph $y > 2x + 1$.

There are two steps to graph an inequality.

1. Graph the line $y = 2x + 1$. Because the inequality symbol is $>$, (instead of $\geq$) the line should be dashed (instead of solid).
2. Next, we need to decide whether to shade the region above $y = 2x + 1$ or below it. We will choose a point to test whether $y > 2x + 1$ is true. As long as the line doesn't cross

$(0,0)$, we will use $(0,0)$ to test, because the number 0 is the easiest number for calculation.

$$y > 2x + 1$$
$$0 \stackrel{?}{>} 2(0) + 1$$
$$0 \stackrel{?}{>} 1$$

Because $0 > 1$ is not true, the point $(0,0)$ is not a solution and should not be shaded. As a result, we shade the region without $(0,0)$.

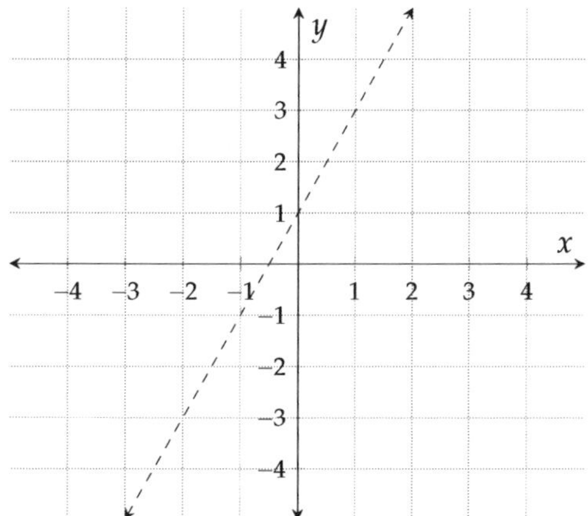

Figure 4.11.16: Step 1 of graphing $y > 2x + 1$

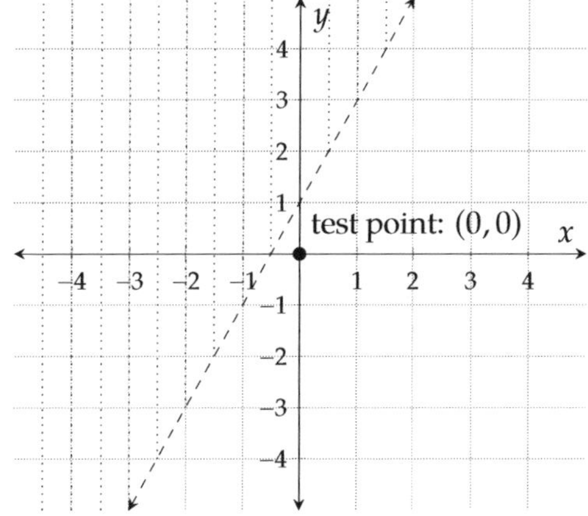

Figure 4.11.17: Step 2 of graphing $y > 2x + 1$

4.11.10 Exercises

1. Sketch the points $(8,2)$, $(5,5)$, $(-3,0)$, $\left(0,-\frac{14}{3}\right)$, $(3,-2.5)$, and $(-5,7)$ on a Cartesian plane.

2. Consider the equation
$y = -\frac{5}{6}x - 4$
Which of the following ordered pairs are solutions to the given equation? There may be more than one correct answer.

□ $(24,-21)$ □ $(-24,16)$ □ $(-30,24)$ □ $(0,-4)$

3.

x	y
0	−4
1	−2
2	0
3	2

Write an equation in the form $y = \ldots$ suggested by the pattern in the table.

4.

x	y
0	3
1	−2
2	−7
3	−12

Write an equation in the form $y = \ldots$ suggested by the pattern in the table.

5. A line's graph is shown below.

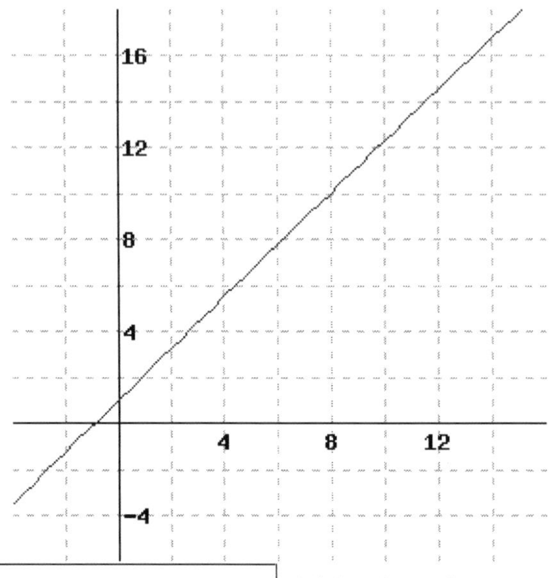

The slope of this line is ______. (If the slope does not exist, enter DNE or NONE.)

6. A line's graph is shown below.

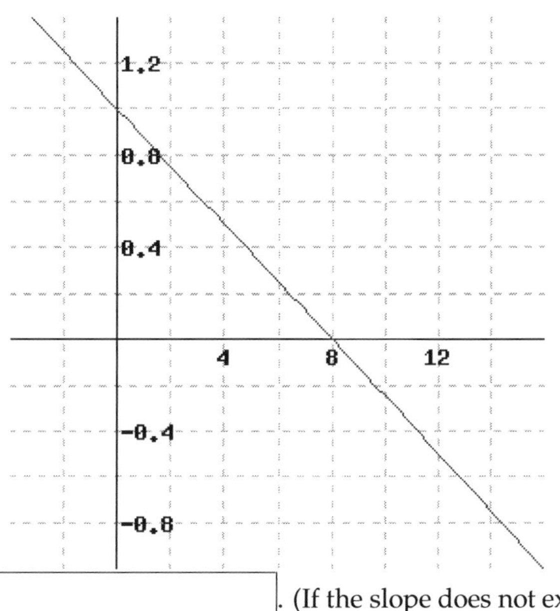

The slope of this line is [ ]. (If the slope does not exist, enter DNE or NONE.)

7. A line passes through the points $(-12, -1)$ and $(8, -16)$. Find this line's slope. If the slope does not exists, you may enter DNE or NONE.

This line's slope is [ ].

8. A line passes through the points $(1, -4)$ and $(-3, -4)$. Find this line's slope. If the slope does not exists, you may enter DNE or NONE.

This line's slope is [ ].

9. A line passes through the points $(-2, -1)$ and $(-2, 5)$. Find this line's slope. If the slope does not exists, you may enter DNE or NONE.

This line's slope is [ ].

10. A line's graph is given.

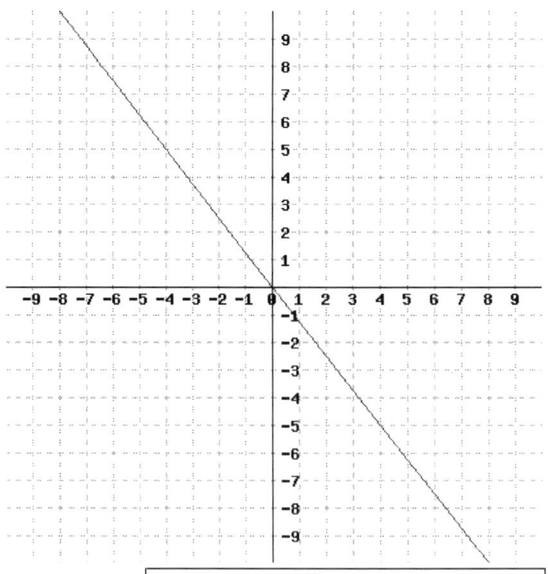

This line's slope-intercept equation is [ ]

11. Find the line's slope and y-intercept.
A line has equation $7x - 6y = -24$.

This line's slope is [ ].

This line's y-intercept is [ ].

12. A line passes through the points $(24, 13)$ and $(0, -8)$. Find this line's equation in point-slope form.
Using the point $(24, 13)$, this line's point-slope form equation is [ ].

Using the point $(0, -8)$, this line's point-slope form equation is [ ].

13. Scientists are conducting an experiment with a gas in a sealed container. The mass of the gas is measured, and the scientists realize that the gas is leaking over time in a linear way. Its mass is leaking by 4.7 grams per minute. Eight minutes since the experiment started, the remaining gas had a mass of 192.7 grams.
Let x be the number of minutes that have passed since the experiment started, and let y be the mass of the gas in grams at that moment. Use a linear equation to model the weight of the gas over time.

a. This line's slope-intercept equation is [ ].

b. 36 minutes after the experiment started, there would be [ ] grams of gas left.

c. If a linear model continues to be accurate, [ ] minutes since the experiment started, all gas in the container will be gone.

14. Find the y-intercept and x-intercept of the line given by the equation

$$8x + 3y = -72$$

If a particular intercept does not exist, enter *none* into all the answer blanks for that row.

	x-value	y-value	Location
y-intercept	______	______	__________
x-intercept	______	______	__________

15. Rewrite $y = 2x + 4$ in standard form.

16. Rewrite $y = -\frac{3}{4}x + 6$ in standard form.

17. Fill out this table for the equation $x = -8$. The first row is an example.

x	y	Points
-8	-3	$(-8, -3)$
______	-2	__________
______	-1	__________
______	0	__________
______	1	__________
______	2	__________

18. A line's graph is given.

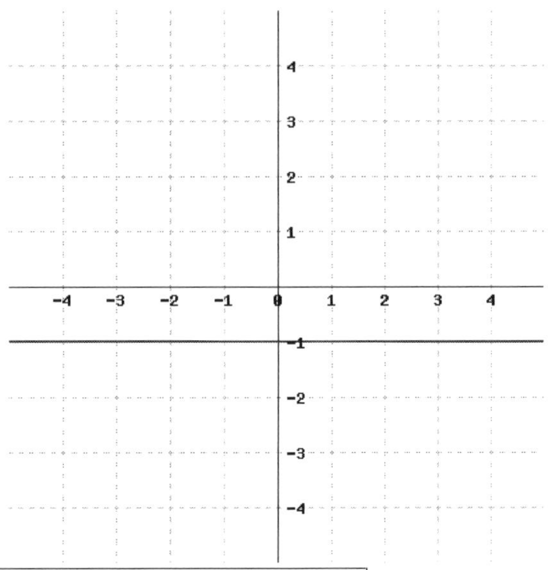

This line's equation is ______

19. Line m passes points $(0, -4)$ and $(0, -10)$.
Line n passes points $(10, -1)$ and $(10, 10)$.
Determine how the two lines are related.
These two lines are

- ○ parallel
- ○ perpendicular

○ neither parallel nor perpendicular

20. Line k's equation is $y = \frac{6}{5}x - 1$.
Line ℓ is perpendicular to line k and passes through the point $(18, -11)$.
Find an equation for line ℓ in both slope-intercept form and point-slope forms.

An equation for ℓ in slope-intercept form is: ________.

An equation for ℓ in point-slope form is: ________.

Index

x-axis, 261
x-coordinate, 261
y-axis, 261
y-coordinate, 261
y-intercept, 303

absolute value, 28
axis
    horizontal, 261
    vertical, 261

Cartesian Coordinate System, 261
constant, 288
coordinate plane, 261
coordinates, 261

equation, 99
evaluate, 81
expression, 80

factor, 83
form
    point-slope, 328
    slope-intercept, 304
    standard, 340

greater-than symbol, 58
greater-than-or-equal-to symbol, 59

inequality, 99
intercept, 303

less-than symbol, 60
linear equation, 99
linear relationship, 288

negative, 2
number line, 2

ordered pair, 261
origin, 261

point-slope form, 328
positive, 2
Properties of Equivalent Equations, 112

quadrant, 261

reduced fraction, 16
relationship
    linear, 288

set notation, 49
slope, 288
    formula, 296
    triangle, 290
slope-intercept form, 304
solution, 100
standard form, 340

term, 83

variable, 79

Made in the USA
San Bernardino, CA
11 January 2018